T0212521

Lecture Notes in Computer Science   10606

Editorial Board

David Hutchison
  Lancaster University, Lancaster, UK
Takeo Kanade
  Carnegie Mellon University, Pittsburgh, PA, USA
Josef Kittler
  University of Surrey, Guildford, UK
Jon M. Kleinberg
  Cornell University, Ithaca, NY, USA
Friedemann Mattern
  ETH Zurich, Zurich, Switzerland
John C. Mitchell
  Stanford University, Stanford, CA, USA
Moni Naor
  Weizmann Institute of Science, Rehovot, Israel
C. Pandu Rangan
  Indian Institute of Technology, Madras, India
Bernhard Steffen
  TU Dortmund University, Dortmund, Germany
Demetri Terzopoulos
  University of California, Los Angeles, CA, USA
Doug Tygar
  University of California, Berkeley, CA, USA
Gerhard Weikum
  Max Planck Institute for Informatics, Saarbrücken, Germany

More information about this series at http://www.springer.com/series/7408

Kim Guldstrand Larsen · Oleg Sokolsky
Ji Wang (Eds.)

Dependable Software Engineering

Theories, Tools, and Applications

Third International Symposium, SETTA 2017
Changsha, China, October 23–25, 2017
Proceedings

 Springer

Editors
Kim Guldstrand Larsen
Aalborg University
Aalborg
Denmark

Oleg Sokolsky 🆔
University of Pennsylvania
Philadelphia, PA
USA

Ji Wang
National University of Defense Technology
Changsha
China

ISSN 0302-9743 ISSN 1611-3349 (electronic)
Lecture Notes in Computer Science
ISBN 978-3-319-69482-5 ISBN 978-3-319-69483-2 (eBook)
https://doi.org/10.1007/978-3-319-69483-2

Library of Congress Control Number: 2017956763

LNCS Sublibrary: SL2 – Programming and Software Engineering

© Springer International Publishing AG 2017
This work is subject to copyright. All rights are reserved by the Publisher, whether the whole or part of the material is concerned, specifically the rights of translation, reprinting, reuse of illustrations, recitation, broadcasting, reproduction on microfilms or in any other physical way, and transmission or information storage and retrieval, electronic adaptation, computer software, or by similar or dissimilar methodology now known or hereafter developed.
The use of general descriptive names, registered names, trademarks, service marks, etc. in this publication does not imply, even in the absence of a specific statement, that such names are exempt from the relevant protective laws and regulations and therefore free for general use.
The publisher, the authors and the editors are safe to assume that the advice and information in this book are believed to be true and accurate at the date of publication. Neither the publisher nor the authors or the editors give a warranty, express or implied, with respect to the material contained herein or for any errors or omissions that may have been made. The publisher remains neutral with regard to jurisdictional claims in published maps and institutional affiliations.

Printed on acid-free paper

This Springer imprint is published by Springer Nature
The registered company is Springer International Publishing AG
The registered company address is: Gewerbestrasse 11, 6330 Cham, Switzerland

Preface

This volume contains the papers presented at the third in the SETTA (the Symposium on Dependable Software Engineering: Theories, Tools and Applications) series of conferences – held during October 23–25, 2017, in Changsha, China. The purpose of SETTA is to provide an international forum for researchers and practitioners to share cutting-edge advancements and strengthen collaborations in the field of formal methods and its interoperability with software engineering for building reliable, safe, secure, and smart systems. The inaugural and second SETTA symposiums were successfully held in Nanjing (2015) and Beijing (2016).

SETTA 2017 attracted 31 good submissions coauthored by researchers from 18 countries. Each submission was reviewed by at least three Program Committee members with help from additional reviewers. The Program Committee discussed the submissions online and decided to accept 19 papers for presentation at the conference. The program also included four invited talks given by Prof. Cliff B. Jones from Newcastle University, Prof. Rupak Majumdar from Max Planck Institute for Software Systems, Prof. Sanjit A. Seshia from the University of California, Berkeley, and Prof. Jean-Pierre Talpin from Inria, Centre de recherche de Rennes-Bretagne-Atlantique.

We would like to express our gratitude to the authors for submitting their papers to SETTA 2017. We are particularly thankful to all members of Program Committee and the additional reviewers, whose hard and professional work in the review process helped us prepare the high-quality conference program. Special thanks go to our four invited speakers for presenting their research at the conference. We would like to thank the Steering Committee for their advice.

Like the previous editions of SETTA, SETTA 2017 had a young SETTA Researchers Workshop held on October 22, 2017. The Second National Conference on Formal Methods and Applications in China was also co-located with SETTA during October 21–22, 2017.

Finally, we thank the conference chair, Prof. Xiangke Liao, publicity chair, Prof. Fu Song, local chair, Prof. Wei Dong, and the local Organizing Committee of Ling Chang, Liqian Chen, Zhenbang Chen, Shanshan Li, Wanwei Liu, Yanjun Wen, and Liangze Yin. We are grateful to Prof. Shuling Wang for taking care of the conference website. We also thank the Chinese Computer Federation, the National Natural Science Foundation of China, and the State Key Laboratory for High Performance Computing of China for their financial support. Furthermore, we would like to especially thank Springer for sponsoring the Best Paper Award.

August 2017

Kim Guldstrand Larsen
Oleg Sokolsky
Ji Wang

Organization

Program Committee

Erika Abraham	RWTH Aachen University, Germany
Farhad Arbab	CWI and Leiden University, The Netherlands
Sanjoy Baruah	Washington University in St. Louis, USA
Michael Butler	University of Southampton, UK
Yuxin Deng	East China Normal University, China
Xinyu Feng	University of Science and Technology of China, China
Goran Frehse	University of Grenoble Alpes-Laboratoire Verimag, France
Martin Fränzle	University of Oldenburg, Germany
Lindsay Groves	Victoria University of Wellington, New Zealand
Dimitar Guelev	Bulgarian Academy of Sciences, Bulgaria
Fei He	Tsinghua University, China
Deepak Kapur	University of New Mexico, USA
Kim Guldstrand Larsen	Aalborg University, Denmark
Axel Legay	IRISA/Inria, France
Xuandong Li	Nanjing University, China
Shaoying Liu	Hosei University, Japan
Zhiming Liu	Southwest University, China
Xiaoguang Mao	National University of Defense Technology, China
Markus Müller-Olm	Westfälische Wilhelms-Universität Münster, Germany
Raja Natarajan	Tata Institute of Fundamental Research, India
Jun Pang	University of Luxembourg, Luxembourg
Shengchao Qin	Teesside University, UK
Stefan Ratschan	Institute of Computer Science, Czech Academy of Sciences, Czech Republic
Sriram Sankaranarayanan	University of Colorado Boulder, USA
Oleg Sokolsky	University of Pennsylvania, USA
Martin Steffen	University of Oslo, Norway
Zhendong Su	University of California, Davis, USA
Cong Tian	Xidian University, China
Tarmo Uustalu	Tallinn University of Technology, Estonia
Chao Wang	University of Southern California, USA
Farn Wang	National Taiwan University, China
Ji Wang	National University of Defense Technology, China
Heike Wehrheim	University of Paderborn, Germany
Michael Whalen	University of Minnesota, USA
Wang Yi	Uppsala University, Sweden
Haibo Zeng	Virginia Tech, USA

Naijun Zhan	Institute of Software, Chinese Academy of Sciences, China
Lijun Zhang	Institute of Software, Chinese Academy of Sciences, China
Qirun Zhang	University of California, Davis, USA

Additional Reviewers

Bu, Lei
Dghaym, Dana
Dokter, Kasper
Gerwinn, Sebastian
Ghassabani, Elaheh
Howard, Giles
Jakobs, Marie-Christine
Katis, Andreas
König, Jürgen
Le, Quang Loc
Le, Ton Chanh
Li, Dan

Mizera, Andrzej
Qamar, Nafees
Qu, Hongyang
Singh, Abhishek Kr
Stewart, Danielle
Traonouez, Louis-Marie
Travkin, Oleg
Wang, Shuling
Xu, Lili
Xu, Zhiwu
Xue, Bai

Abstracts of Invited Talks

General Lessons from a Rely/Guarantee Development

Cliff B. Jones[1], Andrius Velykis[1], and Nisansala Yatapanage[1,2]

[1] School of Computing Science,
Newcastle University, Newcastle upon Tyne, UK
[2] School of Computer Science and Informatics,
De Montfort University, Leicester, UK

Abstract. Decomposing the design (or documentation) of large systems is a practical necessity; this prompts the need for a notion of *compositional* development methods; finding such methods for concurrent software is technically challenging because of the interference that characterises concurrency. This paper outlines the development of a difficult example in order to draw out lessons about such development methods. Although the "rely/guarantee" approach is employed in the example, the intuitions are more general.

Towards Verified Artificial Intelligence

Sanjit A. Seshia

Department of Electrical Engineering and Computer Sciences,
University of California, Berkeley, USA

Abstract. The deployment of artificial intelligence (AI), particularly of systems that learn from data and experience, is rapidly expanding in our society. Verified artificial intelligence (AI) is the goal of designing AI-based systems that have strong, ideally provable, assurances of correctness with respect to mathematically-specified requirements. In this talk, I will consider Verified AI from a formal methods perspective. I will describe five challenges for achieving Verified AI, and five corresponding principles for addressing these challenges. I will illustrate these challenges and principles with examples and sample results, particularly from the domain of autonomous vehicles.

Artificial intelligence (AI) is a term used for computational systems that attempt to mimic aspects of human intelligence (e.g., see [1]). In recent years, the deployment of AI, particularly of systems with the ability to learn from data and experience, has expanded rapidly. AI is now being used in applications ranging from analyzing medical data to driving cars. Several of these applications are safety-critical or mission-critical, and require strong guarantees of reliable operation. Therefore, the question of verification and validation of AI-based systems has begun to demand the attention of the research community. We define *verified artificial intelligence* (AI) as the goal of designing AI-based systems that have strong, ideally provable, assurances of correctness with respect to mathematically-specified requirements.

In this talk, I consider Verified AI from the perspective of formal methods — a field of computer science and engineering concerned with the rigorous mathematical specification, design, and verification of systems [2, 9]. Specifically, I describe five major challenges for achieving Verified AI: (i) environment modeling — dealing with environments that have a plethora of unknowns; (ii) formal specification – how to precisely capture assumptions and requirements on learning-based systems; (iii) system modeling — how to model systems that evolve continuously with new data; (iv) computational engines — new scalable techniques for training, testing, and quantitative verification, and (v) designing AI systems "correct-by-construction". I also describe five corresponding principles for addressing these challenges: (i) introspective environment modeling; (ii) system-level specifications; (iii) abstractions and explanations of learning components; (iv) new randomized and quantitative formal techniques, and (v) the paradigm of formal inductive synthesis to design learning systems in a "correct-by-construction" manner. Taken together, we believe that the principles we suggest can point a way towards the goal of Verified AI.

Initial promising results have been obtained for the control and verification of learning-based systems, including deep learning systems, in the domain of autonomous driving [3, 5–7]. The theory of formal inductive synthesis is described in a recent paper [4]. A more detailed exposition of the ideas presented in this talk can be found in a paper first published in 2016 [8].

References

1. Committee on Information Technology, Automation, and the U.S. Workforce: Information technology and the U.S. workforce: where are we and where do we go from here? http://www.nap.edu/24649
2. Clarke, E.M., Wing, J.M.: Formal methods: state of the art and future directions. ACM Comput. Surv. **28**(4), 626–643 (1996)
3. Dreossi, T., Donzé, A., Seshia, S.A.: Compositional falsification of cyber-physical systems with machine learning components. In: NASA Formal Methods - 9th International Symposium, NFM 2017, pp. 357–372 (2017)
4. Jha, S., Seshia, S.A.: A theory of formal synthesis via inductive learning. Acta Informatica (2017)
5. Li, W., Sadigh, D., Sastry, S.S., Seshia, S.A.: Synthesis for human-in-the-loop control systems. In: Ábrahám, E., Havelund, K. (eds.) TACAS 2014. LNCS, vol. 8413, pp. 470–484. Springer, Berlin (2014)
6. Sadigh, D.: Safe and interactive autonomy: control, learning, and verification. PhD thesis, EECS Department, University of California, Berkeley (2017)
7. Sadigh, D., Sastry, S.S., Seshia, S.A., Dragan, A.D.: Information gathering actions over human internal state. In: Proceedings of the IEEE/RSJ International Conference on Intelligent Robots and Systems (IROS), pp. 66–73 (2016)
8. Seshia, S.A., Sadigh, D., Sastry, S.S.: Towards Verified Artificial Intelligence. ArXiv e-prints (2016)
9. Wing, J.M.: A specifier's introduction to formal methods. IEEE Comput. **23**(9), 8–24 (1990)

Compositional Methods for CPS Design (Keynote Abstract)

Jean-Pierre Talpin

Inria, Centre de recherche de Rennes-Bretagne-Atlantique

Abstract. Logic and proof theory are probably have proved to be the most effective mathematical tools to help modularly engineering correct software, and this since the introduction of type inference, up to fantastic progresses of SAT-SMT solvers and provers.

In this talk, I will focus on past experience of my former project-team ESPRESSO in manipulating such tools in the design, implementation, proof and test-case of Polychrony: an open-source synchronous modeling environment hosted on the Polarsys industry working group of the Eclipse foundation.

I will highlight the compositionality and scalability of its concepts by considering two major case studies (with Polychrony and its commercial implementation, RT-Builder) in developing functional and real-time simulators, in avionics and automotive, and the integration of these concepts in the AADL standard.

Before concluding, I will hint on broader perspectives undertaken by my new project-team, TEA, toward logic-focused, compositional and scalable, models of concurrent, timed, cyber and physical systems, and the challenges to build new ADLs from such concepts.

References

1. Nakajima, S., Talpin, J.-P., Toyoshima, M., Yu, H. (eds.): Cyber-Physical System Design from an Architecture Analysis Viewpoint. Communications of the NII Shonan Meetings. Springer (2017)
2. Lunel, S., Boyer, B., Talpin, J.-P.: Compositional proofs in dynamic differential logic. In: International Conference on Applications of Concurrency to System Design. Springer (2017)
3. Besnard, L., Gautier, T., Le Guernic, P., Guy, C., Talpin, J.-P., Larson, B.R., Borde, E.: Formal semantics of behavior specifications in the architecture analysis and design language standard (extended abstract). High-Level Design, Verification and Test. IEEE (2016)
4. Gautier, T., Le Guernic, P., Talpin, J.-P., Besnard, L.: Polychronous automata. In: Theoretical Aspects of Software Engineering. IEEE (2015)

Joint work with David Berner, Loic Besnard, Adnan Bouakaz, Christian Brunette, Abdoulaye Gamatie, Thierry Gautier, Yann Glouche, Clment Guy, Alexandre Honorat, Kai Hu, Christophe Junke, Kenneth Johnson, Sun Ke, Michael Kerboeuf, Paul Le Guernic, Yue Ma, Hugo Metivier, Van Chan Ngo, Julien Ouy, Klaus Schneider, Sandeep Shukla, Eric Vecchie, Zhibin Yang, Huafeng Yu.
Supported by Nankai University and by Beihang University, NSFC grant 61672074.

5. Ngo, V.C., Talpin, J.-P., Gautier, T., Besnard, L., Le Guernic, P.: Modular translation validation of a full-sized synchronous compiler using off-the-shelf verification tools (abstract). In: International Workshop on Software and Compilers for Embedded Systems, invited presentation. ACM (2015)
6. Talpin, J.-P., Brandt, J., Gemnde, M., Schneider, K., Shukla, S.: Constructive polychronous systems. In: Science of Computer Programming. Elsevier (2014)
7. Yu, H., Joshi, P., Talpin, J.-P., Shukla, S., Shiraishi, S.: Model-based integration for automotive control software. In: Digital Automation Conference, Automotive Special Session. ACM (2015)
8. Bouakaz, A., Talpin, J.-P.: Buffer minimization in earliest-deadline first scheduling of dataow graphs. In: Conference on Languages, Compilers and Tools for Embedded Systems. ACM (2013)
9. Yu, H., Ma, Y., Glouche, Y., Talpin, J.-P., Besnard, L., Gautier, T., Le Guernic, P., Toom, A., Laurent, O.: System-level co-simulation of integrated avionics using polychrony. In: ACM Symposium on Applied Computing. ACM (2011)
10. Talpin, J.-P., Ouy, J., Gautier,T., Besnard, L., Le Guernic, P.: Compositional design of isochronous systems. In: Science of Computer Programming, Special Issue on APGES. Elsevier (2011)

Contents

Modeling and Verification

Formalization

Tools

Invited Talk

General Lessons from a Rely/Guarantee Development

Cliff B. Jones[1(✉)], Andrius Velykis[1], and Nisansala Yatapanage[1,2]

[1] School of Computing Science, Newcastle University,
Newcastle upon Tyne, UK
cliff.jones@ncl.ac.uk
[2] School of Computer Science and Informatics,
De Montfort University, Leicester, UK

Abstract. Decomposing the design (or documentation) of large systems is a practical necessity; this prompts the need for a notion of *compositional* development methods; finding such methods for concurrent software is technically challenging because of the interference that characterises concurrency. This paper outlines the development of a difficult example in order to draw out lessons about such development methods. Although the "rely/guarantee" approach is employed in the example, the intuitions are more general.

1 Introduction

The aim of this paper is to contribute to the discussion about compositional development for concurrent programs. Much of the paper is taken up with the development, from its specification, of a concurrent garbage collector but the important messages are by no means confined to the example and are identified as *lessons*.

1.1 Compositional Methods

To clarify the notion of "compositional" development of concurrent programs, it is worth beginning with some observations about the specification and design of sequential programs. A developer faced with a specification for S might make the design decision to decompose the task using two components that are to be executed sequentially ($S1; S2$); that top-level step can be justified by discharging a proof obligation involving only the specifications of S and $S1/S2$. Moreover, the developer of either of the sub-components need only be concerned with its specification—not that of its sibling nor that of its parent S. This not only facilitates separate development, it also increases the chance that any subsequent modifications are isolated within the boundary of one specified component.

As far as is possible, the advantages of compositional development should be retained for concurrent programs.

© Springer International Publishing AG 2017
K.G. Larsen et al. (Eds.): SETTA 2017, LNCS 10606, pp. 3–22, 2017.
https://doi.org/10.1007/978-3-319-69483-2_1

Lesson 1. *The notion of "compositionality" is best understood by thinking about a development process: specifications of separate components ought genuinely insulate them from one another (and from their context). The ideal is, faced with a specified task (module), propose a decomposition (combinator) and specify any sub-tasks; then prove the decomposition correct wrt (only) the specifications. The same process is then repeated on the sub-tasks.*

Because of the interference inherent in concurrency, this is not easy to achieve and, clearly, (pre/)post conditions will not suffice. However, numerous examples exist to indicate that rely/guarantee conditions (see Sect. 1.2) facilitate the required separation where a designer chooses a decomposition of S into shared-variable sub-components that are to be executed in parallel ($S1 \parallel S2$).

1.2   Rely/Guarantee Thinking

The origin of the rely/guarantee (R/G) work is in [Jon81]. More than 20 theses have developed the original idea including [Stø90, Xu92] that look at progress arguments, [Din00] that moves in the direction of a refinement calculus form of R/G, [Pre01] that provides an Isabelle soundness proof of a slightly restricted form of R/G rules, [Col08] that revisits soundness of general R/G rules, [Pie09] that addresses usability and [Vaf07, FFS07] manage to combine R/G thinking with Separation Logic. Furthermore, a number of separation logic (see below) papers also employ R/G reasoning (e.g. [BA10, BA13]) and [DFPV09, DYDG+10] from separation logic researchers build on R/G. Any reader who is unfamiliar with the R/G approach can find a brief introduction in [Jon96]. (A fuller set of references is contained in [HJC14, JHC15].)

The original way of writing R/G specifications displayed the predicates of a specification delimited by keywords; some of the subsequent papers (notably those concerned with showing the soundness of the system) present specifications as 5-tuples. The reformulation in [HJC14, JHC15] employs a refinement calculus format [Mor90, BvW98] in which it is much more natural to investigate algebraic properties of specifications. Since some of the predicates for the garbage collection example are rather long, the keyword style is adopted in this paper but algebraic properties such as distribution are used as required.

The literature contains many and diverse examples of R/G developments including:

- Susan Owicki's [Owi75] problem of finding the minimum index to an array at which an element can be found satisfying a predicate is tackled using R/G thinking in [HJC14]
- a staple of R/G presentations is a concurrent version of the *Sieve of Eratosthenes* introduced in [Hoa72]—see for example [JHC15]
- parallel "cleanup" operations for the *Fisher/Galler* Algorithm for the *union/find* problem are developed in [CJ00]
- a development of *Simpson's 4-slot algorithm* is given in [JP11]—an even nicer specification using "possible values" (see Sect. 1.3) is contained in [JH16]

The first two represent examples in which the R/G conditions are symmetric in the sense that the concurrent sub-processes have the same specifications; the last two items and the concurrent garbage collector presented below are more interesting because the concurrent processes need different specifications.

Lesson 2. *By using relations to express interference, R/G conditions offer a plausible balance of expressiveness versus tractability—see Sects. 3.2 and 4.*

1.3   Challenges

The extent to which compositionality depends on the expressivity of the specification notation is an issue and the "possible values" notation used below provides an interesting discussion point. Much more telling is the contrast with methods which need the code of sibling processes to reason about interference. For example [Owi75, OG76] not only postpones a final (*Einmischungsfrie*) check until the code of all concurrent processes is to hand—this expensive test has to be repeated when changes are made to any sub-component.

It is useful to distinguish progressively more challenging cases of interference and the impact that the difficulty has on reasoning about correctness:

1. The term "parallel" is often used for threads that share no variables: threads are in a sense entirely independent and only interact in the sense that they overlap in time. Hoare [Hoa72] observes that, in this simple case, the conjunction of the post conditions of the individual threads is an acceptable post condition for their combination.
2. Over-simplifying, this is a basis for concurrent separation logic. CSL [O'H07] and the many related logics are, however, aimed at –and capable of– reasoning about intricate heap based-programs. See also [Par10].
3. It is argued in [JY15] that careful use of abstraction can serve the purpose of reasoning about some forms of separation.
4. The interference in Owicki's example that is referred to in the preceding section is non-trivial because one thread affects a variable used to control repetition in the other thread. It would be possible to reason about the development of this example using "auxiliary" (aka "ghost") variables. The approach in [Owi75] actually goes further in that the code of the combined system is employed in the final *Einmischungsfrei* check. Using the R/G approach in [HJC14], however, the interference is adequately characterised by relation.
5. There are other examples in which relations alone do not appear to be enough. This is true of even the early stages of development of the concurrent garbage collector below. A notation for "possible values" [JP11, HBDJ13, JH16] obviates the need for auxiliary variables in some cases see Sect. 2.1.
6. The question of whether some examples require ghost variables is open and the discussion is resumed in Sect. 4. That their use is tempting in order to simplify reasoning about concurrent processes is attested by the number of proofs that employ them.

Lesson 3. *The use of "ghost" (aka "auxiliary") variables presents a subtle danger to compositional development (cf. Lesson 1). The case against is, however, clear: in the extreme, ghost variables can be used to record complete detail about the environment of a process. Few researchers would succumb to such extreme temptation but minimising the use of ghost variables ought be an objective.*

It might surprise readers who have heard the current authors inveigh against ghost variables that the development in this conference paper does in fact use such a variable. The acknowledgements report a relevant discussion on this point and a journal paper is planned to explore other options.

1.4   Plan of the Paper

The bulk of this paper (Sects. 2, 3, 4 and 5) offers a development of the concurrent garbage collector described in [BA84, vdS87]. At several points, "lessons" are identified. It is these lessons that are actually the main point of the paper and the example is chosen to give credence to these intuitions.

The development uses R/G ideas (often referred to as "R/G thinking") but the lessons have far wider applicability.

It is also worth mentioning that it is intended to write an extended journal version of this paper that will contain a full development from an abstract specification hopefully with proofs checked on Isabelle. That paper will also review various modelling decisions made in the development. Many of twhich were revised (in some cases more than once) e.g. the decision how to model *Heap* was revised several times![1].

2   Preliminary Development

This section builds up to a specification of concurrent garbage collection that is then used as the basis for development in Sects. 3, 4 and 5. The main focus is on the *Collector* but, since this runs concurrently with some form of *Mutator*, some assumptions have to be recorded about the latter.

2.1   Abstract Spec

It is useful to pin down the basic idea of inaccessible addresses (aka "garbage") before worrying about details of heap storage (see Sect. 2.2) and marking (Sect. 3).

Lesson 4. *It is widely accepted that the use of abstract datatypes can clarify key concepts before discussion turns to implementation details. Implementations are then viewed as "reifications" that achieve the same effect as the abstraction. Formal proof obligations are given, for example, in [Jon90].*

---

[1] One plausible alternative is: $Heap = (Addr \times Pos) \xrightarrow{m} Addr$.

Lesson 4 is common place for sequential programs but it actually has even greater force for concurrent program development (where it is perhaps under-employed by many researchers). For example, it is argued in [JY15] that careful use of abstraction can serve the purpose of reasoning about separation. Furthermore, in R/G examples such as [JP11], such abstractions also make it possible to address interference and separation at early stages of design.

The set of addresses ($Addr$) is assumed to be some arbitrary but finite set; it is not to be equated with natural numbers since that would suggest that addresses could have arithmetic operators applied to them.

The abstract states[2] contain two sets of addresses: those that are in use ($busy$) and those that have been collected into a $free$ set.

Σ_0 :: $busy$: $Addr$-**set**
　　　$free$: $Addr$-**set**

where

$inv\text{-}\Sigma_0(mk\text{-}\Sigma_0(busy, free)) \quad \triangle \quad busy \cap free = \{\,\}$

It is, of course, an essential property that the sets $busy/free$ are always disjoint. (VDM types are restricted by datatype invariants and the set Σ_0 only contains values that satisfy the invariant.) There can however be elements of $Addr$ that are in neither set—such addresses are to be considered as "garbage" and the task of a garbage collector is to add such addresses to $free$.

Effectively, the GC process is an infinite loop repeatedly executing the $Collector$ operation whose specification is:

$Collector$
ext wr $free$
　　　rd $busy$
pre true
rely $(busy' - busy) \subseteq free \wedge free' \subseteq free$
guar $free \subseteq free'$
post $(Addr - busy) \subseteq \bigcup \widehat{free}$

The predicate $guar\text{-}Collector$ reassures the designer of $Mutator$ that a chosen $free$ cell will not disappear. The read/write "frames" in a VDM specification provide a shorthand for access and interference: thus $Collector$ actually has an implied guarantee condition that it cannot change $busy$.

The rely condition warns the developer of $Collector$ that the $Mutator$ can consume $free$ addresses. Given this fact, recording a post condition for $Collector$ is not quite trivial. In a sequential setting, it would be correct to write:

$free' = (Addr - busy)$

---

[2] The use of VDM notation should present the reader with no difficulty: it has been widely used for decades and is the subject of an ISO standard; one useful reference is [Jon90].

but the concurrent *Mutator* might be removing addresses from the free set so the best that the *collector* can promise is to place all addresses that are originally garbage into the free set at some point in time. Here is the first use of the "possible values" notation in this paper. In a sequential formulation, *post-Collector* would set the lower bound for garbage collection by requiring that any addresses not reachable (in the initial *hp*) from *roots* would be in the final *free* set. To cope with the fact that a concurrent *Mutator* can acquire addresses from *free*, the correct statement is that all unreachable addresses should appear in some value of *free*. The notation discussed in [JP11, HBDJ13, JH16] for the set of possible values that can be observed by a component is $\widehat{free}$.

Lesson 5. *The "possible values" notation is a useful addition to –at least– R/G.*

2.2   The Heap

This section introduces a model of the heap. The set of addresses that are busy is defined to be those that are reachable from a set of roots by tracing all of the pointers in a heap.

$$\Sigma_1 :: \quad roots \ : \ Addr\text{-}\mathbf{set}$$
$$hp \quad : \ Heap$$
$$free \quad : \ Addr\text{-}\mathbf{set}$$

where

$$inv\text{-}\Sigma_1(mk\text{-}\Sigma_1(roots, hp, free)) \quad \triangle$$
$$\mathbf{dom} \ hp = Addr \ \wedge$$
$$free \cap reach(roots, hp) = \{\} \ \wedge \qquad\qquad\qquad \text{upper bound for GC}$$
$$\forall a \in free \cdot hp(a) = \{[]\}$$

$$Heap = Addr \xrightarrow{m} Node$$

$$Node = \left[Addr\right]^*$$

To smooth the use of this model of *Heap*, $hp(a, i)$ is written for $hp(a)(i)$ and $(a, i) \in \mathbf{dom} \ hp$ has the obvious meaning. When addresses are deleted from nodes, their position is set to the **nil** value.

The second conjunct of the invariant defines the upper bound of garbage collection; the final conjunct requires that free addresses map to empty nodes. The *roots* component of Σ_1 is taken to be constant.

The *child-rel* function extracts the relation over addresses from the heap (i.e. ignoring pointer positions); it drops any **nil** values.

$$child\text{-}rel : Heap \rightarrow (Addr \times Addr)\text{-}\mathbf{set}$$

$$child\text{-}rel(hp) \quad \triangle \quad \{(a, b) \mid a \in \mathbf{dom} \ hp \wedge b \in (\mathbf{elems} \ hp(a)) \cap Addr\}$$

The *reach* function computes the relational image (with respect to its first argument) of the transitive closure of the heap:

$$reach : Addr\text{-}\mathbf{set} \times Heap \rightarrow Addr\text{-}\mathbf{set}$$

$$reach(s, hp) \quad \triangleq \quad \mathbf{rng}\,(s \lhd child\text{-}rel(hp)^{\star})$$

A useful lemma states that, starting from some set s, if there is an element a reachable from s that is not in s, then there must exist a *Node* which contains an address not in s (but notice that $hp(b, j)$ might not be a).

A useful lemma is:

$$\exists a \cdot a \in reach(s, hp) \wedge a \notin s \;\Rightarrow\; \exists (b, j) \in \mathbf{dom}\,hp \cdot b \in s \wedge hp(b, j) \notin s$$

3   Marking

The intuition behind the garbage collection (GC) algorithm in [BA84] is to mark all addresses reachable (over the relation defined by the *Heap*) from *roots*, then sweep any unmarked addresses into *free*.

The state underlying the chosen garbage collector has an additional component to record the addresses that have been marked (the third conjunct of the invariant ensures that all addresses in (*roots* ∪ *free*) are always marked).

$$\Sigma_2 :: \begin{array}{ll} roots & : Addr\text{-}\mathbf{set} \\ hp & : Heap \\ free & : Addr\text{-}\mathbf{set} \\ marked & : Addr\text{-}\mathbf{set} \end{array}$$

$$inv\text{-}\Sigma_2(mk\text{-}\Sigma_2(roots, hp, free, marked)) \quad \triangleq$$
$$\mathbf{dom}\,hp = Addr \,\wedge$$
$$free \cap reach(roots, hp) = \{\,\} \,\wedge \qquad\qquad \text{upper bound for GC}$$
$$(roots \cup free) \subseteq marked \,\wedge$$
$$\forall a \in free \cdot hp(a) = \{[\,]\}$$

3.1   Sequential Algorithm

Garbage collection runs concurrently with a *Mutator* which can acquire *free* addresses and give rise to garbage that is no longer accessible from *roots*. A fully concurrent garbage collector is covered in Sect. 4. This section introduces (in Fig. 1) code that can be viewed as sequential in the sense that the *Mutator* would have to pause; interestingly this same code satisfies specifications for two more challenging concurrent situations (see Sects. 3.2 and 4).

As observed above, the full garbage collector repeatedly iterates the code called here *Collector*. This can be split into three phases. Providing the invariant is respected, *Mark/Sweep* do not depend on how many addresses are marked initially (*Unmark* is there to ensure that garbage is collected in at most two passes) but, thinking of the *Collector* being run intermittently, it is reasonable to start by removing any surplus marks.

$Collector \triangleq (Unmark; Mark; Sweep)$

The main interest is in the marking phase. As shown in Fig. 1, the outer loop propagates a wave of marking over the *hp* relation; it iterates until no new addresses are marked. The inner *Propagate* iterates over all addresses: for each address that is itself marked, all of its children are marked. (Specifications of *Mark-kids* are in Sects. 3.4 and 4.3.)

```
Mark △                      Propagate △
    repeat                      consid ← { };
        mc ← card marked;       do while consid ≠ Addr
        Propagate                   let x ∈ (Addr − consid) in
    until card marked = mc          if x ∈ marked then Mark-kids(x) else skip;
                                    consid ← consid ∪ {x}
                                od
```

Fig. 1. Code for *Mark*

In the case when the code is running with no interference, R/G reasoning is not required and the specification of *Mark* and proof that the code in Fig. 1 satisfies that specification are straightforward. (In fact, they are simplified cases of what follows in Sect. 3.2.) When the same code is considered in the interfering environments in Sects. 3.2 and 4, (differing) R/Gs and, of course, proofs are needed. The elaboration of the R/Gs is particularly interesting.

Lesson 6. *Considering the sequential case is useful because it is then possible to note how the rely condition (nothing changes) and the guarantee condition* (**true**) *need to be changed to handle concurrency.*

3.2   Concurrent GC with Atomic Interference

The complication in the concurrent case is that the *Mutator* can interfere with the marking strategy of the *Collector* by redirecting pointers. This can be accommodated providing the *Mutator* marks appropriately whenever it makes a change.

The development is tackled in two stages: firstly, this section assumes a *Mutator* that atomically both redirects a pointer in a *Node* and marks the new address; Sect. 4 shows that even separating the two steps still allows the *Collector* code of Fig. 1 to achieve the lower bound of marking but the argument is more delicate and indicates an expressive limitation of R/G conditions. The argument to establish the upper bound for marking (and thus the lower bound of garbage collection) is given in Sect. 5.

If the *Mutator* were able to update and mark atomically, specifications and proofs are straightforward; although this atomicity assumption is unrealistic, it is informative to compare this section with Sect. 4. As adumbrated in Sect. 1, the argument is split into a justification of the parallel decomposition (Sect. 3.3) and the decompositions of the *Collector/Mutator* sub-components, addressed in Sects. 3.4 and 3.5 respectively.

3.3   Parallel Decomposition

An R/G specification of the *Collector* is:

Collector
ext wr *free, marked*
    rd *roots, hp*
pre true
rely *free'* $\subseteq$ *free* $\wedge$ *marked* $\subseteq$ *marked'* $\wedge$
    $\forall (a, i) \in$ **dom** *hp* ·
        $hp'(a, i) \neq hp(a, i) \wedge hp'(a, i) \in Addr \;\Rightarrow\; hp'(a, i) \in marked'$
guar *free* $\subseteq$ *free'*
post $(Addr - reach(roots, hp)) \subseteq \bigcup \widehat{free}$ lower bound for GC

Here again, the notation for possible values is used to cover interference.

The final conjunct of the rely condition is the key property that (for now) assumes that the environment (i.e. the *Mutator*) simultaneously marks any change it makes to the heap.[3]

The lower bound of addresses to be collected is one part of the requirement; the upper bound is constrained by the second conjunct of *inv-Σ_2*. The lower bound for garbage collection requires setting an upper bound for marking addresses; this topic is postponed to Sect. 5.

Lesson 7. *Such splitting of what would be an equality in the specification of a sequential component is a common R/G tactic.*

The corresponding specification of the *Mutator* is:

Mutator
ext wr *hp, free, marked*
    rd *roots*
pre true
rely *free* $\subseteq$ *free'*
guar *free'* $\subseteq$ *free* $\wedge$ *marked* $\subseteq$ *marked'* $\wedge$
    $\forall (a, i) \in$ **dom** *hp* ·
        $hp'(a, i) \neq hp(a, i) \wedge hp'(a, i) \in Addr \;\Rightarrow\; hp'(a, i) \in marked'$
post true

---

[3] Strictly, the fact that the *Collector* (in particular, its *Sweep* component) does not have write access to *hp* means that it cannot clean up the nodes in *free* as required by the final conjunct of *inv-Σ_2*. Changing the guarantee conditions is routine but uninformative.

The R/G proof obligation (PO) for concurrent processes requires that each one's guarantee condition implies the rely condition of the other(s); in this case they match identically so the result is immediate.

3.4 Developing the *Collector* Code

As outlined in Sect. 1, what remains to be done is to develop code that satisfies the specification of the *Collector* (in isolation from that of the *Mutator*)—i.e. show that the decomposition of the *Collector* into three phases given in Sect. 3.1 satisfies the *Collector* specification in Sect. 3.3 and then to develop code for *Mark*.

A post condition for a sequential version of *Unmark* could constrain *marked'* to be exactly equal to *roots*∪*free* but, again, interference must be considered. The rely condition indicates that the environment can mark addresses so whatever *Unmark* removes from *marked* could be replaced. The possible values notation is again deployed so that *post-Unmark* requires that, for every address which should not be marked, a possible value of *marked* exists which does not contain the address. However, this post condition alone would permit an implementation of *Unmark* itself first to mark an address and then remove the marking; this erroneous behaviour is ruled out by *guar-Unmark*. The rely condition indicates that the *free* set can also change but, since it can only reduce, this poses no problem.

Unmark

ext wr *marked*
     rd *roots*, *free*

pre true

rely *free'* ⊆ *free*
guar *marked'* ⊆ *marked*
post $\forall a \in (Addr - (roots \cup free)) \cdot \exists m \in \overparen{marked} \cdot a \notin m$

The relaxing of the post condition again uses the idea in Lesson 7.

The post condition for *Mark* also has to cope with the interference absent from a sequential specification and this requires more thought. In the sequential case, *post-Mark* could use a strict equality to require that all reachable nodes are added to *marked* but here the equality is split into a lower and upper bound. The lower bound for marking is crucial to preserve the upper bound of garbage collection (see the second conjunct of *inv-Σ_2*). This lower bound is recorded in the post condition. (The use of *hp'* is, of course, challenging but the post condition is stable [CJ07, WDP10] under the rely condition.) The "loss" (from the equality in the sequential case) of the other containment is compensated for by setting an upper bound for marking (see *no-mog* in Sect. 5).

Mark

ext wr *marked*
    rd *roots, hp, free*

pre true

rely *rely-Collector*
guar $marked \subseteq marked'$
post $reach(marked, hp') \subseteq marked'$

The relaxing of the post condition once again uses the idea in Lesson 7. Similar observations to those for *Unmark* relate to the specification of *Sweep* which, for the concurrent case, becomes:

Sweep

ext wr *free*
    rd *marked*

pre true

rely $free' \subseteq free \land marked \subseteq marked'$
guar $free \subseteq free'$
post $(free' - free) \cap marked = \{\ \} \land$
    $\forall a \in (Addr - marked) \cdot \exists f \in \widehat{free} \cdot a \in f$

The rely and guarantee conditions of *Collector* are distributed (with appropriate weakening/strengthening) over the three sub-components; all of the pre conditions are **true**; so the remaining PO is:

$$post\text{-}Unmark(\sigma, \sigma') \land post\text{-}Mark(\sigma', \sigma'') \land post\text{-}Sweep(\sigma'', \sigma''') \Rightarrow$$
$$post\text{-}Collector(\sigma, \sigma''')$$

The proof is straightforward.

Turning to the decomposition of *Mark* (see Fig. 1), in order to prove *post-Mark*, a specification is needed for *Propagate* that copes with interference:

Propagate

ext wr *marked*
    rd *hp*

pre true

rely *rely-Collector*
guar $marked \subseteq marked'$
post $\bigcup\{\textbf{elems}\ hp'(a) \cap Addr \mid a \in marked\} \subseteq marked' \land$
    $(marked \subset marked' \lor reach(roots, hp') \subseteq marked')$

The first conjunct of the post condition indicates the progress required of the wave of marking; the second triggers further iterations if any marking has occurred.

To prove the lower marking bound (i.e. must mark everything that is reachable from *roots*), we use an argument that composes on the right a relation that expresses the rest of the computation as in [Jon90]: essentially the *to-end* relation states that the remaining iterations of the loop will mark everything reachable from what is already marked:

$$to\text{-}end(\sigma, \sigma') \quad \triangleq \quad reach(marked, hp') \subseteq marked'$$

The PO is:

$$post\text{-}Propagate(\sigma, \sigma') \wedge \sigma.marked \subset \sigma'.marked \wedge to\text{-}end(\sigma', \sigma'') \Rightarrow$$
$$to\text{-}end(\sigma, \sigma'')$$

whose proof is straightforward.

The termination argument follows from there being a limit to the markable elements: a simple upper bound is **dom** hp but there is a tighter one (cf. Section 5).

Then trivially:

$$\sigma.marked = \sigma.roots \wedge to\text{-}end(\sigma, \sigma') \Rightarrow post\text{-}Mark(\sigma, \sigma')$$

Pursuing the decomposition of *Propagate* (again, see Fig. 1) needs a specification of the inner operation:

Mark-kids $(x \colon Addr)$
ext wr *marked*
     rd *hp*
pre true
rely *rely-Collector*
guar *marked* $\subseteq$ *marked'*
post (**elems** $hp'(x) \cap Addr) \subseteq marked'$

In this case, the proof is more conventional and a relation that expresses how far the marking has progressed is composed on the left:

$$so\text{-}far(\sigma, \sigma') \quad \triangleq$$
$$\bigcup \{\mathbf{elems}\ hp(a) \cap Addr \mid a \in (marked \cap consid')\} \subseteq marked'$$

The relevant PO is:

$$so\text{-}far(\sigma, \sigma') \wedge$$
$$consid' \neq Addr \wedge post\text{-}Mark\text{-}kids(\sigma', x, \sigma'') \wedge consid'' = consid' \cup \{x\} \Rightarrow$$
$$so\text{-}far(\sigma, \sigma'')$$

whose discharge is obvious.

The final obligation is to show:

$$so\text{-}far(\sigma, \sigma') \wedge consid' = Addr \Rightarrow post\text{-}Propagate(\sigma, \sigma')$$

The first conjunct of *post-Propagate* is straightforward; the fact that (unless the marking process is complete) some marking must occur in this iteration of *Propagate* follows from the lemma in Sect. 2.2.

3.5   Checking the Interference from *Mutator*

The mutator is viewed as an infinite loop non-deterministically selecting one of *Redirect, Malloc, Zap* as specified below. At this stage, these are viewed as atomic operations so no R/Gs are supplied here:[4] their respective post conditions must be shown to imply *rely-Mark*):

Redirect $(a: Addr, i: \mathbb{N}_1, b: Addr)$

ext wr $hp, marked$

pre $\{a, b\} \subseteq reach(roots, hp) \wedge i \in \mathbf{inds}\, hp(a)$

post $hp' = hp \dagger \{(a, i) \mapsto b\} \wedge marked' = marked \cup \{b\}$

It follows trivially that:

$post\text{-}Redirect(\sigma, \sigma') \;\Rightarrow\; guar\text{-}Mutator(\sigma, \sigma')$

For this atomic case, the code (using multiple assignment) would be:

$< hp(a), marked := hp(a) \dagger \{i \mapsto b\}, marked \cup \{b\} >$

Malloc $(a: Addr, i: \mathbb{N}_1, b: Addr)$

ext wr $hp, free$

pre $a \in reach(roots, hp) \wedge i \in \mathbf{inds}\, hp(a) \wedge b \in free$

post $hp' = hp \dagger \{(a, i) \mapsto b\} \wedge free' = free - \{b\}$

Malloc preserves the invariant because $inv\text{-}\Sigma_2$ insists that free addresses are always marked. It follows trivially that:

$post\text{-}Malloc(\sigma, \sigma') \;\Rightarrow\; guar\text{-}Mutator(\sigma, \sigma')$

Zap $(a: Addr, i: \mathbb{N}_1)$

ext wr hp

pre $a \in reach(roots, hp) \wedge i \in \mathbf{inds}\, hp(a)$

post $hp' = hp \dagger \{(a, i) \mapsto \mathbf{nil}\}$

It again follows trivially that:

$post\text{-}Zap(\sigma, \sigma') \;\Rightarrow\; guar\text{-}Mutator(\sigma, \sigma')$

4   Relaxing Atomicity: Reasoning Using a Ghost Variable

The interesting challenge remaining is to consider the impact of acknowledging that the atomicity assumption in Sect. 3.2 about the mutator is unrealistic. Splitting the atomic assignment (on the two shared variables $hp, marked$) in Sect. 3.5 is delicate. The reader would be excused for thinking that performing

---

[4] The all-important non-atomic case for *Redirect* is covered in Sect. 4.

the marking first would be safe but there is a counter example in the case where the *Collector* executes *Unmark* between the two steps of such an erroneous *Redirect*.

For the general lessons that this example illustrates, the interesting conclusion is that there appears to be no way to maintain full compositionality (i.e. expressing all we need to know about the mutator) with standard rely conditions.

Lesson 8. *Ghost variables can undermine compositionality/separation (cf. Lesson 3). Where they appear to be essential, it would be useful to have a test for this fact.*

The difficulty can be understood by considering the following scenario. *Redirect* can, at the point that it changes $hp(a, i)$ to point to some address b, go to sleep before performing the marking on which the *Collector* of Sect. 3.4 relies. There is in fact no danger since, even if b was not marked, there must be another path to b (see *pre-Redirect* in Sect. 3.5) and the *Collector* should perform the marking when that path (say $hp(c, j)$) is encountered. If, however, that $hp(c, j)$ could be destroyed before the *Collector* gets to c, an incomplete marking would result that could cause live addresses to be collected as garbage. What saves the day is that the *Mutator* cannot make another change without waking up and marking b.[5]

It is precisely this three step argument that pinpoints the limitation of using two state relations in R/G reasoning.

Here, despite all of the reservations expressed in Sect. 1, a ghost variable is employed to complete the justification of the design; thus the state Σ_2 is extended with a ghost variable $tbm: [Addr]$ that can record an address as "to be marked".

Lesson 9. *Lesson 8 asks for a test that would justify the use of ghost variable; the need for a "3-state" justification is such a test.*[6]

4.1   Parallel Decomposition

The rely condition used in Sect. 3.3 is replaced for the non-atomic interference from the mutator by:

$$rely\text{-}Collector : \Sigma_2 \times \Sigma_2 \to \mathbb{B}$$

$$rely\text{-}Collector(\sigma, \sigma') \quad \triangle$$
$$\quad free' \subseteq free \land marked \subseteq marked' \land$$
$$\quad (\forall (a, i) \in hp \cdot$$
$$\quad\quad hp'(a, i) \neq hp(a, i) \land hp'(a, i) \in Addr \Rightarrow$$
$$\quad\quad\quad\quad tbm' = hp'(a, i) \lor hp'(a, i) \in marked') \land$$
$$\quad (tbm \neq \mathbf{nil} \land tbm' \neq tbm \Rightarrow tbm \in marked' \land tbm' = \mathbf{nil})$$

---

[5] This line of argument rules out multiple *Mutator* threads.

[6] A planned journal version of this paper will investigate other options. It is also hoped to compare the approaches with RGITL [STER11].

Here again, the PO of the parallel introduction rule is trivial to discharge because the guarantee condition of the *Mutator* is identical with the rely condition of the *Collector*.

4.2 Developing *Mutator* Code

As indicated in Sect. 1, it still remains to establish that the design of each component satisfies its specification. Looking first at the non-atomic *Mutator* argument, the only real challenge is:[7]

> *Redirect* $(a: Addr, i: \mathbb{N}_1, b: Addr)$
> **ext wr** $hp, marked$
> **pre** $\{a, b\} \subseteq reach(roots, hp) \land i \in \mathbf{inds}\, hp(a)$
> **rely** $hp' = hp$
> **guar** $rely\text{-}Collector$
> **post** $hp' = hp \dagger \{(a, i) \mapsto b\} \land b \in marked'$

Redirect can satisfy this specification by executing the following two atomic steps (but the atomic brackets only surround one shared variable in each case):

> $< hp(a), tbm := hp(a) \dagger \{i \mapsto b\}, b >;$
> $< marked, tbm := marked \cup \{b\}, \mathbf{nil} >$

This not only guarantees *rely-Collector*, but also preserves the following invariant:

$$tbm \neq \mathbf{nil} \Rightarrow$$
$$\exists \{(a, i), (b, j)\} \subseteq \mathbf{dom}\, hp \cdot (a, i) \neq (b, j) \land hp(a, i) = hp(b, j) = tbm$$

4.3 Developing *Collector* Code

Turning to the development of *Collector*, code must be developed relying only on the above interface. The only challenge is the mark phase whose specification is:

> *Mark*
> **ext wr** $marked$
>   **rd** $roots, hp, free$
> **pre true**
> **rely** $rely\text{-}Collector$
> **guar** $marked \subseteq marked'$
> **post** $reach(marked, hp') \subseteq marked'$

The code for *Mark* is still that in Fig. 1—under interference, the post condition of *Propagate* has to be further weakened (from Sect. 3.4) to reflect that,

---

[7] When removing a pointer, no *tbm* is set—see $Zap(a, i)$ in Sect. 3.5; also no *tbm* is needed in the *Malloc* case because $inv\text{-}\Sigma_2$ ensures that any *free* address is marked.

if there is an address in *tbm*, its reach might not yet be marked. Importantly, if the marking is not yet complete, there must have been some node marked in the current iteration:

Propagate

ext wr *marked*
    rd *hp*

pre true

rely *rely-Collector*
guar $marked \subseteq marked'$
post $\bigcup\{\textbf{elems }hp'(a)\cap Addr \mid a \in marked\} \subseteq (marked' \cup (\{tbm'\} \cap Addr)) \land$
    $(marked \subset marked' \lor reach(roots, hp') \subseteq marked')$

Notice that *post-Propagate* implies there can be at most one address whose marking is problematic; this fact must be established using the final conjunct of *rely-Collector*.

The correctness of this loop is interesting—it follows the structure of that in Sect. 3.4 using a *to-end* relation and, in fact, the relation is still:

$$to\text{-}end(\sigma, \sigma') \quad \triangleq \quad reach(marked, hp') \subseteq marked'$$

The PO is now:

$$post\text{-}Propagate(\sigma, \sigma') \land \sigma.marked \subset \sigma'.marked \land to\text{-}end(\sigma', \sigma'') \Rightarrow$$
$$to\text{-}end(\sigma, \sigma'')$$

In comparison with the PO in Sect. 3.4, the difficult case is where $tbm' \neq \textbf{nil}$ (in the converse case the earlier proof would suffice). What needs to be shown is that the stray address in tbm' will be marked. The lemma in Sect. 4.2 ensures there is another path to the address in tbm'; this will be marked if there are further iterations of *Propagate* and these are ensured by the lemma at the end of Sect. 2.2 which, combined with the second conjunct of *post-Propagate*, avoids premature termination.

The code in Fig. 1 shows how *Propagate* uses *Mark-kids* in the inner loop.

Mark-kids $(x : Addr)$

ext wr *marked*
    rd *hp*

pre true

rely *rely-Collector*
guar $marked \subseteq marked'$
post $(\textbf{elems }hp'(x) \cap Addr) \subseteq (marked' \cup (\{tbm'\} \cap Addr)))$

Again, the POs are as for the atomic case, but with:

$$so\text{-}far(\sigma, \sigma') \quad \triangleq$$
$$\bigcup \{\textbf{elems } hp(a) \cap Addr \mid a \in (marked \cap consid')\} \subseteq$$
$$(marked' \cup (\{tbm'\} \cap Addr))$$

5   Lower Limit of GC

Sections 3.2 and 4 address (under different assumptions) the lower bound for marking and thus ensure that no active addresses are treated as garbage. Unless an upper bound for marking is established however, *Mark* could mark every address and no garbage would be collected. The R/G technique of splitting, for example, a set equality into two containments often results in such a residual PO.

Addresses that were garbage in the initial state ($Addr - (reach(roots, hp) \cup free)$) should not be marked (thus any garbage will be collected at the latest after two passes of *Collect*). A predicate "no marked old garbage" can be used for the upper bound of marking:

$$no\text{-}mog : Addr\text{-}\textbf{set} \times Addr\text{-}\textbf{set} \times Heap \times Addr\text{-}\textbf{set} \to \mathbb{B}$$

$$no\text{-}mog(r, f, h, m) \quad \triangleq \quad (Addr - (reach(r, h) \cup f)) \cap m = \{\}$$

The intuitive argument is simple: the *Collector* and *Mutator* only mark things reachable from *roots* and the *Mutator* can change the reachable graph but only links to addresses (from *free* or previously reachable from *roots*) that were never "garbage".

6   Related Work

The nine lessons are the real message of this paper; the (garbage collection) example illustrates and hopefully clarifies the issues for the reader. The current authors believe that examples are essential to drive research.

Many papers exist on garbage collection algorithms, where the verification is usually performed at the code level, e.g. [GGH07,HL10], which both use the PVS theorem prover. In [TSBR08], a copying collector with no concurrency is verified using separation logic. An Owicki-Gries proof of Ben-Ari's algorithm is given in [NE00]; while this examines multiple mutators, the method results in a very large number of POs. The proof of Ben-Ari's algorithm in [vdS87], also using Owicki-Gries, reasons directly at the code level without using abstraction.

Perhaps the closest to our approach is [PPS10], which presents a refinement-based approach for deriving various garbage collection algorithms from an abstract specification. This approach is very interesting and for future work it is worth exploring how the approach given here could be used to verify a similar family of algorithms. It would appear that the rely-guarantee method produces a more compositional proof, as the approach in [PPS10] requires more integrated reasoning about the actions of the Mutator and Collector. Similarly, in [VYB06], a series of transformations is used to derive various concurrent garbage collection algorithms from an initial algorithm.

Acknowledgements. We have benefited from productive discussions with researchers including José Nuno Oliviera and attendees at the January 2017 *Northern Concurrency Working Group* held at Teesside University. In particular, Simon Doherty pointed out that GC is a nasty challenge for any compositional approach because the mutator/collector were clearly thought out together; this is true but looking at an example at the fringe of R/G expressivity has informed the notion of compositional development.

Our colleagues in Newcastle, Leo Freitas and Diego Machado Dias are currently formalising proofs of the lemmas and POs using Isabelle.

The authors gratefully acknowledge funding for their research from EPSRC grant *Taming Concurrency*.

References

[BA84] Ben-Ari, M.: Algorithms for on-the-fly garbage collection. ACM Trans. Program. Lang. Syst. **6**(3), 333–344 (1984)

[BA10] Bornat, R., Amjad, H.: Inter-process buffers in separation logic with rely-guarantee. Formal Aspects Comput. **22**(6), 735–772 (2010)

[BA13] Bornat, R., Amjad, H.: Explanation of two non-blocking shared-variable communication algorithms. Formal Aspects Comput. **25**(6), 893–931 (2013)

[BvW98] Back, R.-J.R., von Wright, J.: Refinement Calculus: A Systematic Introduction. Springer, New York (1998)

[CJ00] Collette, P., Jones, C.B.: Enhancing the tractability of rely/guarantee specifications in the development of interfering operations. In: Plotkin, G., Stirling, C., Tofte, M. (eds.) Proof, Language and Interaction, Chap. 10, pp. 277–307. MIT Press (2000)

[CJ07] Coleman, J.W., Jones, C.B.: A structural proof of the soundness of rely/guarantee rules. J. Logic Comput. **17**(4), 807–841 (2007)

[Col08] Coleman, J.W.: Constructing a Tractable Reasoning Framework upon a Fine-Grained Structural Operational Semantics. Ph.D. thesis, Newcastle University, January 2008

[DFPV09] Dodds, M., Feng, X., Parkinson, M., Vafeiadis, V.: Deny-guarantee reasoning. In: Castagna, G. (ed.) ESOP 2009. LNCS, vol. 5502, pp. 363–377. Springer, Heidelberg (2009). doi:10.1007/978-3-642-00590-9_26

[Din00] Jürgen Dingel. Systematic Parallel Programming. Ph.D. thesis, Carnegie Mellon University (2000). CMU-CS-99-172

[DYDG+10] Dinsdale-Young, T., Dodds, M., Gardner, P., Parkinson, M.J., Vafeiadis, V.: Concurrent abstract predicates. In: D'Hondt, T. (ed.) ECOOP 2010. LNCS, vol. 6183, pp. 504–528. Springer, Heidelberg (2010). doi:10.1007/978-3-642-14107-2_24

[FFS07] Feng, X., Ferreira, R., Shao, Z.: On the relationship between concurrent separation logic and assume-guarantee reasoning. In: Nicola, R. (ed.) ESOP 2007. LNCS, vol. 4421, pp. 173–188. Springer, Heidelberg (2007). doi:10.1007/978-3-540-71316-6_13

[GGH07]   Gao, H., Groote, J.F., Hesselink, W.H.: Lock-free parallel and concurrent
          garbage collection by mark & sweep. Sci. Comput. Program. **64**(3), 341–
          374 (2007)
[HBDJ13] Hayes, I.J., Burns, A., Dongol, B., Jones, C.B.: Comparing degrees of
          non-determinism in expression evaluation. Comput. J. **56**(6), 741–755
          (2013)
[HJC14]   Hayes, I.J., Jones, C.B., Colvin, R.J.: Laws and semantics for rely-
          guarantee refinement. Technical Report CS-TR-1425, Newcastle Univer-
          sity, July 2014
[HL10]   Hesselink, W.H., Lali, M.I.: Simple concurrent garbage collection almost
          without synchronization. Formal Methods Syst. Des. **36**(2), 148–166
          (2010)
[Hoa72]   Hoare, C.A.R.: Towards a theory of parallel programming. In: Operating
          System Techniques, pp. 61–71. Academic Press (1972)
[JH16]   Jones, C.B., Hayes, I.J.: Possible values: exploring a concept for concur-
          rency. J. Logical Algebraic Methods Program. **85**(5, Part 2), 972–984
          (2016). Articles dedicated to Prof. J. N. Oliveira on the occasion of his
          60th birthday
[JHC15]   Jones, C.B., Hayes, I.J., Colvin, R.J.: Balancing expressiveness in for-
          mal approaches to concurrency. Formal Aspects Comput. **27**(3), 475–497
          (2015)
[Jon81]   Jones,C.B.: Development Methods for Computer Programs including
          a Notion of Interference. Ph.D. thesis, Oxford University, June 1981.
          Oxford University Computing Laboratory (now Computer Science) Tech-
          nical Monograph PRG-25
[Jon90]   Jones, C.B.: Systematic Software Development using VDM, 2nd edn.
          Prentice Hall International, Englewood Cliffs (1990)
[Jon96]   Jones, C.B.: Accommodating interference in the formal design of concur-
          rent object-based programs. Formal Methods Syst. Des. **8**(2), 105–122
          (1996)
[JP11]   Jones, C.B., Pierce, K.G.: Elucidating concurrent algorithms via layers
          of abstraction and reification. Formal Aspects Comput. **23**(3), 289–306
          (2011)
[JY15]   Jones, C.B., Yatapanage, N.: Reasoning about separation using abstrac-
          tion and reification. In: Calinescu, R., Rumpe, B. (eds.) SEFM 2015.
          LNCS, vol. 9276, pp. 3–19. Springer, Cham (2015). doi:10.1007/
          978-3-319-22969-0_1
[Mor90]   Morgan, C.: Programming from Specifications. Prentice-Hall, New York
          (1990)
[NE00]   Nieto, L.P., Esparza, J.: Verifying single and multi-mutator garbage col-
          lectors with owicki-gries in Isabelle/HOL. In: Nielsen, M., Rovan, B.
          (eds.) MFCS 2000. LNCS, vol. 1893, pp. 619–628. Springer, Heidelberg
          (2000). doi:10.1007/3-540-44612-5_57
[OG76]   Owicki, S.S., Gries, D.: An axiomatic proof technique for parallel pro-
          grams I. Acta Informatica **6**(4), 319–340 (1976)
[O'H07]   O'Hearn, P.W.: Resources, concurrency and local reasoning. Theoret.
          Comput. Sci. **375**(1–3), 271–307 (2007)
[Owi75]   Owicki, S.: Axiomatic Proof Techniques for Parallel Programs. Ph.D.
          thesis, Department of Computer Science, Cornell University (1975)

[Par10] Parkinson, M.: The next 700 separation logics. In: Leavens, G.T., O'Hearn, P., Rajamani, S.K. (eds.) VSTTE 2010. LNCS, vol. 6217, pp. 169–182. Springer, Heidelberg (2010). doi:10.1007/978-3-642-15057-9_12

[Pie09] Pierce, K.: Enhancing the Useability of Rely-Guaranteee Conditions for Atomicity Refinement. Ph.D. thesis, Newcastle University (2009)

[PPS10] Pavlovic, D., Pepper, P., Smith, D.R.: Formal derivation of concurrent garbage collectors. In: Bolduc, C., Desharnais, J., Ktari, B. (eds.) MPC 2010. LNCS, vol. 6120, pp. 353–376. Springer, Heidelberg (2010). doi:10. 1007/978-3-642-13321-3_20

[Pre01] Nieto, L.P.: Verification of parallel programs with the owicki-gries and rely-guarantee methods in Isabelle/HOL. Ph.D. thesis, Institut für Informatic der Technischen Universitaet München (2001)

[STER11] Schellhorn, G., Tofan, B., Ernst, G., Reif, W.: Interleaved programs and rely-guarantee reasoning with ITL. In: TIME, pp. 99–106 (2011)

[Stø90] Stølen, K.: Development of Parallel Programs on Shared Data-Structures. Ph.D. thesis, Manchester University (1990). UMCS-91-1-1

[TSBR08] Torp-Smith, N., Birkedal, L., Reynolds, J.C.: Local reasoning about a copying garbage collector. ToPLaS 30, 1–58 (2008)

[Vaf07] Vafeiadis, V.: Modular fine-grained concurrency verification. Ph.D. thesis, University of Cambridge (2007)

[vdS87] van de Jan, L.A.: "Algorithms for on-the-fly garbage collection" revisited. Inf. Process. Lett. 24(4), 211–216 (1987)

[VYB06] Vechev, M.T., Yahav, E., Bacon, D.F.: Correctness-preserving derivation of concurrent garbage collection algorithms. In: PLDI, pp. 341–353 (2006)

[WDP10] Wickerson, J., Dodds, M., Parkinson, M.: Explicit stabilisation for modular rely-guarantee reasoning. In: Gordon, A.D. (ed.) ESOP 2010. LNCS, vol. 6012, pp. 610–629. Springer, Heidelberg (2010). doi:10.1007/ 978-3-642-11957-6_32

[Xu92] Xu, Q.: A Theory of State-based Parallel Programming. Ph.D. thesis, Oxford University (1992)

Probabilistic and Statistical Analysis

Polynomial-Time Alternating Probabilistic Bisimulation for Interval MDPs

Vahid Hashemi[1]($^{(\boxtimes)}$), Andrea Turrini[2], Ernst Moritz Hahn[1,2],
Holger Hermanns[1], and Khaled Elbassioni[3]

[1] Saarland University, Saarland Informatics Campus, Saarbrücken, Germany
hashemi@depend.uni-saarland.de
[2] State Key Laboratory of Computer Science, ISCAS, Beijing, China
[3] Masdar Institute of Science and Technology, Abu Dhabi, UAE

Abstract. Interval Markov decision processes (IMDPs) extend classical MDPs by allowing intervals to be used as transition probabilities. They provide a powerful modelling tool for probabilistic systems with an additional variation or uncertainty that relaxes the need of knowing the exact transition probabilities, which are usually difficult to get from real systems. In this paper, we discuss a notion of alternating probabilistic bisimulation to reduce the size of the IMDPs while preserving the probabilistic CTL properties it satisfies from both computational complexity and compositional reasoning perspectives. Our alternating probabilistic bisimulation stands on the competitive way of resolving the IMDP nondeterminism which in turn finds applications in the settings of the controller (parameter) synthesis for uncertain (parallel) probabilistic systems. By using the theory of linear programming, we improve the complexity of computing the bisimulation from the previously known EXPTIME to PTIME. Moreover, we show that the bisimulation for IMDPs is a congruence with respect to two facets of parallelism, namely synchronous product and interleaving. We finally demonstrate the practical effectiveness of our proposed approaches by applying them on several case studies using a prototypical tool.

1 Introduction

Markov Decision Processes (*MDPs*) are a widely and commonly used mathematical abstraction that permits to study properties of real world systems in a rigorous way. The actual system is represented by means of a model subsuming the states the system can be in and the transitions representing how the system evolves from one state to another; the actual properties are encoded as logical formulas that are then verified against the model.

This work is supported by the ERC Advanced Investigators Grant 695614 (POWVER), by the CAS/SAFEA International Partnership Program for Creative Research Teams, by the National Natural Science Foundation of China (Grants No. 61550110506 and 61650410658), by the Chinese Academy of Sciences Fellowship for International Young Scientists, and by the CDZ project CAP (GZ 1023).

© Springer International Publishing AG 2017
K.G. Larsen et al. (Eds.): SETTA 2017, LNCS 10606, pp. 25–41, 2017.
https://doi.org/10.1007/978-3-319-69483-2_2

MDPs are suitable for modelling two core aspects of the behavior of the real world systems: *nondeterminism* and *probability*. A nondeterministic behavior can be introduced to model a behavior of the system that is just partially known (like receiving an asynchronous message, of which it is known it can be received in the current state but no information is available so to quantify its likelihood) or to leave implementation details open. A probabilistic behavior occurs whenever the successor state of the system is not uniquely determined by the current system and the performed action, but depends on a random choice; such a choice can be due to the design of the system, as it is required by the implementation of a distributed consensus protocol with faulty processes [3,14], or by physical properties that need to be taken into account, like transmission errors.

Finding the exact probability values for the transitions is sometimes a difficult task: while probabilities introduced by design can be well known, probabilities modelling physical properties are usually estimated by observing the actual system. This means that the resulting *MDP* is a more or less appropriate abstraction of the real system, depending on how close the estimated probability values are to the actual values; as a consequence, the actual properties of the real system are more or less reflected by the satisfaction of the formulas by the model.

Interval Markov Decision Processes (IMDPs) extend the classical *MDPs* by including uncertainty over the transition probabilities. Instead of a single value for the probability of reaching a specific successor by taking a transition, *IMDPs* allow ranges of possible probability values given as closed intervals of the reals. Thereby, *IMDPs* provide a powerful modelling tool for probabilistic systems with an additional variation or uncertainty concerning the knowledge of exact transition probabilities. They are especially useful to represent realistic stochastic systems that, for instance, evolve in unknown environments with bounded behavior or do not preserve the Markov property.

Since their introduction (under the name of bounded-parameter *MDPs*) [16], *IMDPs* have been receiving a lot of attention in the formal verification community. They are particularly viewed as the appropriate abstraction model for uncertain systems with large state spaces, including continuous dynamical systems, for the purpose of analysis, verification, and control synthesis. Several model checking and control synthesis techniques have been developed [37,38,43] causing a boost in the applications of *IMDPs*, ranging from verification of continuous stochastic systems (e.g., [30]) to robust strategy synthesis for robotic systems (e.g., [32–34,43]).

Bisimulation minimisation is a well-known technique that has been successfully used to reduce the size of a system while preserving the properties it satisfies [5,8,9,23,27]; this helps the task of the property solver, since it has to work on a smaller system. Compositional minimisation permits to minimise the single components of the system before combining them, thus making the task of the minimiser easier and extending its applicability to larger systems. In this paper, we show that this approach is suitable also for *IMDPs*. The contributions of the paper are as follows.

– We define alternating probabilistic bisimulations to compress the *IMDP* model size with respect to the controller synthesis semantics while preserving probabilistic CTL property satisfaction. We show that the compressed models can be computed in polynomial time.
– From the perspective of compositional reasoning, we show that alternating probabilistic bisimulations for *IMDPs* are congruences with respect to two facets of parallelism, namely synchronous product and interleaving.
– We show promising results on a variety of case studies, obtained by proto-typical implementations of all algorithms.

Related work. Related work can be grouped into three categories: uncertain Markov model formalisms, bisimulation minimization, and compositional minimization.

Firstly, from the modelling viewpoint, various probabilistic modelling formalisms with uncertain transitions are studied in the literature. Interval Markov Chains (*IMCs*) [25,28] or abstract Markov chains [13] extend standard discrete-time Markov Chains (*MCs*) with interval uncertainties. They do not feature the non-deterministic choices of transitions. Uncertain *MDPs* [38] allow more general sets of distributions to be associated with each transition, not only those described by intervals. They usually are restricted to *rectangular uncertainty sets* requiring that the uncertainty is linear and independent for any two transitions of any two states. Parametric *MDPs* [17], to the contrary, allow such dependencies as every probability is described as a rational function of a finite set of global parameters. *IMDPs* extend *IMCs* by inclusion of nondeterminism and are a subset of uncertain *MDPs* and parametric *MDPs*.

Secondly, as regards to the bisimulation minimization for uncertain or parametric probabilistic models, works in [18,20,21] explored the computational complexity and approximability of deciding probabilistic bisimulation for *IMDPs* with respect to the cooperative resolution of nondeterminism. In this work, we show that *IMDPs* can be minimized efficiently with respect to the competitive resolution of nondeterminism.

Lastly, from the viewpoint of compositional minimization, *IMCs* [25] and abstract Probabilistic Automata (*PA*) [10,11] serve as specification theories for *MC* and *PA*, featuring satisfaction relation and various refinement relations. In [22], the authors discuss the key ingredients to build up the operations of parallel composition for composing *IMDP* components at run-time. Our paper follows this spirit for alternating probabilistic bisimulation on *IMDPs*.

Structure of the paper. We start with necessary preliminaries in Sect. 2. In Sect. 3, we give the definitions of alternating probabilistic bisimulation for interval *MDP* and discuss their properties. A polynomial time decision algorithm to decide alternating probabilistic bisimulation for *IMDPs* and also compositional reasoning are discussed in Sects. 4 and 5, respectively. In Sect. 6, we demonstrate our approach on some case studies and present promising experimental results. Finally, in Sect. 7 we conclude the paper.

2   Mathematical Preliminaries

For a set X, denote by $\mathrm{Disc}(X)$ the set of discrete probability distributions over X. Intuitively, a discrete probability distribution ρ is a function $\rho\colon X \to \mathbb{R}_{\geq 0}$ such that $\sum_{x\in X} \rho(x) = 1$; for $X' \subseteq X$, we write $\rho(X')$ for $\sum_{x\in X'} \rho(x)$. Given $\rho \in \mathrm{Disc}(X)$, we denote by $\mathrm{Supp}(\rho)$ the set $\{\, x \in X \mid \rho(x) > 0 \,\}$ and by δ_x, where $x \in X$, the *Dirac* distribution such that $\delta_x(y) = 1$ for $y = x$, 0 otherwise. For a probability distribution ρ, we also write $\rho = \{\, (x, p_x) \mid x \in X \,\}$ where p_x is the probability of x.

The lifting $\mathcal{L}(\mathcal{R})$ [31] of a relation $\mathcal{R} \subseteq X \times Y$ is defined as follows: for $\rho_X \in \mathrm{Disc}(X)$ and $\rho_Y \in \mathrm{Disc}(Y)$, $\rho_X\ \mathcal{L}(\mathcal{R})\ \rho_Y$ holds if there exists a *weighting function* $w\colon X \times Y \to [0,1]$ such that (1) $w(x,y) > 0$ implies $x\ \mathcal{R}\ y$, (2) $\sum_{y\in Y} w(x,y) = \rho_X(x)$, and (3) $\sum_{x\in X} w(x,y) = \rho_Y(y)$. When $\mathcal{R}$ is an equivalence relation on X, $\rho_1\ \mathcal{L}(\mathcal{R})\ \rho_2$ holds if for each $\mathcal{C} \in X/\mathcal{R}$, $\rho_1(\mathcal{C}) = \rho_2(\mathcal{C})$ where $X/\mathcal{R} = \{\, [x]_\mathcal{R} \mid x \in X \,\}$ and $[x]_\mathcal{R} = \{\, y \in X \mid y\ \mathcal{R}\ x \,\}$.

For a vector $\boldsymbol{x} \in \mathbb{R}^n$ we denote by $\boldsymbol{x}_i$, its i-th component, and we call $\boldsymbol{x}$ a *weight vector* if $\boldsymbol{x} \in \mathbb{R}^n_{\geq 0}$ and $\sum_{i=1}^n \boldsymbol{x}_i = 1$. Given two vectors $\boldsymbol{x}, \boldsymbol{y} \in \mathbb{R}^n$, their Euclidean inner product $\boldsymbol{x} \cdot \boldsymbol{y}$ is defined as $\boldsymbol{x} \cdot \boldsymbol{y} = \boldsymbol{x}^T \boldsymbol{y} = \sum_{i=1}^n \boldsymbol{x}_i \cdot \boldsymbol{y}_i$. We write $\boldsymbol{x} \leq \boldsymbol{y}$ if $\boldsymbol{x}_i \leq \boldsymbol{y}_i$ for each $1 \leq i \leq n$ and we denote by $\boldsymbol{1} \in \mathbb{R}^n$ the vector such that $\boldsymbol{1}_i = 1$ for each $1 \leq i \leq n$. For a set of vectors $S = \{\boldsymbol{s}^1, \ldots, \boldsymbol{s}^m\} \subseteq \mathbb{R}^n$, we say that $\boldsymbol{s}$ is a *convex combination* of elements of S, if $\boldsymbol{s} = \sum_{i=1}^m \boldsymbol{w}_i \cdot \boldsymbol{s}^i$ for some weight vector $\boldsymbol{w} \in \mathbb{R}^m_{\geq 0}$. For a given set $P \subseteq \mathbb{R}^n$, we denote by $\mathrm{conv}\, P$ the convex hull of P and by $\mathrm{Ext}(P)$ the set of extreme points of P. If P is a polytope in $\mathbb{R}^n$ then for each $1 \leq i \leq n$, the projection $\mathrm{proj}_{\mathbf{e^i}} P$ on the i-th dimension of P is defined as $\mathrm{proj}_{\mathbf{e^i}} P = [\min_i P, \max_i P]$ where $\mathbf{e^i} \in \mathbb{R}^n$ is such that $\mathbf{e}_i^{\mathbf{i}} = 1$ and $\mathbf{e}_j^{\mathbf{i}} = 0$ for each $j \neq i$, $\min_i P = \min\{\, \boldsymbol{x}_i \mid \boldsymbol{x} \in P \,\}$, and $\max_i P = \max\{\, \boldsymbol{x}_i \mid \boldsymbol{x} \in P \,\}$.

2.1   Interval Markov Decision Processes

We now define *Interval Markov Decision Processes* (IMDPs) as an extension of MDPs, which allows for the inclusion of transition probability uncertainties as *intervals*. IMDPs belong to the family of uncertain MDPs and allow to describe a set of MDPs with identical (graph) structures that differ in distributions associated with transitions.

Definition 1 (IMDPs). *An Interval Markov Decision Process (IMDP) $\mathcal{M}$ is a tuple $(S, \bar{s}, \mathcal{A}, \mathsf{AP}, L, I)$, where S is a finite set of states, $\bar{s} \in S$ is the initial state, $\mathcal{A}$ is a finite set of actions, AP is a finite set of atomic propositions, $L\colon S \to 2^{\mathsf{AP}}$ is a labelling function, and $I\colon S \times \mathcal{A} \times S \to \mathbb{I} \cup \{[0,0]\}$ is a total interval transition probability function with $\mathbb{I} = \{\, [l, u] \subseteq \mathbb{R} \mid 0 < l \leq u \leq 1 \,\}$.*

Given $s \in S$ and $a \in \mathcal{A}$, we call $\mathfrak{h}^a_s \in \mathrm{Disc}(S)$ a *feasible distribution* reachable from s by a, denoted by $s \xrightarrow{a} \mathfrak{h}^a_s$, if, for each state $s' \in S$, we have $\mathfrak{h}^a_s(s') \in I(s, a, s')$. We denote the set of feasible distributions for state s and action a by $\mathcal{H}^a_s$, i.e., $\mathcal{H}^a_s = \{\, \mathfrak{h}^a_s \in \mathrm{Disc}(S) \mid s \xrightarrow{a} \mathfrak{h}^a_s \,\}$ and we denote the set of available actions at state $s \in S$ by $\mathcal{A}(s)$, i.e., $\mathcal{A}(s) = \{\, a \in \mathcal{A} \mid \mathcal{H}_s^a \neq \emptyset \,\}$. We assume that $\mathcal{A}(s) \neq \emptyset$ for all $s \in S$.

We define the *size* of $\mathcal{M}$, written $|\mathcal{M}|$, as the number of non-zero entries of I, i.e., $|\mathcal{M}| = |\{\, (s, a, s', \iota) \in S \times \mathcal{A} \times S \times \mathbb{I} \mid I(s, a, s') = \iota \,\}| \in \mathcal{O}(|S|^2 \cdot |\mathcal{A}|)$.

A *path* ξ in $\mathcal{M}$ is a finite or infinite sequence of states $\xi = s_0 s_1 \ldots$ such that for each $i \geq 0$ there exists $a_i \in \mathcal{A}(s_i)$ such that $I(s_i, a_i, s_{i+1}) \in \mathbb{I}$. The i-th state along the path ξ is denoted by $\xi[i]$ and, if the path is finite, we denote by $last(\xi)$ its last state. The sets of all finite and infinite paths in $\mathcal{M}$ are denoted by $Paths^*$ and $Paths$, respectively.

The nondeterministic choices between available actions and feasible distributions present in an *IMDP* are resolved by strategies and natures, respectively.

Definition 2 (Strategy and Nature in IMDPs). *Given an IMDP $\mathcal{M}$, a strategy is a function $\sigma \colon Paths^* \to \mathrm{Disc}(\mathcal{A})$ such that for each path $\xi \in Paths^*$, $\sigma(\xi) \in \mathrm{Disc}(\mathcal{A}(last(\xi))$. A nature is a function $\pi \colon Paths^* \times \mathcal{A} \to \mathrm{Disc}(S)$ such that for each path $\xi \in Paths^*$ and action $a \in \mathcal{A}(s)$, $\pi(\xi, a) \in \mathcal{H}_s^a$ where $s = last(\xi)$.*

The sets of all strategies and all natures are denoted by Σ and Π, respectively.

Given a finite path ξ of an *IMDP*, a strategy σ, and a nature π, the system evolution proceeds as follows: let $s = last(\xi)$. First, an action $a \in \mathcal{A}(s)$ is chosen probabilistically by σ. Then, π resolves the uncertainties and chooses one feasible distribution $\mathfrak{h}_s^a \in \mathcal{H}_s^a$. Finally, the next state s' is chosen according to the distribution $\mathfrak{h}_s^a$, and the path ξ is extended by s'.

A strategy σ and a nature π induce a probability measure over paths as follows. The basic measurable events are the cylinder sets of finite paths, where the *cylinder set* of a finite path ξ is the set $Cyl_\xi = \{\, \xi' \in Paths \mid \xi$ is a prefix of $\xi' \,\}$. The probability $\Pr_{\mathcal{M}}^{\sigma,\pi}$ of a state s' is defined to be $\Pr_{\mathcal{M}}^{\sigma,\pi}[Cyl_{s'}] = \delta_{\bar{s}}(s')$ and the probability $\Pr_{\mathcal{M}}^{\sigma,\pi}[Cyl_{\xi s'}]$ of traversing a finite path $\xi s'$ is defined to be

$$\Pr_{\mathcal{M}}^{\sigma,\pi}[Cyl_{\xi s'}] = \Pr_{\mathcal{M}}^{\sigma,\pi}[Cyl_\xi] \cdot \sum_{a \in \mathcal{A}(last(\xi))} \sigma(\xi)(a) \cdot \pi(\xi, a)(s').$$

Standard measure theoretical arguments ensure that $\Pr_{\mathcal{M}}^{\sigma,\pi}$ extends uniquely to the σ-field generated by cylinder sets.

As an example of *IMDPs*, consider the one depicted in Fig. 1. The set of states is $S = \{s, t, u\}$ with s being the initial one; the set of actions is $\mathcal{A} = \{a, b\}$ while the set of atomic propositions assigned to each state by the labelling function L is represented by the letters in curly brackets near each state. Finally, the transition probability intervals are $I(s, a, t) = [\frac{1}{3}, \frac{2}{3}]$, $I(s, a, u) = [\frac{1}{10}, 1]$, $I(s, b, t) = [\frac{2}{5}, \frac{3}{5}]$, $I(s, b, u) = [\frac{1}{4}, \frac{2}{3}]$, $I(t, a, t) = I(u, b, u) = [1, 1]$, and $I(t, b, t) = I(u, a, u) = [0, 0]$.

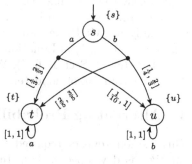

Fig. 1. An example of *IMDP*.

2.2 Probabilistic Computation Tree Logic (PCTL)

There are various ways how to describe properties of *IMDPs*. Here we focus on *probabilistic CTL* (PCTL) [19]. The syntax of PCTL state formulas φ and PCTL path formulas ψ is given by:

$$\varphi := a \mid \neg\varphi \mid \varphi_1 \wedge \varphi_2 \mid \mathsf{P}_{\bowtie p}(\psi)$$

$$\psi := \mathsf{X}\varphi \mid \varphi_1 \mathsf{U} \varphi_2 \mid \varphi_1 \mathsf{U}^{\leq k} \varphi_2$$

where $a \in \mathsf{AP}$, $p \in [0,1]$ is a rational constant, $\bowtie \in \{\leq, <, \geq, >\}$, and $k \in \mathbb{N}$.

The semantics of a PCTL formula with respect to *IMDPs* is very similar to the classical PCTL semantics for *MDPs*: they coincide on all formulas except for $\mathsf{P}_{\bowtie p}(\psi)$, where they may differ depending on how the nondeterminism is resolved. Formally, for the formulas they agree on, given a state s and a state formula φ, the satisfaction relation $s \models \varphi$ is defined as follows:

$$\begin{array}{lll} s \models a & & \text{if } a \in L(s); \\ s \models \neg\varphi & & \text{if it is not the case that } s \models \varphi, \text{ also written } s \not\models \varphi; \\ s \models \varphi_1 \wedge \varphi_2 & & \text{if } s \models \varphi_1 \text{ and } s \models \varphi_2. \end{array}$$

Given an infinite path $\xi = s_1 s_2 \ldots$ and a path formula ψ, the satisfaction relation $\xi \models \psi$ is defined as follows:

$$\begin{array}{lll} \xi \models \mathsf{X}\varphi & & \text{if } s_2 \models \varphi; \\ \xi \models \varphi_1 \mathsf{U}^{\leq k} \varphi_2 & & \text{if there exists } i \leq k \text{ such that } s_i \models \varphi_2 \\ & & \text{and } s_j \models \varphi_1 \text{ for every } 1 \leq j < i; \\ \xi \models \varphi_1 \mathsf{U} \varphi_2 & & \text{if there exists } k \in \mathbb{N} \text{ such that } \xi \models \varphi_1 \mathsf{U}^{\leq k} \varphi_2. \end{array}$$

Regarding the state formula $\mathsf{P}_{\bowtie p}(\psi)$, its semantics depends on the way the nondeterminism is resolved for the probabilistic operator $\mathsf{P}_{\bowtie p}(\psi)$. When quantifying both types of nondeterminism universally, the corresponding satisfaction relation $s \models \mathsf{P}_{\bowtie p}(\psi)$ is defined as follows:

$$s \models \mathsf{P}_{\bowtie p}(\psi) \text{ if } \forall \sigma \in \Sigma : \forall \pi \in \Pi : \mathrm{Pr}_s^{\sigma,\pi}\left[Paths_\psi\right] \bowtie p \qquad (\forall)$$

where $Paths_\psi = \{\xi \in Paths \mid \xi \models \psi\}$ denotes the set of infinite paths satisfying ψ. It is easy to show that the set $Paths_\psi$ is measurable for any path formula ψ, hence its probability can be computed and compared with p. When the *IMDP* is actually an *MDP*, i.e., all intervals are single values, then the satisfaction relation $s \models \mathsf{P}_{\bowtie p}(\psi)$ in Equation $(\forall)$ coincides with the corresponding definition for *MDPs* (cf. [2, Sect. 10.6.2]). We explain later how the semantics differs for a different resolution of nondeterminism for strategy and nature.

3 Alternating Probabilistic Bisimulation for *IMDPs*

This section revisits required main results on probabilistic bisimulation for *IMDPs*, as developed in [20]. In the setting of this paper, we consider alternating probabilistic bisimulation which stems from the competitive resolution of nondeterminisms in *IMDPs*. In the competitive semantics, the strategy and nature are playing in a game *against* each other; therefore, they are resolved *competitively*. This semantics is very natural in the context of controller synthesis for systems with uncertain probabilities or in the context of parameter synthesis for parallel systems.

In this paper, in order to resolve the stochastic nondeterminism we focus on the dynamic approach [24,42], i.e., independently at each computation step as it is easier to work with algorithmically and can be seen as a relaxation of the static approach that is often intractable [4,7,12,16].

To this end, we consider the *controller synthesis* semantics to resolve the two sources of *IMDP* nondeterminisms and discuss the resultant alternating probabilistic bisimulation. Note that there is another variant of alternating probabilistic bisimulation based on the *parameter synthesis* semantics [20]. However, the alternating bisimulations relations resulting from these two semantics coincide [20, Theorem 4].

In the controller synthesis semantics, we search for a strategy σ such that for any nature π, a fixed property φ is satisfied. This corresponds to the satisfaction relation $\models_{(\exists\sigma\forall)}$ in PCTL, obtained from $\models$ by replacing the rule ($\forall$) with

$$s \models_{(\exists\sigma\forall)} \mathsf{P}_{\bowtie p}(\psi) \text{ if } \exists\sigma \in \Sigma : \forall\pi \in \Pi : \mathrm{Pr}_s^{\sigma,\pi}\left[Paths_\psi\right] \bowtie p. \qquad (\exists\sigma\forall)$$

As regards to bisimulation, the competitive setting is not common. We define a bisimulation similar to the alternating bisimulation of [1] applied to non-stochastic two-player games. For a decision $\rho \in \mathrm{Disc}(\mathcal{A})$ of σ, let $s \xrightarrow{\rho} \mu$ denote that μ is a possible successor distribution, i.e., there are decisions μ_a of π for each $a \in \mathrm{Supp}(\rho)$ such that $\mu = \sum_{a\in\mathcal{A}} \rho(a) \cdot \mu_a$.

Definition 3. *Given an IMDP $\mathcal{M}$, let $\mathcal{R} \subseteq S \times S$ be an equivalence relation. We say that $\mathcal{R}$ is an* alternating probabilistic $(\exists\sigma\forall)$-bisimulation *if for any $(s,t) \in \mathcal{R}$ we have that $L(s) = L(t)$ and for each $\rho_s \in \mathrm{Disc}(\mathcal{A}(s))$ there exists $\rho_t \in \mathrm{Disc}(\mathcal{A}(t))$ such that for each $t \xrightarrow{\rho_t} \mu_t$ there exists $s \xrightarrow{\rho_s} \mu_s$ such that $\mu_s \, L(\mathcal{R}) \, \mu_t$. We write $s \sim_{(\exists\sigma\forall)} t$ whenever $(s,t) \in \mathcal{R}$ for some alternating probabilistic $(\exists\sigma\forall)$-bisimulation $\mathcal{R}$.*

The exact alternation of quantifiers might be counter-intuitive at first sight. Note that it exactly corresponds to the situation in non-stochastic games [1]. The defined bisimulation preserves the PCTL logic with respect to the $\models_{(\exists\sigma\forall)}$ semantics.

Theorem 4. *For states $s \sim_{(\exists\sigma\forall)} t$ and any PCTL formula φ, we have $s \models_{(\exists\sigma\forall)} \varphi$ if and only if $t \models_{(\exists\sigma\forall)} \varphi$.*

As a concluding remark, it is worthwhile to note that Definition 3 can be seen as the conservative extension of probabilistic bisimulation for (state-labelled) MDPs. To see that, assume the set of uncertainty for every transition is a singleton. Since there is only one choice for the nature, the role of nature can be safely removed from the definitions.

4   A PTIME Decision Algorithm for Bisimulation Minimization

Computation of the alternating probabilistic bisimulation $\sim_{(\exists\sigma\forall)}$ for *IMDPs* follows the standard partition refinement approach [6,15,26,35]. However, the core part is finding out whether two states "violate the definition of bisimulation". This verification routine amounts to check that s and t have the same set of *strictly minimal polytopes* detailed as follows.

For $s \in S$ and $a \in \mathcal{A}(s)$, recall that $\mathcal{H}_s^a$ denotes the polytope of feasible successor distributions over *states* with respect to taking the action a in the state s. By $\mathcal{P}_{\mathcal{R}}^{s,a}$, we denote the polytope of feasible successor distributions over *equivalence classes* of $\mathcal{R}$ with respect to taking the action a in the state s. Given an interval $[l, u]$, let $\inf[l, u] = l$ and $\sup[l, u] = u$. For $\mu \in \text{Disc}(S/\mathcal{R})$ we set $\mu \in \mathcal{P}_{\mathcal{R}}^{s,a}$ if, for each $\mathcal{C} \in S/\mathcal{R}$, we have $\mu(\mathcal{C}) \in I(s, a, \mathcal{C})$ where

$$I(s, a, \mathcal{C}) = \left[\min\left\{1, \sum_{s' \in \mathcal{C}} \inf I(s, a, s')\right\}, \min\left\{1, \sum_{s' \in \mathcal{C}} \sup I(s, a, s')\right\} \right].$$

It is not difficult to see that each $\mathcal{P}_{\mathcal{R}}^{s,a}$ can be represented as an $\mathcal{H}$-polytope. To simplify our presentation, we shall fix an order over the equivalence classes in $S/\mathcal{R}$. By doing so, any distribution $\rho \in \text{Disc}(S/\mathcal{R})$ can be seen as a vector $\boldsymbol{v} \in \mathbb{R}_{\geq 0}^n$ such that $\boldsymbol{v}_i = \rho(\mathcal{C}_i)$ for each $1 \leq i \leq n$, where $n = |S/\mathcal{R}|$ and $\mathcal{C}_i$ is the i-th equivalence class in the order. For the above discussion, $\rho \in \mathcal{P}_{\mathcal{R}}^{s,a}$ if and only if $\rho(\mathcal{C}_i) \in [\boldsymbol{l}_i^{s,a}, \boldsymbol{u}_i^{s,a}]$ for any $1 \leq i \leq n$ and $\rho \in \text{Disc}(S/\mathcal{R})$, where $\boldsymbol{l}^{s,a}$ and $\boldsymbol{u}^{s,a}$ are vectors such that $\boldsymbol{l}_i^{s,a} = \min\{1, \sum_{s' \in \mathcal{C}_i} \inf I(s, a, s')\}$ and $\boldsymbol{u}_i^{s,a} = \min\{1, \sum_{s' \in \mathcal{C}_i} \sup I(s, a, s')\}$ for each $1 \leq i \leq n$. Therefore, $\mathcal{P}_{\mathcal{R}}^{s,a}$ corresponds to an $\mathcal{H}$-polytope defined by $\{\boldsymbol{x}^{s,a} \in \mathbb{R}^n \mid \boldsymbol{l}^{s,a} \leq \boldsymbol{x}^{s,a} \leq \boldsymbol{u}^{s,a}, \boldsymbol{1} \cdot \boldsymbol{x}^{s,a} = 1\}$.

Definition 5 (Strictly minimal polytopes). *Given an IMDP $\mathcal{M}$, a state s, an equivalence relation $\mathcal{R} \subseteq S \times S$, and a set $\{\mathcal{P}_{\mathcal{R}}^{s,a} \mid a \in \mathcal{A}(s)\}$ where for each $a \in \mathcal{A}(s)$, for given $\boldsymbol{l}^{s,a}, \boldsymbol{u}^{s,a} \in \mathbb{R}^n$, $\mathcal{P}_{\mathcal{R}}^{s,a}$ is the convex polytope $\mathcal{P}_{\mathcal{R}}^{s,a} = \{\boldsymbol{x}^{s,a} \in \mathbb{R}^n \mid \boldsymbol{l}^{s,a} \leq \boldsymbol{x}^{s,a} \leq \boldsymbol{u}^{s,a}, \boldsymbol{1} \cdot \boldsymbol{x}^{s,a} = 1\}$, a polytope $\mathcal{P}_{\mathcal{R}}^{s,a}$ is called strictly minimal, if for no $\rho \in \text{Disc}(\mathcal{A}(s) \setminus \{a\})$, we have $\mathcal{P}_{\mathcal{R}}^{s,\rho} \subseteq \mathcal{P}_{\mathcal{R}}^{s,a}$ where $\mathcal{P}_{\mathcal{R}}^{s,\rho}$ is defined as $\mathcal{P}_{\mathcal{R}}^{s,\rho} = \{\boldsymbol{x}^{s,\rho} \in \mathbb{R}^n \mid \boldsymbol{x}^{s,\rho} = \sum_{b \in \mathcal{A}(s) \setminus \{a\}} \rho(b) \cdot \boldsymbol{x}^{s,b} \wedge \boldsymbol{x}^{s,b} \in \mathcal{P}_{\mathcal{R}}^{s,b}\}$.*

Checking violation of a given pair of states amounts to check if the states have the same set of strictly minimal polytopes. Formally,

Lemma 6 (cf. [20]). *Given an IMDP $\mathcal{M}$ and $s, t \in S$, we have $s \sim_{(\exists\sigma\forall)} t$ if and only if $L(s) = L(t)$ and $\{\mathcal{P}_{\sim_{(\exists\sigma\forall)}}^{s,a} \mid a \in \mathcal{A} \text{ and } \mathcal{P}_{\sim_{(\exists\sigma\forall)}}^{s,a} \text{ is strictly minimal}\} = \{\mathcal{P}_{\sim_{(\exists\sigma\forall)}}^{t,a} \mid a \in \mathcal{A} \text{ and } \mathcal{P}_{\sim_{(\exists\sigma\forall)}}^{t,a} \text{ is strictly minimal}\}.$*

The expensive procedure in the analysis of the worst case time complexity of computing the coarsest alternating probabilistic bisimulation $\sim_{(\exists\sigma\forall)}$, as described in [20], is to check the strict minimality of a polytope $\mathcal{P}_{\mathcal{R}}^{s,a}$ for $a \in \mathcal{A}(s)$. This decision problem has been shown to be exponentially verifiable via a reduction to a system of linear (in)equalities in EXPTIME. In this paper, we give a poly-nomial time routine to verify the strict minimality of a polytope which in turn enables a polynomial time decision algorithm to decide $\sim_{(\exists\sigma\forall)}$. To this aim, we use the following equivalent form of the Farkas' Lemma [39].

Lemma 7. *Let $A \in \mathbb{R}^{m\times n}$, $b \in \mathbb{R}^m$ and $c \in \mathbb{R}^n$. Then, $Ax \leq b$ implies $c \cdot x \leq d$ if and only if there exists $y \in \mathbb{R}^m_{\geq 0}$ such that $A^T y = c$ and $b \cdot y \leq d$.*

This variant of Farkas' Lemma leads us to establish the main result of the paper. Formally,

Theorem 8. *Given an IMDP $\mathcal{M}$, a state $s \in S$, an equivalence relation $\mathcal{R} \subseteq S \times S$ and a set $\{\mathcal{P}^{s,a}_{\mathcal{R}} \mid a \in \mathcal{A}(s)\}$ defined as in Definition 5, checking whether for each $a \in \mathcal{A}(s)$, the polytope $\mathcal{P}^{s,a}_{\mathcal{R}}$ is strictly minimal, is in $\boldsymbol{P}$.*

Proof. Let $\mathcal{A}(s) = \{a_0, a_1, \ldots, a_m\}$, $n = |S/\mathcal{R}|$, and $P_i = \mathcal{P}^{s,a_i}_{\mathcal{R}}$ for $0 \leq i \leq m$. We describe the verification routine to check the strict minimality of P_0; the same routine applies to the other polytopes. We consider the converse of the strict min-imality problem which asks to decide whether there exist $\lambda_1, \lambda_2, \ldots, \lambda_m \in \mathbb{R}_{\geq 0}$ such that $\sum_{i=1}^m \lambda_i = 1$ and $\sum_{i=1}^m \lambda_i P_i \subseteq P_0$. We show that the latter problem can be casted as an LP via Farkas' Lemma 7. To this aim, we alternatively reformulate the converse problem as *"do there exist $\lambda_1, \lambda_2, \ldots, \lambda_m \in \mathbb{R}_{\geq 0}$ with $\sum_{i=1}^m \lambda_i = 1$, such that $x^i \in P_i$ for each $1 \leq i \leq m$ implies $\sum_{i=1}^m \lambda_i x^i \in P_0$?"*.

For every fixed $\lambda_1, \lambda_2, \ldots, \lambda_m \in \mathbb{R}_{\geq 0}$ with $\sum_{i=1}^m \lambda_i = 1$, the implication *"$(\forall 1 \leq i \leq m : x^i \in P_i) \implies \sum_{i=1}^m \lambda_i x^i \in P_0$"* can be written as the conjunction of $2n$ conditions:

$$\bigwedge_{i=1}^m l^i \leq x^i \leq u^i \wedge \bigwedge_{i=1}^m 1 \cdot x^i = 1 \implies \sum_{i=1}^m \lambda_i x^i_k \geq l^0_k \tag{1}$$

$$\bigwedge_{i=1}^m l^i \leq x^i \leq u^i \wedge \bigwedge_{i=1}^m 1 \cdot x^i = 1 \implies \sum_{i=1}^m \lambda_i x^i_k \leq u^0_k \tag{2}$$

for all $1 \leq k \leq n$. (Note that the condition $1 \cdot \sum_{i=1}^m \lambda_i x^i = 1$ is trivially satisfied if $1 \cdot x^i = 1$ for all $1 \leq i \leq m$.) Each of the conditions (1) and (2), by Farkas' Lemma, is equivalent to the feasibility of a system of inequalities; for instance, for a given k, (1) is true if and only if there exist vectors $\mu^{k,i}, \nu^{k,i} \in \mathbb{R}^n_{\geq 0}$ and scalars $\theta^{k,i}, \eta^{k,i} \in \mathbb{R}_{\geq 0}$ for each $1 \leq i \leq m$ satisfying:

$$\mu^{k,i} - \nu^{k,i} + \theta^{k,i} 1 - \eta^{k,i} 1 = -\lambda_i e^k \qquad \forall 1 \leq i \leq m \tag{3}$$

$$\sum_{i=1}^m \left(u^i \cdot \mu^{k,i} - l^i \cdot \nu^{k,i} + \theta^{k,i} - \eta^{k,i} \right) \leq -l^0_k \tag{4}$$

Similarly, for a given k, (2) is true if and only if there exist vectors $\widehat{\mu}^{k,i}, \widehat{\nu}^{k,i} \in \mathbb{R}^n_{\geq 0}$ and scalars $\widehat{\theta}^{k,i}, \widehat{\eta}^{k,i} \in \mathbb{R}_{\geq 0}$ for each $1 \leq i \leq m$ satisfying:

$$\widehat{\mu}^{k,i} - \widehat{\nu}^{k,i} + \widehat{\theta}^{k,i} 1 - \widehat{\eta}^{k,i} 1 = \lambda_i e^k \qquad \forall 1 \leq i \leq m \tag{5}$$

$$\sum_{i=1}^m (u^i \cdot \widehat{\mu}^{k,i} - l^i \cdot \widehat{\nu}^{k,i} + \widehat{\theta}^{k,i} - \widehat{\eta}^{k,i}) \leq u^0_k \tag{6}$$

Algorithm 1: BISIMULATION($\mathcal{M}$)

Input: A relation $\mathcal{R}$ on $S \times S$
Output: A probabilistic bisimulation $\mathcal{R}$

1 begin
2     $\mathcal{R} \leftarrow \{\, (s,t) \in S \times S \mid L(s) = L(t) \,\}$;
3   repeat
4        $\mathcal{R}' \leftarrow \mathcal{R}$;
5   forall $s \in S$ do
6           $D \leftarrow \emptyset$;
7   forall $t \in [s]_\mathcal{R}$ do
8   if VIOLATE($s,t,\mathcal{R}$) then
9                 $D \leftarrow D \cup \{t\}$;
10   split $[s]_\mathcal{R}$ in $\mathcal{R}$ into D and $[s]_\mathcal{R} \setminus D$;
11   until $\mathcal{R} = \mathcal{R}'$;
12   return $\mathcal{R}$;

Procedure 2: VIOLATE($s,t,\mathcal{R}$)

Input: States s,t and relation $\mathcal{R}$
Output: Checks if $s \sim_\mathcal{R} t$

1 begin
2     $S, T \leftarrow \emptyset$;
3   forall $a \in \mathcal{A}$ do
4   if $\mathcal{P}_\mathcal{R}^{s,a}$ is strictly minimal then
5           $S \leftarrow S \cup \{\mathcal{P}_\mathcal{R}^{s,a}\}$;
6   if $\mathcal{P}_\mathcal{R}^{t,a}$ is strictly minimal then
7           $T \leftarrow T \cup \{\mathcal{P}_\mathcal{R}^{t,a}\}$;
8   return $S \neq T$;

Fig. 2. Alternating probabilistic bisimulation algorithm for interval MDPs

Thus, the converse problem we are aiming to solve reduces to checking the existence of vectors $\boldsymbol{\mu}^{k,i}, \boldsymbol{\nu}^{k,i}, \widehat{\boldsymbol{\mu}}^{k,i}, \widehat{\boldsymbol{\nu}}^{k,i} \in \mathbb{R}_{\geq 0}^n$ and scalars $\lambda_i, \theta^{k,i}, \eta^{k,i}, \widehat{\theta}^{k,i}, \widehat{\eta}^{k,i} \in \mathbb{R}_{\geq 0}$ for each $1 \leq i \leq m$ satisfying (3)–(6) and $\sum_{i=1}^m \lambda_i = 1$. That amounts to solve an LP problem, which is known to be in **P**. □

As stated earlier, in order to compute $\sim_{(\exists \sigma \forall)}$ we follow the standard partition refinement approach formalized by the procedure BISIMULATION in Fig. 2. Namely, we start with $\mathcal{R}$ being the complete relation and iteratively remove from $\mathcal{R}$ pairs of states that violate the definition of bisimulation with respect to $\mathcal{R}$. Clearly the core part of the algorithm is to check if two states "violate the definition of bisimulation". The violation of bisimilarity of s and t with respect to $\mathcal{R}$, which is addressed by the procedure VIOLATE, is checked by verifying if states s and t have the same set of strictly minimal polytopes. As a result of Theorem 8, this verification routine can be checked in polynomial time. As regards the computational complexity of Algorithm 1, let $|S| = n$ and $|\mathcal{A}| = m$. The procedure VIOLATE in Fig. 2 is called at most n^3 times. The procedure VIOLATE is then linear in m and in the complexity of checking strict minimality of $\mathcal{P}_\mathcal{R}^{s,a}$ and $\mathcal{P}_\mathcal{R}^{t,a}$, which is in $\mathcal{O}(|\mathcal{M}|^{\mathcal{O}(1)})$. Putting all these together, we get the following result.

Theorem 9. *Given an IMDP $\mathcal{M}$, computing $\sim_{(\exists \sigma \forall)}$ belongs to $\mathcal{O}(|\mathcal{M}|^{\mathcal{O}(1)})$.*

5 Compositional Reasoning

In order to study the compositional minimization, that is, to split a complex IMDP as parallel composition of several simpler IMDPs and then to use the bisimulation as a means to reduce the size of each of these IMDPs before performing the model checking for a given PCTL formula φ, we have to extend the notion of bisimulation from one IMDP to a pair of IMDPs; we do this by following the usual construction (see, e.g., [6,40]). Given two IMDPs $\mathcal{M}_1$ and

$\mathcal{M}_2$, we say that they are *alternating probabilistic* $(\exists\sigma\forall)$-*bisimilar*, denoted by $\mathcal{M}_1 \sim_{(\exists\sigma\forall)} \mathcal{M}_2$, if there exists an alternating probabilistic $(\exists\sigma\forall)$-bisimulation on the disjoint union of $\mathcal{M}_1$ and $\mathcal{M}_2$ such that $\bar{s}_1 \sim_{(\exists\sigma\forall)} \bar{s}_2$. We can now establish the first property needed for the compositional minimization, that is, transitivity of $\sim_{(\exists\sigma\forall)}$:

Theorem 10. *Given three IMDPs $\mathcal{M}_1$, $\mathcal{M}_2$, and $\mathcal{M}_3$, whenever $\mathcal{M}_1 \sim_{(\exists\sigma\forall)} \mathcal{M}_2$ and $\mathcal{M}_2 \sim_{(\exists\sigma\forall)} \mathcal{M}_3$, then $\mathcal{M}_1 \sim_{(\exists\sigma\forall)} \mathcal{M}_3$.*

For the second property needed by the compositional minimization, that is, that $\sim_{(\exists\sigma\forall)}$ is preserved by the parallel composition operator, we first have to introduce such an operator; to this end, we consider a slight adaption of synchronous product of $\mathcal{M}_1$ and $\mathcal{M}_2$ as introduced in [22]. Such a synchronous product makes use of a subclass of the Segala's (simple) probabilistic automata [40,41], called *action agnostic probabilistic automata* [22], where each automaton has as set of actions the same singleton set $\{f\}$, that is, all transitions are labelled by the same external action f: an *(action agnostic) probabilistic automaton* (PA) is a tuple $\mathcal{P} = (S, \bar{s}, \text{AP}, L, D)$, where S is a set of *states*, $\bar{s} \in S$ is the *start* state, AP is a finite set of *atomic propositions*, $L\colon S \to 2^{\text{AP}}$ is a *labelling function*, and $D \subseteq S \times \text{Disc}(S)$ is a *probabilistic transition relation*.

Definition 11. *Given two IMDPs $\mathcal{M}_1$ and $\mathcal{M}_2$, we define the* synchronous product *of $\mathcal{M}_1$ and $\mathcal{M}_2$ as $\mathcal{M}_1 \otimes \mathcal{M}_2 := \text{F}(\text{UF}(\mathcal{M}_1) \otimes \text{UF}(\mathcal{M}_2))$ where*

- *the unfolding mapping $\text{UF}\colon \text{IMDP} \to \text{PA}$ is a function that maps a given IMDP $\mathcal{M} = (S, \bar{s}, \mathcal{A}, \text{AP}, L, I)$ to the PA $\mathcal{P} = (S, \bar{s}, \text{AP}, L, D)$ where $D = \{(s, \mu) \mid s \in S, \exists a \in \mathcal{A}(s) : \mu \in \text{Ext}(\mathcal{H}_s^a) \wedge \mathcal{H}_s^a \text{ is a strictly minimal polytope}\}$;*
- *the folding mapping $\text{F}\colon \text{PA} \to \text{IMDP}$ transforms a PA $\mathcal{P} = (S, \bar{s}, \text{AP}, L, D)$ into the IMDP $\mathcal{M} = (S, \bar{s}, \{f\}, \text{AP}, L, I)$ where, for each $s, t \in S$, $I(s, f, t) = \text{proj}_{e^t} \text{conv} \{\mu \mid (s, \mu) \in D\}$;*
- *the synchronous product of two PAs $\mathcal{P}_1$ and $\mathcal{P}_2$, denoted by $\mathcal{P}_1 \otimes \mathcal{P}_2$, is the probabilistic automaton $\mathcal{P} = (S, \bar{s}, \text{AP}, L, D)$ where $S = S_1 \times S_2$, $\bar{s} = (\bar{s}_1, \bar{s}_2)$, $\text{AP} = \text{AP}_1 \cup \text{AP}_2$, for each $(s_1, s_2) \in S$, $L(s_1, s_2) = L_1(s_1) \cup L_2(s_2)$, and $D = \{((s_1, s_2), \mu_1 \times \mu_2) \mid (s_1, \mu_1) \in D_1 \text{ and } (s_2, \mu_2) \in D_2\}$, where $\mu_1 \times \mu_2$ is defined for each $(t_1, t_2) \in S_1 \times S_2$ as $(\mu_1 \times \mu_2)(t_1, t_2) = \mu_1(t_1) \cdot \mu_2(t_2)$.*

As stated earlier, Definition 11 is slightly different from its counterpart in [22]. As a matter of fact, due to the competitive semantics for resolving the nondeterminism, only actions whose uncertainty set is a strictly minimal polytope play a role in deciding the alternating bisimulation relation $\sim_{(\exists\sigma\forall)}$. In particular, for the compositional reasoning keeping state actions whose uncertainty set is not strictly minimal induces spurious behaviors and therefore, influences on the soundness of the parallel operator definition. In order to avoid such redundancies, we can either preprocess the *IMDPs* before composing by removing state actions whose uncertainty set is not strictly minimal or restricting the unfolding mapping UF to unfold a given *IMDP* while ensuring that all extreme transitions in the resultant probabilistic automaton correspond to extreme points of strictly minimal polytopes in the original *IMDP*. For the sake of simplicity, we choose the latter.

Theorem 12. *Given three IMDPs $\mathcal{M}_1$, $\mathcal{M}_2$, and $\mathcal{M}_3$, if $\mathcal{M}_1 \sim_{(\exists \sigma \forall)} \mathcal{M}_2$, then $\mathcal{M}_1 \otimes \mathcal{M}_3 \sim_{(\exists \sigma \forall)} \mathcal{M}_2 \otimes \mathcal{M}_3$.*

We have considered so far the parallel composition via synchronous production, which is working by the definition of folding collapsing all labels to a single transition. Here we consider the other extreme of the parallel composition: interleaving only.

Definition 13. *Given two IMDPs $\mathcal{M}_l$ and $\mathcal{M}_r$, we define the* interleaved composition *$\mathcal{M}_l \wedge\!\!\!\wedge \mathcal{M}_r$ of $\mathcal{M}_l$ and $\mathcal{M}_r$ as the IMDP $\mathcal{M} = (S, \bar{s}, \mathcal{A}, AP, L, I)$ where*

- $S = S_l \times S_r$;
- $\bar{s} = (\bar{s}_l, \bar{s}_r)$;
- $\mathcal{A} = (\mathcal{A}_l \times \{l\}) \cup (\mathcal{A}_r \times \{r\})$;
- $AP = AP_l \cup AP_r$;
- *for each* $(s_l, s_r) \in S$, $L(s_l, s_r) = L_l(s_l) \cup L_r(s_r)$; *and*
-

$$I((s_l, s_r), (a, i), (t_l, t_r)) = \begin{cases} I_l(s_l, a, t_l) & \text{if } i = l \text{ and } t_r = s_r, \\ I_r(s_r, a, t_r) & \text{if } i = r \text{ and } t_l = s_l, \\ [0, 0] & \text{otherwise.} \end{cases}$$

Theorem 14. *Given three IMDPs $\mathcal{M}_1$, $\mathcal{M}_2$, and $\mathcal{M}_3$, if $\mathcal{M}_1 \sim_{(\exists \sigma \forall)} \mathcal{M}_2$, then $\mathcal{M}_1 \wedge\!\!\!\wedge \mathcal{M}_3 \sim_{(\exists \sigma \forall)} \mathcal{M}_2 \wedge\!\!\!\wedge \mathcal{M}_3$*

6   Case Studies

We implemented in a prototypical tool the proposed bisimulation minimization algorithm and applied it to several case studies. The bisimulation algorithm is tested on several PRISM [29] benchmarks extended to support also intervals in the transitions. For the evaluation, we have used a machine with a 3.6 GHz Intel i7-4790 with 16 GB of RAM of which 12 assigned to the tool; the timeout has been set to 30 mins. Our tool reads a model specification in the PRISM input language and constructs an explicit-state representation of the state space. Afterwards, it computes the quotient using the algorithm in Fig. 2.

Table 1 shows the performance of our prototype on a number of case studies taken from the PRISM website [36], where we have replaced some of the probabilistic choices with intervals. Despite using an explicit representation for the model, the prototype is able to manage cases studies in the order of millions of states and transitions (columns "Model", "$|S|$", and "$|I|$"). The time in seconds required to compute the bisimulation relation and the size of the corresponding quotient *IMDP* are shown in columns "$t_\sim$", "$|S_\sim|$", and "$|I_\sim|$". In order to improve the performance of the tool, we have implemented optimizations, such as caching equivalent LP problems, which improve the runtime of our prototype. Because of this, we saved to solve several LP problems in each tool run, thereby avoiding the potentially costly solution of LP problems from becoming a bottleneck. However, the more refinements are needed, the more time is required to

Table 1. Experimental evaluation of the bisimulation computation

Model	$|S|$	$|I|$	$t_\sim$ (s)	$|S_\sim|$	$|I_\sim|$
Consensus-Shared-Coin-3	5 216	13 380	1	787	1 770
Consensus-Shared-Coin-4	43 136	144 352	3	2 189	5 621
Consensus-Shared-Coin-5	327 936	1 363 120	26	5 025	14 192
Consensus-Shared-Coin-6	2 376 448	11 835 456	238	10 173	30 861
Crowds-5-10	111 294	261 444	1	107	153
Crowds-5-20	2 061 951	7 374 951	20	107	153
Crowds-5-30	12 816 233	61 511 033	149	107	153
Crowds-5-40	–MO–				
Mutual-Exclusion-PZ-3	2 368	8 724	4	475	1 632
Mutual-Exclusion-PZ-4	27 600	136 992	70	3 061	13 411
Mutual-Exclusion-PZ-5	308 800	1 930 160	534	12 732	65 661
Mutual-Exclusion-PZ-6	3 377 344	25 470 144	–TO–		
Dining-Phils-LR-nofair-3	956	3 048	1	172	509
Dining-Phils-LR-nofair-4	9 440	40 120	14	822	3 285
Dining-Phils-LR-nofair-5	93 068	494 420	622	5 747	29 279
Dining-Phils-LR-nofair-6	917 424	5 848 524	–TO–		

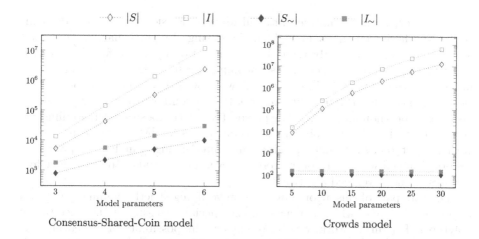

Consensus-Shared-Coin model   Crowds model

Fig. 3. Effectiveness of bisimulation minimization on model reduction

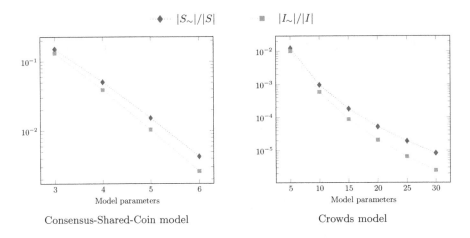

Fig. 4. State and transition reduction ratio by bisimulation minimization

complete the minimization, since several new LP problems need to be solved. The plots in Fig. 3 show graphically the number of states and transitions for the Consensus and Crowds experiments, where for the latter we have considered more instances than the ones reported in Table 1. As we can see, the bisimulation minimization is able to reduce considerably the size of the $IMDP$, by several orders of magnitude. Additionally, this reduction correlates positively with the number of model parameters as depicted in Fig. 4.

7   Concluding Remarks

In this paper, we have analyzed interval Markov decision processes under controller synthesis semantics in a dynamic setting. In particular, we provided an efficient compositional bisimulation minimization approach for $IMDP$s with respect to the competitive semantics, encompassing both the controller and parameter synthesis semantics. In this regard, we proved that alternating probabilistic bisimulation for $IMDP$s with respect to the competitive semantics can be decided in polynomial time. From perspective of compositional reasoning, we showed that alternating probabilistic bisimulations for $IMDP$s are congruences with respect to synchronous product and interleaving. Finally, we presented results obtained with a prototype tool on several case studies to show the effectiveness of the developed algorithm.

The core part of this algorithm relies on verifying strictly minimal polytopes in polynomial time, which depends on the special structure of the uncertainty polytopes. For future work, we aim to explore the possibility of preserving this computational efficiency for MDPs with richer formalisms for uncertainties such as likelihood or ellipsoidal uncertainties.

References

1. Alur, R., Henzinger, T.A., Kupferman, O., Vardi, M.Y.: Alternating refinement relations. In: Sangiorgi, D., de Simone, R. (eds.) CONCUR 1998. LNCS, vol. 1466, pp. 163–178. Springer, Heidelberg (1998). doi:10.1007/BFb0055622
2. Baier, C., Katoen, J.-P.: Principles of Model Checking. The MIT Press, Cambridge (2008)
3. Ben-Or, M.: Another advantage of free choice: completely asynchronous agreement protocols (extended abstract). In: PODC, pp. 27–30 (1983)
4. Benedikt, M., Lenhardt, R., Worrell, J.: LTL model checking of interval Markov chains. In: Piterman, N., Smolka, S.A. (eds.) TACAS 2013. LNCS, vol. 7795, pp. 32–46. Springer, Heidelberg (2013). doi:10.1007/978-3-642-36742-7_3
5. Böde, E., Herbstritt, M., Hermanns, H., Johr, S., Peikenkamp, T., Pulungan, R., Rakow, J., Wimmer, R., Becker, B.: Compositional dependability evaluation for STATEMATE. ITSE 35(2), 274–292 (2009)
6. Cattani, S., Segala, R.: Decision algorithms for probabilistic bisimulation. In: Brim, L., Křetínský, M., Kučera, A., Jančar, P. (eds.) CONCUR 2002. LNCS, vol. 2421, pp. 371–386. Springer, Heidelberg (2002). doi:10.1007/3-540-45694-5_25
7. Chatterjee, K., Sen, K., Henzinger, T.A.: Model-checking ω-regular properties of interval Markov chains. In: Amadio, R. (ed.) FoSSaCS 2008. LNCS, vol. 4962, pp. 302–317. Springer, Heidelberg (2008). doi:10.1007/978-3-540-78499-9_22
8. Chehaibar, G., Garavel, H., Mounier, L., Tawbi, N., Zulian, F.: Specification, verification of the PowerScale® bus arbitration protocol: An industrial experiment with LOTOS. In: FORTE, pp. 435–450 (1996)
9. Coste, N., Hermanns, H., Lantreibecq, E., Serwe, W.: Towards performance prediction of compositional models in industrial GALS designs. In: Bouajjani, A., Maler, O. (eds.) CAV 2009. LNCS, vol. 5643, pp. 204–218. Springer, Heidelberg (2009). doi:10.1007/978-3-642-02658-4_18
10. Delahaye, B., Katoen, J.-P., Larsen, K.G., Legay, A., Pedersen, M.L., Sher, F., Wąsowski, A.: Abstract probabilistic automata. In: Jhala, R., Schmidt, D. (eds.) VMCAI 2011. LNCS, vol. 6538, pp. 324–339. Springer, Heidelberg (2011). doi:10.1007/978-3-642-18275-4_23
11. Delahaye, B., Katoen, J.-P., Larsen, K.G., Legay, A., Pedersen, M.L., Sher, F., Wasowski, A.: New results on abstract probabilistic automata. In: ACSD, pp. 118–127 (2011)
12. Delahaye, B., Larsen, K.G., Legay, A., Pedersen, M.L., Wąsowski, A.: Decision problems for interval Markov Chains. In: Dediu, A.-H., Inenaga, S., Martín-Vide, C. (eds.) LATA 2011. LNCS, vol. 6638, pp. 274–285. Springer, Heidelberg (2011). doi:10.1007/978-3-642-21254-3_21
13. Fecher, H., Leucker, M., Wolf, V.: *Don't Know* in probabilistic systems. In: Valmari, A. (ed.) SPIN 2006. LNCS, vol. 3925, pp. 71–88. Springer, Heidelberg (2006). doi:10.1007/11691617_5
14. Fischer, M.J., Lynch, N.A., Paterson, M.S.: Impossibility of distributed consensus with one faulty process. J. ACM 32(2), 374–382 (1985)
15. Gebler, D., Hashemi, V., Turrini, A.: Computing behavioral relations for probabilistic concurrent systems. In: Remke, A., Stoelinga, M. (eds.) Stochastic Model Checking. Rigorous Dependability Analysis Using Model Checking Techniques for Stochastic Systems. LNCS, vol. 8453, pp. 117–155. Springer, Heidelberg (2014). doi:10.1007/978-3-662-45489-3_5

16. Givan, R., Leach, S.M., Dean, T.L.: Bounded-parameter Markov decision processes. Artif. Intell. **122**(1–2), 71–109 (2000)
17. Hahn, E.M., Han, T., Zhang, L.: Synthesis for PCTL in parametric markov decision processes. In: Bobaru, M., Havelund, K., Holzmann, G.J., Joshi, R. (eds.) NFM 2011. LNCS, vol. 6617, pp. 146–161. Springer, Heidelberg (2011). doi:10.1007/978-3-642-20398-5_12
18. Hahn, E.M., Hashemi, V., Hermanns, H., Turrini, A.: Exploiting robust optimization for interval probabilistic bisimulation. In: Agha, G., Van Houdt, B. (eds.) QEST 2016. LNCS, vol. 9826, pp. 55–71. Springer, Cham (2016). doi:10.1007/978-3-319-43425-4_4
19. Hansson, H., Jonsson, B.: A logic for reasoning about time and reliability. Formal Asp. Comput. **6**(5), 512–535 (1994)
20. Hashemi, V., Hatefi, H., Krčál, J.: Probabilistic bisimulations for PCTL model checking of interval MDPs. In: SynCoP, pp. 19–33. EPTCS (2014)
21. Hashemi, V., Hermanns, H., Song, L., Subramani, K., Turrini, A., Wojciechowski, P.: Compositional bisimulation minimization for interval Markov decision processes. In: Dediu, A.-H., Janoušek, J., Martín-Vide, C., Truthe, B. (eds.) LATA 2016. LNCS, vol. 9618, pp. 114–126. Springer, Cham (2016). doi:10.1007/978-3-319-30000-9_9
22. Hashemi, V., Hermanns, H., Turrini, A.: Compositional reasoning for interval Markov decision processes. http://arxiv.org/abs/1607.08484
23. Hermanns, H., Katoen, J.-P.: Automated compositional Markov chain generation for a plain-old telephone system. SCP **36**(1), 97–127 (2000)
24. Iyengar, G.N.: Robust dynamic programming. Math. Oper. Res. **30**(2), 257–280 (2005)
25. Jonsson, B., Larsen, K.G.: Specification and refinement of probabilistic processes. In: LICS, pp. 266–277 (1991)
26. Kanellakis, P.C., Smolka, S.A.: CCS expressions, finite state processes, and three problems of equivalence. I&C, pp. 43–68 (1990)
27. Katoen, J.-P., Kemna, T., Zapreev, I., Jansen, D.N.: Bisimulation minimisation mostly speeds up probabilistic model checking. In: Grumberg, O., Huth, M. (eds.) TACAS 2007. LNCS, vol. 4424, pp. 87–101. Springer, Heidelberg (2007). doi:10.1007/978-3-540-71209-1_9
28. Kozine, I., Utkin, L.V.: Interval-valued finite Markov chains. Reliable Comput. **8**(2), 97–113 (2002)
29. Kwiatkowska, M., Norman, G., Parker, D.: PRISM 4.0: verification of probabilistic real-time systems. In: Gopalakrishnan, G., Qadeer, S. (eds.) CAV 2011. LNCS, vol. 6806, pp. 585–591. Springer, Heidelberg (2011). doi:10.1007/978-3-642-22110-1_47
30. Lahijanian, M., Andersson, S.B., Belta, C.: Formal verification and synthesis for discrete-time stochastic systems. IEEE Tr. Autom. Contr. **60**(8), 2031–2045 (2015)
31. Larsen, K.G., Skou, A.: Bisimulation through probabilistic testing (preliminary report). In: POPL, pp. 344–352 (1989)
32. Luna, R., Lahijanian, M., Moll, M., Kavraki, L.E.: Asymptotically optimal stochastic motion planning with temporal goals. In: Akin, H.L., Amato, N.M., Isler, V., van der Stappen, A.F. (eds.) Algorithmic Foundations of Robotics XI. STAR, vol. 107, pp. 335–352. Springer, Cham (2015). doi:10.1007/978-3-319-16595-0_20
33. Luna, R., Lahijanian, M., Moll, M., Kavraki, L.E.: Fast stochastic motion planning with optimality guarantees using local policy reconfiguration. In: ICRA, pp. 3013–3019 (2014)
34. Luna, R., Lahijanian, M., Moll, M., Kavraki, L.E.: Optimal and efficient stochastic motion planning in partially-known environments. In: AAAI, pp. 2549–2555 (2014)

35. Paige, R., Tarjan, R.E.: Three partition refinement algorithms. SIAM J. Comput. **16**(6), 973–989 (1987)
36. PRISM model checker. http://www.prismmodelchecker.org/
37. Puggelli, A.: Formal Techniques for the Verification and Optimal Control of Probabilistic Systems in the Presence of Modeling Uncertainties. Ph.D. thesis, EECS Department, University of California, Berkeley (2014)
38. Puggelli, A., Li, W., Sangiovanni-Vincentelli, A.L., Seshia, S.A.: Polynomial-time verification of PCTL properties of MDPs with convex uncertainties. In: Sharygina, N., Veith, H. (eds.) CAV 2013. LNCS, vol. 8044, pp. 527–542. Springer, Heidelberg (2013). doi:10.1007/978-3-642-39799-8_35
39. Schrijver, A.: Theory of Linear and Integer Programming. Wiley, New York (1998)
40. Segala, R.: Modeling and Verification of Randomized Distributed Real-Time Systems. Ph.D. thesis, MIT (1995)
41. Segala, R.: Probability and nondeterminism in operational models of concurrency. In: Baier, C., Hermanns, H. (eds.) CONCUR 2006. LNCS, vol. 4137, pp. 64–78. Springer, Heidelberg (2006). doi:10.1007/11817949_5
42. Sen, K., Viswanathan, M., Agha, G.: Model-checking markov chains in the presence of uncertainties. In: Hermanns, H., Palsberg, J. (eds.) TACAS 2006. LNCS, vol. 3920, pp. 394–410. Springer, Heidelberg (2006). doi:10.1007/11691372_26
43. Wolff, E.M., Topcu, U., Murray, R.M.: Robust control of uncertain Markov decision processes with temporal logic specifications. In: CDC, pp. 3372–3379 (2012)

Better Automated Importance Splitting
for Transient Rare Events

Carlos E. Budde[1,2]([⊠]), Pedro R. D'Argenio[2,3]([⊠]), and Arnd Hartmanns[1]([⊠])

[1] University of Twente, Enschede, The Netherlands
a.hartmanns@utwente.nl
[2] Universidad Nacional de Córdoba, Córdoba, Argentina
[3] Saarland University, Saarbrücken, Germany

Abstract. Statistical model checking uses simulation to overcome the state space explosion problem in formal verification. Yet its runtime explodes when faced with rare events, unless a rare event simulation method like importance splitting is used. The effectiveness of importance splitting hinges on nontrivial model-specific inputs: an importance function with matching splitting thresholds. This prevents its use by non-experts for general classes of models. In this paper, we propose new method combinations with the goal of fully automating the selection of all parameters for importance splitting. We focus on transient (reachability) properties, which particularly challenged previous techniques, and present an exhaustive practical evaluation of the new approaches on case studies from the literature. We find that using RESTART simulations with a compositionally constructed importance function and thresholds determined via a new *expected success* method most reliably succeeds and performs very well. Our implementation within the MODEST TOOLSET supports various classes of formal stochastic models and is publicly available.

1 Introduction

Nuclear reactors, smart power grids, automated storm surge barriers, networked industrial automation systems: We increasingly rely on critical technical systems and infrastructures whose failure would have drastic consequences. It is imperative to perform a quantitative evaluation in the design phase based on a formal stochastic model, e.g. on extensions of continuous-time Markov chains (CTMC), stochastic Petri nets (SPN), or fault trees. Only if the probability of failure can be shown to be sufficiently low can the system design be implemented. Calculating such probabilities—which may be on the order of 10^{-19} or lower—is challenging: For finite-state Markov chains or probabilistic timed automata (PTA [23]), probabilistic model checking can numerically approximate the desired probabilities, but the state space explosion problem limits it to small models. For other models, in particular those involving events governed by general continuous probability distributions, model checking techniques only exist for specific subclasses with limited scalability [26] or merely compute probability bounds [14].

© Springer International Publishing AG 2017
K.G. Larsen et al. (Eds.): SETTA 2017, LNCS 10606, pp. 42–58, 2017.
https://doi.org/10.1007/978-3-319-69483-2_3

Statistical model checking (SMC [17,33]), i.e. using Monte Carlo simulation with formal models, has become a popular alternative for large models and formalisms not amenable to model checking. It trades memory for runtime: memory usage is constant but the number of simulation runs explodes with the desired precision. When an event's true probability is 10^{-19}, for example, we may want to be confident that the error of our estimation is at most on the order of 10^{-20}. *Rare event simulation* (RES [28]) methods have been developed to attack this problem. They increase the number of simulation runs that reach the rare event and adjust the statistical evaluation accordingly. The main RES methods are *importance sampling* and *importance splitting*. The former modifies probabilities in the model to make the event more likely. The challenge lies in finding a good such *change of measure*. Importance splitting instead performs more simulation runs, which however may start from a non-initial state and end early. Here, the challenge is to find an *importance function* that assigns to each state a value indicating how "close" it is to the rare event. More (partial) runs will be started from states with higher importance. Additionally, depending on the concrete splitting method used, *thresholds* (the subset of importance values at which to start new runs) and splitting *factors* (how many new runs to generate at each threshold) need to be chosen. The performance of RES varies drastically with the choices made for these parameters. The quality of a choice of parameters highly depends on the model at hand; making good choices requires an expert in the system domain, the modelling formalism, and the selected RES method.

Aligning RES with the spirit of (statistical) model checking as a "push-button" approach requires methods to automatically select (usually) good parameters. These methods must not negate the memory usage advantages of SMC. Between importance sampling and splitting, the latter appears more amenable to automatic approaches that work well across modelling formalisms (CTMC, PTA, etc.). We previously proposed a compositional method to automatically construct an importance function [4]. Its compositionality is the key to low memory usage. Our FIG tool [3] for RES of input-output stochastic automata (IOSA [8]) implements this method together with the RESTART splitting algorithm [30], thresholds computed via a sequential Monte Carlo (SEQ) approach [3,6], and a single fixed splitting factor specified by the user for all thresholds. Experimental results [3] show that FIG works well for steady-state measures, but less so for transient properties. In particular, runtime varies significantly between tool invocations due to different thresholds being computed by SEQ, and the optimal splitting factor varies significantly between different models.

Our contributions. In this paper, we investigate several alternative combinations of splitting and threshold/factor selection algorithms with the goal of improving the automation, robustness and performance of importance splitting for RES in SMC. We keep the compositional method for automatic importance function construction as implemented in FIG. Aside from RESTART, we consider the fixed effort [9] and fixed success [25,28] splitting methods (Sect. 3). While RESTART was proposed for steady-state measures and only later extended to transient properties [31], the latter two were designed for estimating probabilities of transient events in the first place. For threshold selection, we specify a new "expected success" (EXP) technique as an alternative to SEQ (Sect. 4). EXP selects thresholds *and* an individual splitting factor for each threshold, removing the need for the user to manually select a global splitting factor. We implemented all techniques in the modes simulator of the MODEST TOOLSET [16]. They can be freely combined, and work for all the formalisms supported by modes—including CTMC, IOSA, and deterministic PTA. Our final and major contribution is an extensive experimental evaluation (Sect. 5) of the various combinations on six case studies.

Related work. A thorough theoretical and empirical comparison of variants of RESTART is presented in [9], albeit in a non-automated setting. Approaching the issue of automation, Jégourel et al. [19,20] use a layered restatement of the formula specifying the rare event to build an importance function for use with adaptive multilevel splitting [5], the predecessor of SEQ. Garvels et al. [10] derive importance functions for finite Markov chains from knowledge about their steady-state behaviour. For SPN, Zimmermann and Maciel [34] provide a monolithic method, though limited to a restricted class of models and throughput measures [35]. Importance *sampling* has been automated for SPN [27] restricted to Markovian firing delays and a global parameterisation of the transition intensities [35]. The difficulties of automating importance sampling are also illustrated in [18]: the proposed automatic change of measure guarantees a variance reduction, yet is proved for stochastic behaviour described by integrable products of exponentials and uniforms only. We do not aim at provable improvements in specific settings, but focus on general models and empirically study which methods work best in practice. We are not aware of other practical methods for, or comparisons of, automated splitting approaches on general models.

2   Preliminaries

We write $\{| \ldots |\}$ for multisets, in contrast to sets written as $\{ \ldots \}$. $\mathbb{N}$ is the set of natural numbers $\{ 0, 1, \ldots \}$ and $\mathbb{N}^+ = \mathbb{N} \setminus \{ 0 \}$. In our algorithms, operation S.remove() returns and removes an element from the set S. The element may be picked according to any policy (e.g. uniformly at random, in FIFO order, etc.).

2.1   Simulation Models

We develop RES approaches that can work for any stochastic formalism combining discrete and continuous state. We thus use an abstract notion of models:

Definition 1. *A (simulation) model* M *is a discrete-time Markov process whose states consist of a discrete and (optionally) a continuous part. It has a fixed initial state that can be obtained as* M.initial()*. Operation* M.step(s) *samples a path from state s and returns the path's next state after one time step.*

Example 1. A CTMC M_{ctmc} is a continuous-time stochastic process. We can cast it as a simulation model M_{sim} by using the number of transitions taken as the (discrete) time steps of M_{sim}. Thus, given a state s of M_{ctmc}, M_{sim}.step(s) returns the first state s' of M_{ctmc} encountered after taking a single transition from s on a sample path. In effect, we follow the embedded discrete-time Markov chain. Only if the event of interest refers to time do we also need to keep track of the global elapsed (continuous) time as part of the states of M_{sim}.

We require models to be Markov processes. For formalisms with memory, e.g. due to general continuous probability distributions, we encode the memory (e.g. values and expiration times of clocks in the case of IOSA) in the state space. We compute probabilities for transient properties, or more precisely the probability to reach a set of target states while avoiding another disjoint set of states:

Definition 2. *A transient property $\phi \in S \rightarrow \{$ true, false, undecided $\}$ for a model with state space S maps target states to true, states to be avoided to false, and all other states to undecided. We require that the probability of reaching a state where ϕ returns true or false is 1 from a model's initial state.*

To determine whether a sample path satisfies ϕ, evaluate ϕ sequentially for every state on the path and return the first outcome $\neq$ *undecided*. Standard SMC/ Monte Carlo simulation generates a large number n of sample paths to estimate the probability p of the transient property as $\hat{p} \stackrel{\text{def}}{=} \frac{n_{true}}{n}$, where n_{true} is the number of paths that satisfied ϕ, and reports a confidence interval around the estimate for a specified confidence level. This corresponds to estimating the value of the until formula $P_{=?}(\neg \textit{avoid}\, U\, \textit{target})$ in a logic like PCTL (as used in e.g. PRISM [22]) for state formulas *avoid* and *target*. Time-bounded until $U_{\leq b}$ is encoded by tracking the elapsed time t_{global} in states and including $t_{global} > b$ in *avoid*.

2.2 Ingredients of Importance Splitting

Importance splitting increases the simulation effort for states "close" to the target set. Closeness is represented by an *importance function* $f_I \in S \rightarrow \mathbb{N}$ that maps each state to its importance in $\{0, \ldots, \max f_I\}$. To simplify our presentation, we assume that $f_I(\text{M.initial}()) = 0$, $(\phi(s_{target}) = \textit{true}) \Rightarrow f_I(s_{target}) = \max f_I$, and if $s' := \text{M.next}(s)$, then $|f_I(s) - f_I(s')| \leq 1$. These assumptions can easily be removed. The performance, but not the correctness, of all importance splitting methods hinges on the quality of the importance function. Traditionally, it is specified ad hoc for each model domain by a RES expert [9, 28, 30]. Methods to automatically compute one [10, 19, 20, 34] are usually specialised to a specific formalism or a particular model structure, potentially providing guaranteed efficiency improvements. We build on the method of [4] that is applicable to any stochastic compositional model with a partly discrete state space. It does not provide mathematical guarantees of performance improvements, but is aimed at generality and providing "usually good" results with minimal user input.

Compositional f_I. A compositional model is a parallel composition of components $M = M_1 \parallel \ldots \parallel M_n$. Each component can be seen as a model on its own, but the components may interact, usually via some synchronisation/handshaking mechanism. We write the projection of state s of M to the discrete local variables of component M_i as $s|_i$. The compositional method works as follows:

1. Convert the target set formula *target* to negation normal form (NNF) and associate each literal $target^j$ with the component $M(target^j)$ whose local state variables it refers to. Literals must not refer to multiple components.
2. Explore the *discrete part* of the state space of each component M_i. For each $target^j$ with $M_i = M(target^j)$, use reverse breadth-first search to compute the local minimum distance $f_i^j(s|_i)$ of each state $s|_i$ to a state satisfying $target^j$.
3. In the syntax of the NNF of *target*, replace every occurrence of $target^j$ by $f_i^j(s|_i)$ with i such that $M_i = M(target^j)$, and every Boolean operator $\wedge$ or $\vee$ by $+$. Use the resulting formula as the importance function $f_I(s)$.

Full implementation details can be found in [3]. Other operators can be used in place of $+$, e.g. max or multiplication. Aside from the choice of operator, with $+$ as default since it works well for most models, the procedure requires no user input. It takes into account both the structure of the target set formula and the structure of the state space. Memory usage is determined by the number of discrete local states (required to be finite) over all components. Typically, component state spaces are small even when the composed state space explodes.

Levels, Thresholds and Factors. Given a model and importance function f_I, importance splitting could spawn more simulation runs whenever the current sample path moves from a state with importance i to one with importance $j > i$. Using the compositional approach, the probability of visiting a state with a higher importance is often close to 1 for many of the i, so splitting on every increment would lead to excessively many (partial) runs and high runtime. Importances are hence partitioned into a set of intervals called *levels*. This results in a *level function* $f_L \in S \to \mathbb{N}$ where, again, the initial state is on level 0 and all target states are on the highest level max f_L. We refer to the boundary between the highest importance of level $l-1$ and the lowest importance i of level l as the *threshold* T_l, identified by i. Some splitting methods are further parameterised by the "amount of splitting" at each threshold or the "effort" at each level; we use *splitting factor* and *effort* functions f_S resp. f_E in $\mathbb{N} \to \mathbb{N}^+$ for this purpose.

3 Splitting Methods

We now briefly describe, from a practical perspective, the three different approaches to importance splitting that we implemented and evaluated.

3.1 RESTART

Originally discovered in 1970 [2] and popularised by J. Villén-Altamirano and M. Villén-Altamirano [30], the RESTART importance splitting method was designed

Input: model M, level function f_L, splitting factors f_S, transient property ϕ

1   $S := \{| \langle \text{M.initial}(), 0 \rangle |\}, \hat{p} := 0$ // *start with the initial state from level 0*
2   **while** $S \neq \varnothing$ **do**   // *perform main and child runs (RESTART loop)*
3   $\quad$ $\langle s, l \rangle := S.\text{remove}(), l_{create} := l$ // *get next split and store creation level*
4   $\quad$ **while** $\phi(s) = undecided$ **do**   // *run until property decided (simulation loop)*
5   $\quad\quad$ $s := \text{M.step}(s)$ // *simulate up to next change in discrete state*
6   $\quad\quad$ **if** $f_L(s) < l_{create}$ **then break**   // *moved below creation level: kill run*
7   $\quad\quad$ **else if** $f_L(s) > l$ **then**   // *moved one level up: split run*
8   $\quad\quad\quad$ $l := f_L(s), S := S \cup \{| \langle s, l \rangle, \ldots (f_S(l) - 1 \text{ times}) \ldots, \langle s, l \rangle |\}$
9   $\quad$ **if** $\phi(s)$ **then** $\hat{p} := \hat{p} + 1 / \prod_{i=1}^{l} f_S(l)$ // *update result if we hit the rare event*
10 **return** $\hat{p}$

Algorithm 1. The RESTART method for importance splitting

for steady-state measures and later extended to transient properties [31]. It works by performing one main simulation run from the initial state. As soon as any run crosses a threshold from below, new child runs are started from the first state in the new level l (the run is split). Their number is given by l's splitting factor: $f_S(l) - 1$ child runs are started, resulting in $f_S(l)$ runs that continue after splitting. Each run is tagged with the level on which it is created. When a run crosses a thresh-

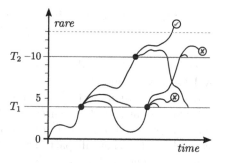

Fig. 1. RESTART

old from above into a level below its creation level, it ends (the run is killed). A run also ends when it reaches an *avoid* or *target* state. We state RESTART formally as Algorithm 1. Figure 1 illustrates its behaviour. The horizontal axis is the model's time steps while the vertical direction shows the current state's importance. *target* states are marked ✓and *avoid* states are marked ✗. We have three levels with thresholds at importances 3 to 4 and 9 to 10. f_S is $\{1 \mapsto 3, 2 \mapsto 2\}$.

The result of a RESTART run—consisting of a main and several child runs—is the weighted number of runs that reach *target*. Each run's weight is 1 divided by the product of the splitting factors of all levels. The result is thus a positive rational number. Note that this is in contrast to standard Monte Carlo simulation, where each run is a Bernoulli trial with outcome 0 or 1. This affects the statistical analysis on which the confidence interval over multiple runs is built. RESTART is carefully designed s.t. the mean of the results of many RESTART runs is an unbiased estimator for the true probability of the transient property [32].

3.2   Fixed Effort

In contrast to RESTART, each run of the *fixed effort* method [9,11] performs a fixed number $f_E(l)$ of partial runs on each level l. Each of these ends when it

Input: model M, level function f_L, effort function f_E, transient property ϕ

```
1  L := { 0 ↦ [ S := { M.initial() }, n := 0, up := 0 ] }        // set up data for level 0
2  for l from 0 to max f_L do          // iterate over all levels from initial to target
3     for i from 1 to f_E(l) do        // perform sub-runs on level (fixed effort loop)
4        s :∈ L(l).S, L(l).n := L(l).n + 1      // pick from the level's initial states
5        while φ(s) = undecided do     // run until φ is decided (simulation loop)
6           s := M.step(s)              // simulate up to next change in discrete state
7           if f_L(s) > l then          // moved one level up: end sub-run
8              L(l).up := L(l).up + 1          // level-up run for current level
9              L(f_L(s)).S := L(f_L(s)).S ∪ { s }     // initial state for next level
10             break
11       if φ(s) then L(l).up := L(l).up + 1  // hit rare event (highest level only)
12    if L(l).up = 0 then return 0          // we cannot reach the target any more
13 return ∏_{i=0}^{max f_L} L(l).up/L(l).n   // multiply conditional level-up prob. estimates
```

Algorithm 2. The fixed effort method for importance splitting

either crosses a threshold from below into level $l + 1$, encounters a *target* state, or encounters an *avoid* state. We count the first two cases as $n^l{}_{up}$. In the first case, the new state is stored in a set of initial states for level $l + 1$. When all partial runs for level l have ended, the algorithm moves to level $l + 1$, starting the next round of partial runs from the previously collected initial states of the new level. This behaviour is illustrated in Fig. 2 (with $f_E(l) = 5$ for all levels) and formally stated as Algo-

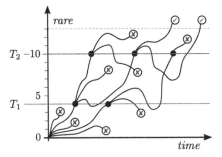

Fig. 2. Fixed effort

rithm 2. The initial state of each partial run can be chosen randomly, or in a round-robin fashion among the available initial states [9]. When a fixed effort run ends, the fraction of partial runs started in level l that moved up is an approximation of the conditional probability of reaching level $l + 1$ given that level l was reached. Since *target* states exist only on the highest level, the overall result is thus simply the product of the fraction $n^l{}_{up}/f_E(l)$ for all levels l, i.e. a rational number in the interval $[0, 1]$. The average of the result of many fixed effort runs is again an unbiased estimator for the probability of the transient property [11].

The fixed effort method is specifically designed for transient properties. Its advantage is predictability: each run involves at most $\sum_{l=0}^{max f_L} f_E(l)$ partial runs. Like RESTART needs splitting levels via function f_S, fixed effort needs the effort function f_E that determines the number of partial runs for each level.

3.3 Fixed Success

Fixed effort intuitively controls the simulation effort by adjusting the estimator's imprecision. The fixed success method [1,25] turns this around: its parameters control the imprecision, but the effort then varies. Instead of launching a fixed number of partial runs per level, fixed success keeps launching such runs until $f_E(l)$ of them have reached the next level (or a *target* state in case of the highest level). Illustrated in Fig. 3 (with $f_E(l) = 4$ for all levels), the algorithmic steps are as in Algorithm 2 except for two changes: First, the **for** loop in line 3 is replaced by a **while** loop with condition $L(l).up < f_E(l)$. Second, the final return statement in line 13 uses a different estimator: instead of $\prod_{i=0}^{\max f_L} \frac{L(l).up}{L(l).n}$, we have to return $\prod_{i=0}^{\max f_L} \frac{L(l).up-1}{L(l).n-1}$. This is due to the underlying negative binomial distribution; see [1] for details. The method thus requires $f_E(l) \geq 2$ for all levels l.

From the automation perspective, the advantage of fixed success is that it self-adapts to the (a priori unknown) probability of levelling up: if that probability is low for some level, more partial runs will be generated on it, and vice-versa. However, the desired number of successes still needs to be specified. 20 is suggested as a starting point in [1], but for a specific setting already. A disadvantage of fixed success is that it is not guaranteed to terminate: If the model, importance function

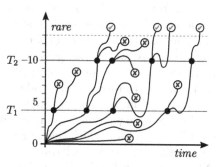

Fig. 3. Fixed success

and thresholds are such that, with positive probability, it may happen that all initial states found for some level lie in a bottom strongly connected component without *target* states, then the (modified) loop of line 3 of the algorithm diverges. We have not encountered this situation in our experiments, though.

4 Determining Thresholds and Factors

To determine the splitting levels/thresholds, we implement and compare two approaches: the sequential Monte Carlo (SEQ) method from [3] and a new technique that tries to ensure a certain expected number of runs that level up.

4.1 Sequential Monte Carlo

Our first approach is inspired by the sequential Monte Carlo splitting technique [6]. It works in two alternating phases: First, n simulation runs determine the importances that can be reached from the current level , keeping track of the state of maximum importance for each run. We sort these states by ascending importance and pick the importance of the one at position $n - k$, i.e. the $(n - k)$-th n-quantile of importances, as the start of the next level. This means that

Input: model M, importance function f_I, transient property ϕ, $n \in \mathbb{N}^+$

1 $f_L := f_I$, $f_E := \{\, l \mapsto n \mid l \in \{\, 0, \ldots, \max f_I \,\} \,\}$
2 $m := 0$, $e := 0$, $p_{up} := \{\, l \mapsto 0 \mid l \in \{\, 0, \ldots, \max f_I \,\} \,\}$
3 **while** $p_{up}(\max f_I) = 0$ **do**   // *roughly estimate the level-up probabilities*
4 $\quad\lfloor$ $m := m + 1$, $L :=$ level data computed in one fixed effort run (Algorithm 2)
5 $\quad\ \ $ **for** l **from** 0 **to** $\max f_I$ **do** $p_{up}(l) := p_{up}(l) + \frac{1}{m}(L(l).up/L(l).n - p_{up}(l))$

6 **for** l **from** 0 **to** $\max f_I$ **do**   // *turn level-up probabilities into splitting factors*
7 $\quad\lfloor$ $split := 1/p_{up}(l) + e$, $F(l) := \lfloor split + 0.5 \rfloor$, $e := split - F(l)$

8 **return** F // *if $F(l) > 1$, then l is a threshold and $F(l)$ the splitting factor*

Algorithm 3. The expected success method for threshold and factor selection

as parameter k grows, the width of the levels decreases and the probability of moving from one level to the next increases. In the second phase, the algorithm randomly selects k new initial states that lie just above the newfound threshold via more simulation runs. This extra phase is needed to obtain new *reachable* states because we cannot *generate* them directly as in the setting of [6]. We then proceed to the next round to compute the next threshold from the new initial states. Detailed pseudocode is shown as Algorithm 5 in [3]. The result is a sequence of importances characterising a level function.

This SEQ algorithm only determines the splitting levels. It does not decide on splitting factors, which the user must select if they wish to run RESTART. FIG and modes request a fixed splitting factor g and then run SEQ with $k = n/g$. When used with fixed effort and fixed success, we set $k = n/2$ and use a user-specified effort value e for all levels. A value for n must also be specified; by default $n = 1000$. The degree of automation offered by SEQ is clearly not satisfactory. Furthermore, we found in previous experiments with FIG that the levels computed by different SEQ runs differed significantly, leading to large variations in RESTART performance [3]. Combined with mediocre results for transient properties, this was the main trigger for the work we present in this paper.

SEQ may get stuck in the same way as fixed success. We encountered this with our *wlan* case study of Sect. 5. Our tool thus restarts SEQ after a 30 s timeout; on the *wlan* model, it then always succeeded with at most two retries.

4.2 Expected Success

To replace SEQ, we propose a new approach based on the rule-of-thumb that one would like the expected number of runs that move up on each level to be 1. This rule is called "balanced growth" by Garvels [11]. The resulting procedure, shown as Algorithm 3, is conceptually much simpler than SEQ: We first perform fixed effort runs, using constant effort n and each importance as a level, until the rare event is encountered. We extract the approximations of the conditional level-up probabilities computed inside the fixed effort runs, averaging the values if we need multiple runs (line 5). After that, we set the factor for each importance to one divided by the (very rough) estimate of the respective conditional probability computed in the first phase. Since splitting factors are natural numbers, we round each factor, but carry the rounding error to the next importance. In this way, even if the exact splitting factors would all be close to 1, we get a rounded splitting factor of 2 for some of the importances. The result is a mapping from importances to splitting factors, characterising both the level function f_L—every importance with a factor $\neq 1$ starts a new level—and the splitting function f_S. We call this procedure the *expected success* (ES) method. Aside from the choice of n (we use a default of $n = 256$, which has worked well in all experiments), it provides full automation with RESTART. To use it with fixed effort, we need a user-specified base effort value e, and then set f_E to $\{\, l \mapsto e \cdot f_S(l) \mid l \in \{0,\ldots,\max f_L\}\,\}$ resulting in a *weighted fixed effort* approach. Note that our default of $n = 256$ is much lower than the default of $n = 1000$ for SEQ. This is because SEQ performs simple simulation runs where ES performs fixed effort runs, each of which provides more information about the behaviour of the model.

We also experimented with expected numbers of runs that move up of 2 and 4, but these always lead to dismal performance or timeouts due to too many splits or partial runs in our experiments, so we do not consider them any further.

5   Experimental Evaluation

The goal of our work was to find a RES approach that provides consistently good performance at a maximal degree of automation. We thus implemented compositional importance function generation, the splitting methods described in Sect. 3, and the threshold calculation methods of Sect. 4 in the modes simulator of the MODEST TOOLSET [14]. This allowed us to study CTMC queueing models, network protocols modelled as PTA, and a more complex fileserver setting modelled as stochastic timed automata (STA [14]) using a single tool.

5.1   Case Studies

tandem: Tandem queueing networks are standard benchmarks in probabilistic model checking and RES [10–12, 24, 29]. We consider the case from [4] with all exponentially distributed interarrival times (a CTMC). The arrival rate into the first queue q_1 (initially empty) is 3 and its service rate is 2. After that, packets move into the second queue q_2 (initially containing one packet), to be processed at rate 6. The model has one parameter C, the capacity of each queue. We estimate the value of the transient property $P_{=?}(q_2 > 0 \ \mathsf{U} \ q_2 = C)$, i.e. of the second queue becoming full without having been empty before.

openclosed: Our second CTMC has two *parallel* queues [13], both initially empty: an *open queue* q_o, receiving packets at rate 1 from an external source, and a *closed queue* q_c that receives internal packets. One server processes packets from both queues: packets from q_o are processed at rate 4 while q_c is empty; otherwise, packets from q_c are served at rate 2. The latter packets are put back into another internal queue, which are independently moved back to q_c at rate $\frac{1}{2}$. We study the system as in [3] with a single packet in internal circulation, i.e. an M/M/1 queue with server breakdowns, and the capacity of q_o as parameter. We estimate $P_{=?}(\neg\, reset\ \mathsf{U}\ lost)$: the probability that q_o overflows before a packet is processed from q_o or q_c such that the respective queue becomes empty again.

breakdown: The final queueing system that we consider [21] as a CTMC consists of ten sources of two types, five of each, that produce packets at rate $\lambda_1 = 3$ (type 1) or $\lambda_2 = 6$ (type 2), periodically break down with rate $\beta_1 = 2$ resp. $\beta_2 = 4$ and get repaired with rate $\alpha_1 = 3$ resp. $\alpha_2 = 1$. The produced packets are collected in a single queue, attended to by a server with service rate $\mu = 100$, breakdown rate $\gamma = 3$ and repair rate $\delta = 4$. Again, and as in [4], we parameterise the model by the queue's capacity, here denoted K, and estimate $P_{=?}(\neg\, reset\ \mathsf{U}\ buf = K)$: starting from a single packet in the queue, what is the probability for the queue to overflow before it becomes empty?

brp: We also study two PTA examples from [15]. The first is the bounded retransmission protocol, another classic benchmark in formal verification. We use parameter M to determine the actual parameters N (the number of chunks to transmit), MAX (the retransmission bound) and TD (the transmission delay) by way of $\langle N, MAX, TD \rangle = \langle 16 \cdot 2^M, 4 \cdot M, 4 \cdot 2^M \rangle$. We thus consider the large instances $\langle 32, 4, 8 \rangle$, $\langle 64, 8, 16 \rangle$ and $\langle 128, 12, 32 \rangle$. To avoid nondeterminism, TD is both lower and upper bound for the delay. We estimate $P_{=?}(true\ \mathsf{U}\ s_{nok} \wedge i > \frac{N}{2})$, i.e. the probability that the sender eventually reports unsuccessful transmission after more than half of the chunks have been sent successfully.

wlan: Our second PTA model is of IEEE 802.11 wireless LAN with two stations. In contrast to [15] and the original PRISM case study, we use the timing parameters from the standard (leading to a model too large for standard probabilistic model checkers) and a stochastic semantics of the PTA (scheduling events as soon as possible and resolving all other nondeterminism uniformly). The parameter is K, the maximum backoff counter value. We estimate $P_{=?}(true\ \mathsf{U}\ bc_1 = bc_2 = K)$, the probability that both stations' backoff counters reach K.

fileserver: Our last case study combines exponentially and uniformly distributed delays. It is an STA model of a file server where some files are archived and require significantly more time to retrieve. Introduced in [14], we change the archive access time from nondeterministic to continuously uniform over the same interval. Model parameter C is the server's queue size. We estimate the time-bounded probability of queue overflow: $P_{=?}(true\ \mathsf{U}_{\leq 1000}\ queue = C)$.

We consider several queueing systems since these are frequently used benchmarks for RES [10–13,21,24,29]. The CTMC could easily be modified to use general distributions and our techniques and tools would still work the same.

5.2   Experimental Setup

The experiments for the *tandem* and *wlan* models were performed on a four-core Intel Core i5-6600T (2.7/3.5 GHz) system running 64-bit Windows 10 v1607 x64 using three simulation threads. All other experiments ran on a six-core Intel Xeon

Table 1. Model data and performance results

model/param		$\hat{p}$	n_I	SMC	RESTART					fixed effort			-weighted			fixed success		
					2	4	8	16	ES	16	64	256	8	16	128	8	32	128
tandem	8	5.6E−6	22	70	3	1	1	11	1	1	1	1	1	1	1	1	1	1
	12	1.9E−8	30	—	45	1	10	190	1	5	4	3	3	2	1	6	2	2
	16	7.1E−11	38	—	—	3	177	588	2	18	8	6	11	6	4	18	7	5
	20	3.0E−13	46	—	—	5	—	—	4	124	23	14	84	21	12	59	17	12
open-	20	3.9E−8	155	—	2	142	3	2	1	5	3	2	6	4	2	5	3	3
closed	30	8.8E−12	235	—	5	—	21	7	1	19	9	9	46	19	6	24	8	8
	40	2.0E−15	315	—	19	—	89	15	3	105	24	17	360	72	14	133	19	20
	50	4.6E−19	395	—	74	—	—	85	4	404	45	33	—	167	38	284	47	34
break-	40	4.6E−4	193	46	7	7	8	11	4	10	10	16	15	13	7	11	9	15
down	80	3.7E−7	353	—	33	24	29	40	23	73	51	61	194	112	44	87	52	54
	120	3.0E−10	513	—	80	59	67	97	104	397	149	173	687	283	139	312	182	136
	160	2.4E−13	673	—	316	109	121	175	583	794	377	290	—	—	335	999	421	313
brp	1	3.5E−7	2k	—	—	—	413	86	21	110	36	33	856	435	226	27	21	50
	2	5.8E−13	6k	—	—	—	—	—	81	—	423	184	—	—	—	208	141	235
	3	9.0E−19	16k	—	—	—	—	—	216	—	—	—	—	—	—	—	420	569
wlan	4	2.2E−5	14k	376	—	—	—	—	—	57	38	31	120	131	221	44	36	39
	5	1.6E−7	23k	—	—	—	—	—	—	457	177	121	784	855	809	139	153	164
file-	50	3.9E−11	156	—	125	88	61	57	27	572	137	75	—	435	79	—	—	140
server	100	4.8E−23	306	—	—	—	—	229	319	—	—	765	—	—	851	—	—	—

E5-2620v3 (2.4/3.2 GHz, 12 logical processors) system with Mono 5.2 on 64-bit Debian v4.9.25 using five simulation threads each for two separate experiments running concurrently. We used a timeout of 600 s for the *tandem*, *openclosed* and *brp* models and 1200 s for the others. Simulations were run until the half-width of the 95 % normal confidence interval was at most 10 % of the currently estimated mean.[1] By this use of a relative width, precision automatically adapted to the rareness of the event. We also performed SMC/Monte Carlo simulation as a comparison baseline (labelled "SMC" in results), where modes uses the Agresti-Coull approximation of the binomial confidence interval. For each case study and parameterisation, we evaluated the following combinations of methods:

- RESTART with thresholds selected via SEQ and a fixed splitting factor $g \in \{2, 4, 8, 16\}$ (labelled "RESTART g"), using $n = 512$ and $k = n/g$ for SEQ;
- RESTART with thresholds and splitting factors determined by the ES method (labelled "RESTART ES") and the default $n = 256$ for ES;
- fixed effort with SEQ ($n = 512$, $k = n/2$) and effort $e \in \{16, 64, 256\}$;
- weighted fixed effort with ES (labelled "-weighted") as described in Sect. 4.2 using base effort $e \in \{8, 16, 128\}$ since all weights are ≥ 2;
- fixed success with SEQ as before ($n = 512$, $k = n/2$) and the required number of successes for each level being either 8, 32 or 128.

We did not consider ES in cases where the splitting factors it computes would not be used (such as with "unweighted" fixed effort or fixed success). The default

---

[1] We rely on the standard CLT assumption for large enough sample sizes; to this end, we do not stop before we obtain at least one sample > 0 and at least 50 samples.

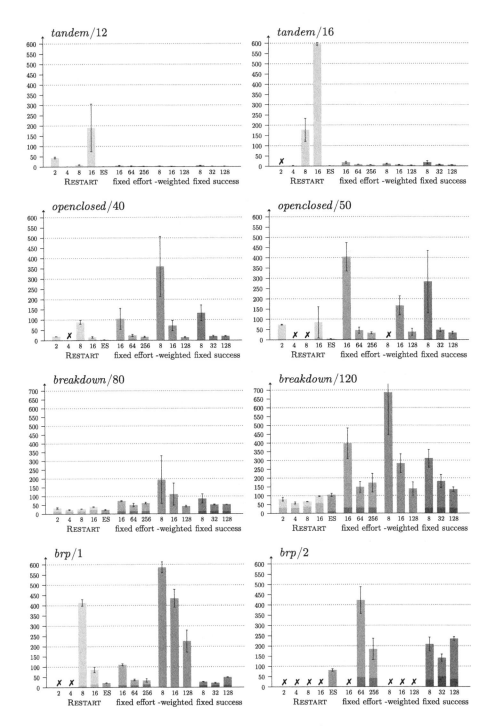

Fig. 4. Selected performance results compared (runtimes in seconds)

of using addition to replace $\wedge$ and $\vee$ in the compositional importance function (cf. Sect. 2.2) worked well except for *wlan*, where we used max instead. RESTART with SEQ and a user-specified splitting factor is the one approach used in FIG [3].

5.3 Results

We provide an overview of the performance results for all model instances in Table 1. We report the averages of three runs of each experiment to account for fluctuations due to the inherent randomisation in the simulation and especially in the threshold selection algorithms. Column $\hat{p}$ lists the average of all (up to 45) individual estimates for each instance. All estimates were consistent, including SMC in the few cases where it did not time out. To verify that the compositional importance function construction does not lead to high memory usage, we list the total number of states that it needs to store in column n_I. These numbers are consistently low; even on the two PTA cases, they are far below the total number of states of the composed state spaces. The remaining columns report the total time, in seconds, that each approach took to compute the importance function, perform threshold selection, and use the respective splitting method to estimate the probability of the transient rare event. Dashes mark timeouts.

We show some interesting cases graphically with added details in Fig. 4. ✗ marks timeouts. Each bar's darker part is the time needed to compute the importance function and thresholds. The lighter part is the time for the actual RES. The former, which is almost entirely spent in threshold selection, is much lower for ES than for SEQ. The error bars show the standard deviation between the convergence times of the three runs that we performed for each experiment.

Our experiments first confirm previous observations made with FIG: The performance of RESTART depends not only on the importance function, but also very much on the thresholds and splitting factor. Out of $g \in \{2, 4, 8, 16\}$, there was no single optimal splitting factor that worked well for all models. RESTART with ES usually performed best, being drastically faster than any other method in many cases. This is a very encouraging result since RESTART with ES is also the one approach that requires no more user-selected parameters. We thus selected it as the default for modes. The *wlan* case is the only one where this default, and in fact none of the RESTART-based methods, terminated within our 1200 s time bound. All of the splitting methods specifically designed for transient properties, however, worked for *wlan*, with fixed success performing best. They also work reasonably well on the other cases, but we see that their performance depends on the chosen effort parameter. In contrast to the splitting factors for RESTART, though, we can make a clear recommendation for this choice: larger effort values rather consistently result in better performance.

6 Conclusion

We investigated ways to improve the automation and performance of importance splitting to perform rare event simulation for general classes of stochastic models. For this purpose, we studied and implemented three existing splitting methods and two threshold selection algorithms, one from a previous tool and one new. Our implementation in the MODEST TOOLSET is publicly available at www.modestchecker.net. We performed extensive experiments, resulting in the only *practical* comparison of RESTART and other methods that we are aware of.

Our results show that we have found a *fully* automated rare event simulation approach based on importance splitting that performs very well: automatic compositional importance functions together with RESTART and the expected success method. It is also easier to implement than our previous approach with SEQ in FIG, and finally pushes automated importance splitting for general models into the realm of very rare events with probabilities down to the order of 10^{-23}.

As future work, we would like to more deeply investigate models with few points of randomisation such as the PTA examples that proved to be the most challenging for our methods, and combine RES with the lightweight scheduler sampling technique [7] to properly handle models that include nondeterminism.

Acknowledgements. We are grateful to José Villén-Altamirano for very helpful discussions that led to our eventual design of the expected success method.

This work is supported by the 3TU.BSR project, ERC grant 695614 (POWVER), the NWO SEQUOIA project, and SeCyT-UNC projects 05/BP12 and 05/B497.

References

1. Amrein, M., Künsch, H.R.: A variant of importance splitting for rare event estimation: Fixed number of successes. ACM Trans. Model. Comput. Simul. **21**(2), 13:1–13:20 (2011)
2. Bayes, A.J.: Statistical techniques for simulation models. Aust. Comput. J. **2**(4), 180–184 (1970)
3. Budde, C.E.: Automation of Importance Splitting Techniques for Rare Event Simulation. Ph.D. thesis, Universidad Nacional de Córdoba, Córdoba, Argentina (2017)
4. Budde, C.E., D'Argenio, P.R., Monti, R.E.: Compositional construction of importance functions in fully automated importance splitting. In: VALUETOOLS (2016)
5. Cérou, F., Guyader, A.: Adaptive multilevel splitting for rare event analysis. Stoch. Anal. Appl. **25**(2), 417–443 (2007)
6. Cérou, F., Moral, P.D., Furon, T., Guyader, A.: Sequential Monte Carlo for rare event estimation. Stat. Comput. **22**(3), 795–808 (2012)
7. D'Argenio, P.R., Hartmanns, A., Legay, A., Sedwards, S.: Statistical approximation of optimal schedulers for probabilistic timed automata. In: Ábrahám, E., Huisman, M. (eds.) IFM 2016. LNCS, vol. 9681, pp. 99–114. Springer, Cham (2016). doi:10.1007/978-3-319-33693-0_7
8. D'Argenio, P.R., Lee, M.D., Monti, R.E.: Input/Output stochastic automata. In: Fränzle, M., Markey, N. (eds.) FORMATS 2016. LNCS, vol. 9884, pp. 53–68. Springer, Cham (2016). doi:10.1007/978-3-319-44878-7_4
9. Garvels, M.J.J., Kroese, D.P.: A comparison of RESTART implementations. In: Winter Simulation Conference, WSC, pp. 601–608 (1998)
10. Garvels, M.J.J., van Ommeren, J.C.W., Kroese, D.P.: On the importance function in splitting simulation. Eur. Trans. Telecommun. **13**(4), 363–371 (2002)

11. Garvels, M.J.J.: The splitting method in rare event simulation. Ph.D. thesis, University of Twente, Enschede, The Netherlands (2000)
12. Glasserman, P., Heidelberger, P., Shahabuddin, P., Zajic, T.: A large deviations perspective on the efficiency of multilevel splitting. IEEE Trans. Autom. Control **43**(12), 1666–1679 (1998)
13. Glasserman, P., Heidelberger, P., Shahabuddin, P., Zajic, T.: Multilevel splitting for estimating rare event probabilities. Oper. Res. **47**(4), 585–600 (1999)
14. Hahn, E.M., Hartmanns, A., Hermanns, H.: Reachability and reward checking for stochastic timed automata. In: ECEASST 70 (2014)
15. Hartmanns, A., Hermanns, H.: A Modest approach to checking probabilistic timed automata. In: QEST, pp. 187–196. IEEE Computer Society (2009)
16. Hartmanns, A., Hermanns, H.: The Modest Toolset: an integrated environment for quantitative modelling and verification. In: Ábrahám, E., Havelund, K. (eds.) TACAS 2014. LNCS, vol. 8413, pp. 593–598. Springer, Heidelberg (2014). doi:10.1007/978-3-642-54862-8_51
17. Hérault, T., Lassaigne, R., Magniette, F., Peyronnet, S.: Approximate probabilistic model checking. In: Steffen, B., Levi, G. (eds.) VMCAI 2004. LNCS, vol. 2937, pp. 73–84. Springer, Heidelberg (2004). doi:10.1007/978-3-540-24622-0_8
18. Jegourel, C., Larsen, K.G., Legay, A., Mikučionis, M., Poulsen, D.B., Sedwards, S.: Importance sampling for stochastic timed automata. In: Fränzle, M., Kapur, D., Zhan, N. (eds.) SETTA 2016. LNCS, vol. 9984, pp. 163–178. Springer, Cham (2016). doi:10.1007/978-3-319-47677-3_11
19. Jegourel, C., Legay, A., Sedwards, S.: Importance splitting for statistical model checking rare properties. In: Sharygina, N., Veith, H. (eds.) CAV 2013. LNCS, vol. 8044, pp. 576–591. Springer, Heidelberg (2013). doi:10.1007/978-3-642-39799-8_38
20. Jegourel, C., Legay, A., Sedwards, S.: An effective heuristic for adaptive importance splitting in statistical model checking. In: Margaria, T., Steffen, B. (eds.) ISoLA 2014. LNCS, vol. 8803, pp. 143–159. Springer, Heidelberg (2014). doi:10.1007/978-3-662-45231-8_11
21. Kroese, D.P., Nicola, V.F.: Efficient estimation of overflow probabilities in queues with breakdowns. Perform. Eval. **36**, 471–484 (1999)
22. Kwiatkowska, M., Norman, G., Parker, D.: PRISM 4.0: verification of probabilistic real-time systems. In: Gopalakrishnan, G., Qadeer, S. (eds.) CAV 2011. LNCS, vol. 6806, pp. 585–591. Springer, Heidelberg (2011). doi:10.1007/978-3-642-22110-1_47
23. Kwiatkowska, M.Z., Norman, G., Segala, R., Sproston, J.: Automatic verification of real-time systems with discrete probability distributions. Theor. Comput. Sci. **282**(1), 101–150 (2002)
24. L'Ecuyer, P., Demers, V., Tuffin, B.: Rare events, splitting, and quasi-Monte Carlo. ACM Trans. Model. Comput. Simul. **17**(2) (2007)
25. LeGland, F., Oudjane, N.: A sequential particle algorithm that keeps the particle system alive. In: EUSIPCO, pp. 1–4. IEEE (2005)
26. Paolieri, M., Horváth, A., Vicario, E.: Probabilistic model checking of regenerative concurrent systems. IEEE Trans. Softw. Eng. **42**(2), 153–169 (2016)
27. Reijsbergen, D., de Boer, P.-T., Scheinhardt, W., Haverkort, B.: Automated rare event simulation for stochastic Petri nets. In: Joshi, K., Siegle, M., Stoelinga, M., D'Argenio, P.R. (eds.) QEST 2013. LNCS, vol. 8054, pp. 372–388. Springer, Heidelberg (2013). doi:10.1007/978-3-642-40196-1_31
28. Rubino, G., Tuffin, B. (eds.): Rare Event Simulation Using Monte Carlo Methods. Wiley (2009)
29. Villén-Altamirano, J.: Rare event RESTART simulation of two-stage networks. Eur. J. Oper. Res. **179**(1), 148–159 (2007)

30. Villén-Altamirano, M., Villén-Altamirano, J.: RESTART: a method for accelerating rare event simulations. In: Queueing, Performance and Control in ATM (ITC-13), pp. 71–76. Elsevier (1991)
31. Villén-Altamirano, M., Villén-Altamirano, J.: RESTART: a straightforward method for fast simulation of rare events. In: WSC, pp. 282–289. ACM (1994)
32. Villén-Altamirano, M., Villén-Altamirano, J.: Analysis of restart simulation: theoretical basis and sensitivity study. Eur. Trans. Telecommun. **13**(4), 373–385 (2002)
33. Younes, H.L.S., Simmons, R.G.: Probabilistic verification of discrete event systems using acceptance sampling. In: Brinksma, E., Larsen, K.G. (eds.) CAV 2002. LNCS, vol. 2404, pp. 223–235. Springer, Heidelberg (2002). doi:10.1007/3-540-45657-0_17
34. Zimmermann, A., Maciel, P.: Importance function derivation for RESTART simulations of Petri nets. In: RESIM 2012, pp. 8–15 (2012)
35. Zimmermann, A., Reijsbergen, D., Wichmann, A., Canabal Lavista, A.: Numerical results for the automated rare event simulation of stochastic Petri nets. In: RESIM, pp. 1–10 (2016)

On the Criticality of Probabilistic Worst-Case Execution Time Models

Luca Santinelli[1]([✉]) and Zhishan Guo[2]

[1] ONERA, Toulouse, France
`luca.santinelli@onera.fr`
[2] Missouri University of Science and Technology, Rolla, USA
`guozh@mst.edu`

Abstract. Probabilistic approaches to timing analysis derive probability distributions to upper bound task execution time. The main purpose of probability distributions instead of deterministic bounds, is to have more flexible and less pessimistic worst-case models. However, in order to guarantee safe probabilistic worst-case models, every possible execution condition needs to be taken into account.

In this work, we propose probabilistic representations which is able to model every task and system execution conditions, included the worst-cases. Combining probabilities and multiple conditions offers a flexible and accurate representation that can be applied with mixed-critical task models and fault effect characterizations on task executions. A case study with single- and multi-core real-time systems is provided to illustrate the completeness and versatility of the representation framework we provide.

1 Introduction

Nowadays real-time systems are mostly implemented with multi-core and many-core commercial-off-the-shelf platforms. Cache memories, branch predictors, communication buses/networks and other features present in such implementations allow increasing average performance; nonetheless, they make predictability much harder to guarantee.

Task execution times are affected by the large variety of system execution conditions which can happen at runtime. Also, the numerous interferences within real-time systems may result into significant variations of tasks execution times. This acerbates with multi- and many-core real-time systems implementations. In essence, task execution time resembles to a random variable where the value depends on different outcomes.

Timing analysis seeks upper bounds to the task execution time, and the predictability required by real-time systems can be granted. Classically, the bounds are deterministic Worst-Case Execution Times (WCET) which are single values able to upper bound the time needed to finish execution. In order to be safe, WCETs have to account for any case/execution condition possible, including the highly improbable pathological cases such as faults. Deterministic WCETs

© Springer International Publishing AG 2017
K.G. Larsen et al. (Eds.): SETTA 2017, LNCS 10606, pp. 59–74, 2017.
https://doi.org/10.1007/978-3-319-69483-2_4

could be very pessimistic with respect to task actual execution times, and could lead to resource under-utilization.

Probabilistic worst-case models generalize WCETs with worst-case distributions, probabilistic WCETs (pWCETs), which upper bounds any possible task execution behavior. pWCET representations focus on probability of occurrence of worst-case conditions and abstract them into multiple worst-case values with their correspondent probability of happening. The challenge with probabilistic timing analysis is guaranteeing the pWCET safety.

System faults have a non negligible impact on worst-case models; although as pathological and improbable cases, they have to be considered with timing analysis. Task models have to embed fault manifestations and fault effects in order to provide upper bounds to every possible condition the real-time system can experience at runtime.

With multi- and many-core implementations, it comes the opportunity to combine different applications on the same platform which may have different criticality levels, e.g. safety-critical, mission-critical, non-critical, designating the level of assurance needed against failure. The very low acceptable failure rates (e.g. 10^{-9} failures per hour) for high criticality applications imply the need for significantly more rigorous and costly development as well as verification processes than required by low criticality applications. The gradual transformation of real-time systems into Mixed Criticality (MC) systems demands for timing analysis ables to cope with multiple criticality levels for tasks.

State of the Art: First papers on probabilistic timing modeling describe the worst-case execution time of tasks with random variables, using either discrete [2,22] or continuous [15] distributions. Since Edgar and Burns [10], several papers have worked on obtaining safe and reliable probabilistic Worst-Case Execution Time (pWCET) estimates [7,13,14].

Among probabilistic timing analysis approaches, it is possible to distinguish between Static Probabilistic Timing Analysis (SPTA) and Measurement-Based Probabilistic Timing Analysis (MBPTA). SPTA methods analyze the software and use a model of the hardware behavior to derive an estimate of pWCET; SPTA is applicable when some part of the system or the environment have been artificially time randomized, [3,8]. MBPTA approaches rely on the Extreme Value Theory (EVT) for computing pWCET estimates out of measured behaviors [6,11,16]. Figure 1 depicts key elements for MBPTA which accepts input measurements of task execution times under specific execution conditions and applies to those the EVT for inferring pWCET estimates.

Fault modeling and fault management intertwines with timing analysis as faults introduce latencies to the task execution behavior which have to be embedded into the task models. As examples, in [19] backups are executed for fault tolerance and recovering form task errors caused by hardware or software faults; in [25] it is proposed an algorithm to abort and restart a task in case of conflicts in shared resources accesses. In [21], an attempt of including faults into MBPTA and apply fault-based task models for schedulability and sensitivity analysis of real-time systems.

Research on MC scheduling focuses upon the Vestal task model which assigns multiple WCET estimates to each task, [23]. This is motivated by the fact that different tools for WCET estimates may be more or less conservative than one another, [4,24] as reviews. To the best of our knowledge, [12] is a first attempt of a probabilistic MC task modeling with WCET estimates associated to the probability/confidence of being WCET bound.

Contribution: This work proposes a probabilistic representation framework for real-time tasks which composes of multiple probabilistic worst-case models, each estimating the worst-case of a specific possible execution condition that both tasks and the system can encounter.

The probabilistic task model proposed can be applied to the MC problem, since probabilities and multiple pWCETs can characterize the criticality modes as different task conditions as well as different confidences. Besides, the probabilistic representation can be used for characterize the effects that faults have on the tasks executions, proving to be flexible and safe. A case study is presented for validating the probabilistic models and illustrating its effectiveness in modeling different conditions/criticalities.

Organization of the paper: Section 2 presents the probabilistic background applied in this work. Section 3 describes the probabilistic worst-case model proposed and based on multiple execution conditions possible. Section 4 details the MC task representation derived from the the probabilistic worst-case model. The impact of faults on task models are also considered. Section 5 shows probabilistic MC task models from three different real-time test cases; Sect. 6 is for conclusions and future works.

2   System Modeling

A real-time task consists of a sequence of recurring jobs, each has to complete execution by a given deadline.

In a periodic task system, a task is described with the 4-tuple (O_i, T_i, D_i, C_i), where O_i is the offset or starting time that specifies the time instant at which the first job of τ_i is released. T_i is the period as the temporal separation between two successive jobs of τ_i. D_i is the deadline which defines the time interval $[O_i + j \cdot T_i, O_i + j \cdot T_i + D_i)$ in which task execution has to take place (the j-th job of τ_i). C_i is the WCET defining the processing requirements for each job.

The scheduling policy decides the task execution ordering, possibly with preemption or migration between cores; schedulability analysis of task models guarantees the respect of the timing constraints (system predictability) checking if there are enough resources for the tasks to finish executions by their deadlines.

In this work, we consider a set $\Gamma = \{\tau_1, \ldots, \tau_n\}$ of n periodic *probabilistic tasks* τ_i as such their worst-case execution time is modeled with pWCETs. The probabilistic modeling framework proposed applies to either single-core, multi-core or many-core processors.

2.1   pWCETs and WCET Thresholds

The pWCET $\mathcal{C}_i$ of a task τ_i is defined as the worst-case estimate distribution that upper-bounds any possible execution time the task can exhibit [8]. $\mathcal{C}_i$ generalize deterministic WCET estimates C_i by including multiple values, each with the probability of being the worst-case of the task execution time[1].

Assuming the pWCET $\mathcal{C}_i$ as continuous distribution, the probability distribution function (pdf) $\mathsf{pdf}_{\mathcal{C}_i}$ describes the probability of happening of certain events from the random variable $\mathcal{C}_i$; it is such that $P(C_1 \le \mathcal{C}_i \le C_2) = \int_{C_1}^{C_2} \mathsf{pdf}_{\mathcal{C}_i}(C)dC$ and $\int_0^\infty \mathsf{pdf}_{\mathcal{C}_i}(C)dC = 1$.

$\mathsf{cdf}_{\mathcal{C}_i}$ denotes the cumulative distribution function (cdf) representation of $\mathcal{C}_i$, $\mathsf{cdf}_{\mathcal{C}_i}(C) = P(\mathcal{C}_i \le C) = \int_0^C \mathsf{pdf}_{\mathcal{C}_i}(x)$ while the inverse cumulative distribution function (icdf) $\mathsf{icdf}_{\mathcal{C}_i}(C)$ outlines the exceedence thresholds. $\mathsf{icdf}_{\mathcal{C}_i}(C) = P(\mathcal{C}_i \ge C)$ is the probability of having execution time greater than C, $\mathsf{icdf}_{\mathcal{C}_i}(C) = 1 - \int_0^C \mathsf{pdf}_{\mathcal{C}_i}(x)$.

In case of discrete distributions pWCETs, it is $\mathsf{pdf}_{\mathcal{C}_i}(C) = P(\mathcal{C}_i = C)$, $\mathsf{cdf}_{\mathcal{C}_i}(C) = \sum_0^C \mathsf{pdf}_{\mathcal{C}_i}(x)$ and $\mathsf{icdf}_{\mathcal{C}_i}(C) = 1 - \sum_0^C \mathsf{pdf}_{\mathcal{C}_i}(x)$.

WCET Thresholds From $\mathcal{C}_i$, it is possible to define *WCET thresholds* $\langle C_{i,j}, p_{i,j} \rangle$ where the value $C_{i,j}$ is associated to the probability $p_{i,j}$ of being the WCET for τ_i. $p_{i,j} \overset{def}{=} \mathsf{icdf}_{\mathcal{C}_i}(C_{i,j})$ quantifies the *confidence* on $C_{i,j}$ of being the task worst-case execution time and $1 - p_{i,j}$ is the probability of respecting $C_{i,j}$. Depending on the granularity of the pWCET, it would be possible to define WCET thresholds at probability of 10^{-3}, 10^{-6}, 10^{-9}, etc.

Worst-Case Distribution Independence. Assuming $\mathcal{C}_i$ to be the probabilistic worst-case distribution estimate of τ_i, it means that in $\mathcal{C}_i$ there have already been included all the possible interferences (and their effects as latencies) that τ_i suffers, [5]. This is the definition of statistical independences between tasks, i.e. the task execution distribution does not change in presence or not of interferences – conditional probability.

For example, the pWCET distribution of task τ_i in presence of τ_j, equivalently while executing together with τ_j (the conditional distribution $\mathsf{pdf}_{\mathcal{C}_i | \mathcal{C}_j}$) does not change from the case where τ_i runs in isolation ($\mathsf{pdf}_{\mathcal{C}_i}$) since all the interferences from τ_j have been taken into account in the worst-case distribution bound and in order to guarantee it. It is $\mathsf{pdf}_{\mathcal{C}_i | \mathcal{C}_j} = \mathsf{pdf}_{\mathcal{C}_i}$ which corresponds to the definition of statistical independence between pWCETs and thus tasks. Previous independence condition holds with all the tasks in Γ and for all the system execution conditions possible, i.e. worst-case distributions guarantee independence between between tasks.

---

[1] In the following, calligraphic letters are used to represent distributions while non-calligraphic letters are for scalars or deterministic values.

3   Probabilistic Worst-Case Representations

Whenever correctly applied, the EVT produces a continuous distribution which is a safe estimation of the worst-case behavior of the task. The EVT guarantees that if certain hypotheses are verified, from the actual measured behavior it is possible to infer rare events, where the worst-case execution time lie [6]. The outcome of the MBPTA/EVT is the pWCET estimate C_i.

Figure 1 shows the basics of MBPTA with the EVT applied to measurements of task execution time – average execution time – for inferring pWCET estimation C_i. The task under observation is τ_i while the rest of the real-time application $\Gamma \setminus \tau_i$ contributes to the interferences on τ_i. Besides, the specific measurement execution condition applied for the task executions $s^k = f(I, Env, Map, \ldots)$ and the measurements themselves define the task's actual behavior.

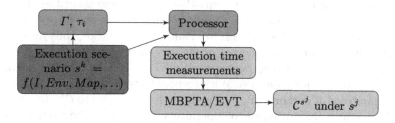

Fig. 1. MBPTA where pWCETs are inferred with the EVT specific execution conditions $s^k = f(I, Env, Map, \ldots)$ applied for measurements.

The guarantees that the EVT provides worst-case task models strongly depend on what has been measured, e.g. the execution conditions for the measurements, the confidence or representativity of the measurements.

A trace of execution time measurements accounts for some of the interfering conditions and inputs (to the system and tasks) which happen at runtime. The pWCET estimate C_i from the EVT embeds those system conditions and others which have not been measured (the so called rare events which are costly to observe by measurements), i.e. the EVT is able to infer some of the unknowns from the known measurements. Unfortunately, not all the unknowns can be estimated with the only use of the EVT.

An execution scenario $s^j = f(I, Env, Map, \ldots)$ abstracts the execution conditions the system (and the task) subdue to. s^j represents instances of the inputs (for tasks and system) I, of the environment Env, of the task mapping and scheduling policy Map, etc.; s^k is a function of I, Env, Map and more. For a real-time system, there exist a finite set S of all the possible execution scenarios, $S = \{s^1, s^2, \ldots, s^n\}$, since inputs, environment conditions, mapping, etc. are finite.

The same reasoning would be applicable to SPTA with different conditions $s^j = f(I, Env, Map, \ldots)$ possible for tasks.

3.1   Worst-Case Bounding

The *absolute* pWCET $\mathcal{C}_i$ is the worst-case distribution that upper bounds every task execution time obtained under any possible execution scenario $s^j \in S$. The absolute pWCET $\mathcal{C}_i$ is safe if it upper bounds every task execution time under any execution scenario.

Given $s^j \in S$, the pWCET $\mathcal{C}_i^{s^j}$ comes from the measurements taken under s^j and the EVT applied to them (Fig. 1). $\mathcal{C}_i^{s^j}$ is the pWCET specific to s^j, the *relative* pWCET; the relative pWCET $\mathcal{C}_i^{s^j}$ is safe if it upper bounds any task execution time under s^j.

Measurement representativity is a fundamental requirement for guaranteeing both absolute and relative pWCETs. We hereby focus on representativity as the measurement capability of well characterizing multiple execution conditions (worst-cases included) like in [1,17]. Differently than those works, we do not consider artificially randomized systems that aim at increasing the chances of measuring the worst-case. We believe that the input representativity can be built from an enhanced knowledge of the system and of its scenarios, thus *from a study of the system, its S and the coverage of the execution conditions.*

From the partial ordering between pWCETs [9], it is possible defining a notion of *dominance* for scenarios. With respect to task τ_i, given s^r and s^t from S, s^r dominates s^t if and only if $\mathcal{C}_i^{s^r}$ is greater than or equal to $\mathcal{C}_i^{s^t}$, $\mathcal{C}_i^{s^r} \succeq \mathcal{C}_i^{s^t}$, $\succeq$ being the "greater than or equal to" operator which defines the partial ordering between distributions [9].

It is also possible defining the notion of *equivalence* between scenarios. Given s^k and s^j from S, with respect to task τ_i s^k is equivalent to s^j if and only if there exist values in the support of $\mathcal{C}_i^{s^k}$ and $\mathcal{C}_i^{s^j}$ for which $\mathcal{C}_i^{s^k} \succeq \mathcal{C}_i^{s^j}$ and there exist other values in the support of $\mathcal{C}_i^{s^k}$ and $\mathcal{C}_i^{s^j}$ for which $\mathcal{C}_i^{s^j} \succeq \mathcal{C}_i^{s^k}$.

For a set of equivalent scenarios $S^j = \{s^j, s^k, \ldots, s^t\} \subseteq S$ ($s^k, \ldots, s^t$ equivalent to s^j), it is possible defining the scenario s^{j*} that dominates all the scenarios in S^j. It would be such that $\mathcal{C}_i^{s^{j*}} \overset{def}{=} max_{s^j \in S^j}\{\mathcal{C}_i^{s^j}\}$, while with the icdf representation, it would be $\mathrm{icdf}_{\mathcal{C}_i^{s^{j*}}}(C) \overset{def}{=} max_{s^j \in S^j}\{\mathrm{icdf}_{\mathcal{C}_i^{s^j}}(C)\}$. s^{j*} is not a real scenario, but it dominates all the $s^k \in S^j$.

Worst-Case Set. The *Worst-Case Set* task representation is the collection of all the pWCET from S; $\overline{\mathcal{C}}_i$ is the Worst-Case Set representation as a set of pWCET estimates such that:

$$\overline{\mathcal{C}}_i \overset{def}{=} (\mathcal{C}_i^{s^1}, \mathcal{C}_i^{s^2}, \ldots, \mathcal{C}_i^{s^n}). \tag{1}$$

With partial ordering between relative pWCETs it is possible ordering scenarios and get $S = \{s^1, s^2, s^3, \ldots, s^k\}$ such that $\mathcal{C}^{s^k} \succeq \mathcal{C}^{s^{k-1}} \succeq \ldots \succeq \mathcal{C}^{s^1}$; the Worst-Case Set becomes:

$$\overline{\mathcal{C}}_i \overset{def}{=} (\mathcal{C}_i^{s^1}, \mathcal{C}_i^{s^2}, \ldots, \mathcal{C}_i^{s^k}), \tag{2}$$

with s^k the worst-case scenario for τ_i, $s^{worst} \equiv s^k$.

Although, we hereby focus on MBPTA, the Worst-Case Set representation applies to both MBPTA and SPTA with multiple execution conditions possible.

Worst-Case Set and Dominance. Although with actual real-time systems it is reasonable to assume a finite number of measurement scenarios, enumerating them all remains a complex problem. With dominance between scenarios, it would be possible neglecting the dominated scenarios in order to ease the task representation from Eqs. (1) and (2). Moreover, with equivalence between scenarios it would be possible to assume the correspondent dominating scenario $s^{*,j}$ to represent all the equivalent scenarios S^j.

From Eqs. (1) and (2), fewer dominating scenarios S^* could be considered to represent the task execution behavior. $S^* = \{s^r, s^j, s^k\} \subseteq S$ is such that s^r dominates some scenarios in S, s^j dominates other scenarios as well as s^r and s^k dominates all the scenarios. The Worst-Case Set becomes:

$$\overline{\mathcal{C}}_i \overset{def}{=} (\mathcal{C}_i^{s^r}, \mathcal{C}_i^{s^j}, \mathcal{C}_i^{s^k}), \tag{3}$$

as a less complex probabilistic representation to the task executions; Eq. (3) remains a safe representation for the task behavior since the worst-cases s^k and $\mathcal{C}_i^{s^k}$ are included.

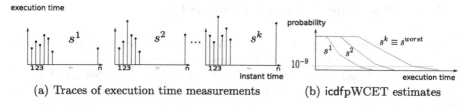

(a) Traces of execution time measurements   (b) icdfpWCET estimates

Fig. 2. Multiple scenarios $S = \{s^1, s^2, \ldots, s^k \equiv s^{worst}\}$, each with a trace of execution time measurements and pWCET estimate.

Figure 2 depicts $S = \{s^1, s^2, \ldots, s^k \equiv s^{worst}\}$, each scenario s^j with a trace of execution time measurements; the resulting pWCETs $\mathcal{C}_i^{s^k}$ are illustrated with the partial ordering guaranteed by $\succeq$.

At this stage, S is assumed to be known; future work will investigate how to obtain the different scenarios and how to guarantee the existence of worst-cases among them.

3.2 Probabilistic Task Models

Combining the orthogonal information of WCET thresholds (with probability associated) and execution scenarios from the Worst-Case Set, Eqs. (1), (2) or (3), there are possible the *inter-scenario* and *intra-scenario* representations.

Inter-scenario Representation. The inter-scenario representation characterizes task behavior across scenarios. Given an exceeding probability p and the
WCET threshold at that exceeding probability for each scenario $s^j \in S$ (equivalently $s^j \in S^*$), it is $\langle \overline{C}_i, p \rangle$ such that:

$$\overline{C}_i \stackrel{def}{=} (C_i^{s^1}, C_i^{s^2}, \dots, C_i^{s^k}) \tag{4}$$

is the set of WCET thresholds such that $\langle C_i^j, p \rangle \; \forall s^j \in S$. As an example, it is
possible picking $p = 10^{-9}$ with $\langle C_i^j, 10^{-9} \rangle \; \forall s^j \in S$. Equation (4) is the inter-
scenario representation for the task worst-case execution time.

Intra-scenario Representation. The intra-scenario representation describes
the task behavior focusing on a specific scenario. For a given scenario $s^j \in
S$ (equivalently $s^j \in S^*$) and a set of exceeding thresholds probabilities
$(p_1, p_2, \dots p_m)$ it is:

$$\hat{C}_i^{s^j} \stackrel{def}{=} (\langle C_{1,i}^{s^j}, p_1 \rangle, \langle C_{2,i}^{s^j}, p_2 \rangle, \dots, \langle C_{m,i}^{s^j}, p_m \rangle). \tag{5}$$

Equation (5) is the intra-scenario representation for the task worst-case execution time on a specific scenario with all the meaningful WCET thresholds and
exceeding probabilities.

4   Task Modeling Through Criticalities

Each pWCET estimations composing the Worst-Case Set representation implicitly carries confidence (as safety) of being the absolute task pWCET; execution
scenarios may be more or less safe in defining pWCET estimates and WCET
thresholds. For example, s^1 from Eq. (2) provides the least confident absolute
pWCET as $C_i^{s^1}$: $C_i^{s^1}$ is the least safe absolute pWCET for τ_i; s^2 provides slightly
more confidence that $C_i^{s^2}$ is the absolute pWCET for τ_i: $C_i^{s^2}$ is relatively more
safe than $C_i^{s^1}$. Going on with the scenarios within S, the safety increases up
until s^k which is the worst-case scenario and $C_i^{s^k}$ is the only 100% safe absolute
pWCET for τ_i; $C_i^{s^k}$ is the safest among the pWCETs.

The MC task model makes use of multiple WCETs for characterizing the
task behavior; such bounds results from different timing analysis tools as well as
different criticality requirements that task can respect at runtime. For example,
in the two-criticality-level case, each task is designated as being of either higher
(HI) or lower (LO) criticality, and two WCETs are specified for each HI-criticality
task: a LO-WCET determined by a less pessimistic tool or a less demanding
safety requirements (e.g. mission-critical or non-critical), and a larger HI-WCET
determined by a more conservative tool or more safety-critical requirements.

For real-time systems, safety and criticality have a strong relationship so that
they can be interchanged whenever applied for timing analysis and schedulability
analysis: a safe pWCET is the worst-case models which can apply with high
critical modes.

Least critical Scenario LO-critical. In case of s^1 from Eq. (2), the task has the least execution time, thus the least dominating relative pWCET $\mathcal{C}_i^{s^1}$. $\mathcal{C}_i^{s^1}$ upper bounds any (and only) possible execution time resulting from s^1; it is the last safe absolute pWCET, equivalently the least critical LO-criticality. $\mathcal{C}_i^{s^1}$ is applied to characterize LO-criticality requirements of τ_1 and the LO-criticality functional mode.

Critical Scenarios MI-critical. From S ordered by dominance, Eq. (2), s^2 dominates s^1 because under s^2 the task suffers execution times bigger than under s^1. Considering $\mathcal{C}_i^{s^2}$ as absolute pWCET, it would be slightly safer than $\mathcal{C}_i^{s^1}$, but it is not safe enough to upper bound the other $s^j \in S$. $\mathcal{C}_i^{s^2}$ is the middle criticality (MI-criticality) characterization for τ_i.

With s^3, it is $\mathcal{C}_i^{s^3}$ dominating $\mathcal{C}_i^{s^2}$ and $\mathcal{C}_i^{s^1}$ since s^3 produces larger execution times than s^1 and s^2. Thus, $\mathcal{C}_i^{s^3}$ would be safer than $\mathcal{C}_i^{s^2}$ as absolute pWCET. Also $\mathcal{C}_i^{s^3}$ is a MI-criticality characterization for τ_i but more critical than $\mathcal{C}_i^{s^2}$.

We distinguish between MI-2-criticality and MI-3-criticality, respectively for s^2 and s^3, and $\mathcal{C}_i^{s^3} \succeq \mathcal{C}_i^{s^2}$. Other intermediate criticality levels can be defined from $s^j \in (S \setminus s^k)$.

Most Critical Scenario, HI-critical. The pWCET $\mathcal{C}_i^{s^k}$ from s^k is the safest absolute pWCET. $\mathcal{C}_i^{s^k}$ is also the HI-criticality bound to the task behavior. $\mathcal{C}_i^{s^k} \equiv \mathcal{C}_i^{s^{worst}}$ represents the worst conditions and is the most conservative upper bound for τ_i to be applied in the highest critical modes.

From $\mathcal{C}_i^{s^1}$ it would be possible to extract $\langle C_i(\text{LO}), 10^{-9} \rangle$. We name $C_i(\text{LO})$ the LO-critical WCET threshold as it results from the least safe pWCET model $\mathcal{C}_i^{s^1} \equiv \mathcal{C}_i^{\text{LO}}$. $C_i(\text{LO})$ is the LO-criticality WCET threshold with a confidence of 10^{-9}.

s^j, with $1 < j < k$ form Eq. (2), is a MI-j-criticality scenario that upper bounds all the scenarios s^r such that $r \leq j$; $\mathcal{C}_i^{s^j} \equiv \mathcal{C}_i^{\text{MI}-j}$ is the MI-j-criticality pWCET. $\mathcal{C}_i^{s^j} \succeq \mathcal{C}_i^{s^{j-1}}$ and $\langle C_i(\text{MI} - j), 10^{-9} \rangle$ is such that $C_i(\text{MI} - j) \geq C_i(\text{MI} - j - 1)$; $C_i(\text{MI} - j)$ is the MI-j-criticality WCET threshold with a confidence of 10^{-9}.

s^k is the worst-case scenario and $\mathcal{C}_i^{s^k}$ is the absolute pWCET for τ_i. $\mathcal{C}_i^{s^k} \equiv \mathcal{C}_i^{\text{HI}}$ is the HI-criticality pWCET and s^k is the HI-criticality scenario for the worst conditions. From $\mathcal{C}_i^{\text{HI}}$, it is $\langle C_i(\text{HI}), 10^{-9} \rangle$ such that $C_i(\text{HI})$ is the HI-criticality WCET threshold. $\mathcal{C}_i^{\text{HI}} \succeq \mathcal{C}_i^{\text{MI}-j}$ and $C_i(\text{HI}) \geq C_i(\text{MI} - j)$.

From the difference in safety/criticality between s^{worst}, s^3, s^2 and s^1 execution conditions, it is $\mathcal{C}_i^{\text{HI}} \succeq \mathcal{C}_i^{\text{MI}} \succeq \mathcal{C}_i^{\text{LO}}$. Also, $C_i(\text{HI}) \geq C_i(\text{MI}) \geq C_i(\text{LO})$ for the same probability p from respectively $\mathcal{C}_i^{\text{HI}}$, $\mathcal{C}_i^{\text{MI}}$ and $\mathcal{C}_i^{\text{LO}}$. How much they differ depends on the relationship between the scenarios and the impact that the scenarios have on the execution times of tasks. $p = 10^{-9}$ is chosen arbitrarily, but the probabilistic modeling proposed can make use of any probability, depending on the confidence requirements.

The MC Worst-Case Set representation for τ_i is:

$$\overline{C}_i \stackrel{def}{=} (\mathcal{C}_i^{\text{LO}}, \ldots, \mathcal{C}_i^{\text{MI}-j}, \ldots, \mathcal{C}_i^{s^k}). \tag{6}$$

For the intra- and inter-scenario perspective, adding criticality levels to Eqs. (4) and (5) it is $\overline{C}_i \stackrel{def}{=} (C(\text{LO})_i, \ldots, C(\text{HI})_i)$ for the inter-scenario MC representation $\langle \overline{C}_i, p \rangle$ at probability p and $\hat{C}(\updownarrow)_i \stackrel{def}{=} (\langle C(\updownarrow)_{1,i}, p_1 \rangle, \langle C(\updownarrow)_{2,i}, p_2 \rangle, \ldots, \langle C(\updownarrow)_{m,i}, p_m \rangle)$ for the intra-scenario MC representation at the criticality level l and probabilities $p_1, p_2, \ldots p_m$.

MC Probabilistic Task Model. With three criticality levels, the MC task model based on the Worst-Case Set is:

$$\tau_i = ([\overline{C}_i, \langle \overline{C}_i, 10^{-9} \rangle, (\hat{C}_i^{\text{LO}}, \hat{C}_i^{\text{MI}}, \hat{C}_i^{\text{HI}})], T_i, D_i), \tag{7}$$

where $\overline{C}_i = (\mathcal{C}_i^{\text{LO}}, \mathcal{C}_i^{\text{MI}}, \mathcal{C}_i^{\text{HI}})$ and $\langle \overline{C}_i, p \rangle = (C_i^{\text{LO}}, C_i^{\text{MI}}, C_i^{\text{HI}})$. The intra-scenario representation is such that $\hat{C}_i^{\text{LO}} = (\langle C_{1,i}^{\text{LO}}, 10^{-3} \rangle, \langle C_{2,i}^{\text{LO}}, 10^{-6} \rangle, \langle C_{3,i}^{\text{LO}}, 10^{-9} \rangle)$ for the LO-safety, $\hat{C}_i^{\text{MI}} = (\langle C_{1,i}^{\text{MI}}, 10^{-3} \rangle, \langle C_{2,i}^{\text{MI}}, 10^{-6} \rangle, \langle C_{3,i}^{\text{MI}}, 10^{-9} \rangle)$ for the MI-safety scenario and $\hat{C}_i^{\text{HI}} = (\langle C_{1,i}^{\text{HI}}, 10^{-3} \rangle, \langle C_{2,i}^{\text{HI}}, 10^{-6} \rangle, \langle C_{3,i}^{\text{HI}}, 10^{-9} \rangle)$ for the HI-safety scenario.

The MC task model is essentially asserting that depending on the conditions for the timing analysis applied it is possible to have more or less guarantees on the pWCET and the WCET thresholds estimates. Only by considering most of the possibilities (necessarily the dominating ones) the MC worst-case models are safe. The MC task model can be generalized to multiple criticality levels and different probabilities in order to better cope with the requirements.

Worst-Case Sets Independence. It is necessary to investigate the statistical independence between criticality levels and pWCETs for the MC Worst-Case Set representation. It has already been showed that there exist independence between absolute pWCETs, thus between HI-criticality representations $\mathcal{C}_i^{\text{HI}}$ and $C_i(\text{HI})$ from s^k.

Supposing to have s^j representing a criticality level other than HI-criticality, what happens to the pWCET estimates of τ_i and τ_k? Under s^j, the conditional probability $\text{pdf}_{\mathcal{C}_i^{s^j} | \mathcal{C}_k^{s^j}}$ equals $\text{pdf}_{\mathcal{C}_i^{s^j}}$ (equivalently $\text{pdf}_{\mathcal{C}_k^{s^j} | \mathcal{C}_i^{s^j}}$ equals $\text{pdf}_{\mathcal{C}_k^{s^j}}$) because all the effects from s^j have been included into the relative pWCETs $\mathcal{C}_i^{s^j}$ and $\mathcal{C}_k^{s^j}$. This assures tasks independence with the same scenario.

With s^r dominating s^j for both τ_i and τ_k, what happens to $\mathcal{C}_i^{s^r}$ and $\mathcal{C}_k^{s^j}$? It is $\text{pdf}_{\mathcal{C}_i^{s^r} | \mathcal{C}_i^{s^j}} = \text{pdf}_{\mathcal{C}_i^{s^r}}$, since all the effects of s^j and τ_k on τ_i have been already taken into account by $\mathcal{C}_i^{s^r}$. This guarantees the independence between τ_i and τ_k under s^r and s^j, respectively for τ_i and τ_k.

Worst-Case Set representations guarantee tasks independence and independence between criticality levels which will ease task combination and schedulability analysis.

4.1 Fault Occurrence and Effect

Faults can be modeled with the probability $f(t, T)$ of a fault occurring; $f(t, T)$ is the probability of fault in a system component by time t, $failure \leq t$, given that the component was still functional at the end of the previous interval $t - T$, $failure > t - T$. T is the scrubbing period, i.e. the time interval between two consecutive fault detection to avoid error accumulation.

Several probability distributions are used to model failure times [18]. One commonly used is the log-normal failure distribution:

$$f(t, T) = \frac{\mathsf{cdf}_{norm}\frac{ln(t)-\mu}{\sigma} - \mathsf{cdf}_{norm}(\frac{ln(t-T)-\mu}{\sigma})}{1 - \mathsf{cdf}_{norm}(\frac{ln(t-T)-\mu}{\sigma})}, \tag{8}$$

where cdf_{norm} the cumulative density function of the normal distribution. The mean and standard deviation parameters of such distribution can be computed from the Mean Time To Failure (MTTF) such that $\mu = ln(MTTF^2/\sqrt{var_{MTTF} + MTTF^2})$ and $\sigma = \sqrt{ln(1 + var_{MTTF}/MTTF^2)}$. In Eq. (8), $f(t, T)$ depends on the actual time t and the scrubbing period T.

Fault and Criticalities. Faults (either transient or permanent) translate into penalties δ (latencies) to the task execution time which depends on the time t the fault happens, $\delta(t)$. With $C(t)$ the expected task execution time at time t, in presence of fault it would be $C(t) + \delta(t)$ the task execution time accounting for the fault penalty on task computations. With a measurement-based approach, fault effects on task execution can be measured and directly embedded into traces of execution time measurements; then, with EVT it is possible infer pWCET estimates which upper bounds faulty execution conditions.

Different scenarios are possible with respect to faults. By considering non-faulty conditions (fault never happening), it is scenario s^{NF} that describes the task behavior. Here, the execution times observed exploit only the task functional behavior due to the absence of faults. For s^{NF} it exists $C_i^{s^{NF}}$; s^{NF} is the LO-criticality scenario with $C_i^{\mathrm{LO}} \equiv C_i^{NF}$, $C(\mathrm{LO})$ and $\hat{C}_i(\mathrm{LO})$ representing it.

It could also exist s^{FW} which assumes that the worst fault condition manifests at runtime; C_i^{FW} is the pWCET estimate for s^{FW}. s^{FW} is the wort-case scenario where the task executes always under the most critical conditions; $C_i^{\mathrm{HI}} \equiv C_i^{FW}$, $C_i(\mathrm{HI})$ and $\hat{C}_i(\mathrm{HI})$ represents it.

In between these two extreme scenarios, it exist a set of possible faulty scenarios where faults are not as extreme as s^{FW} and s^{NF}, nonetheless they happen and affect the normal task behavior. For example, it could exist s^{F1} which is the MI-1-criticality scenario with $C_i^{\mathrm{MI}-1} \equiv C_i^{s^{F1}}$, $C_i(\mathrm{MI} - 1)$ and $\hat{C}_i(\mathrm{MI} - 1)$; $C_i(\mathrm{MI} - 1) \geq C_i(\mathrm{LO})$ and $C_i^{\mathrm{MI}-1} \succeq C_i^{\mathrm{LO}}$. It could also exist s^{F2} as the MI-2-criticality scenario with $C_i^{\mathrm{MI}-2} \equiv C_i^{s^{F2}}$, $C(\mathrm{MI} - 2)$ and $\hat{C}_i(\mathrm{MI} - 2)$; $C_i(\mathrm{MI} - 2) \geq C_i(\mathrm{MI} - 1)$ and $C_i^{\mathrm{MI}-2} \succeq C_i^{\mathrm{MI}-1}$.

Specific to faults and faulty scenarios, it is $S = \{s^{NF} \equiv s^{\mathrm{LO}}, s^{F1} \equiv s^{\mathrm{MI}-1}, s^{F2} \equiv s^{\mathrm{MI}-2}, \dots, s^{FW} \equiv s^{\mathrm{MI}}\}$ with the task MC Worst-Case Set given by $\overline{C}_i = (C_i^{s^{\mathrm{LO}}}, C_i^{s^{\mathrm{MI}-1}}, C_i^{s^{\mathrm{MI}-2}}, \dots, C_i^{s^{\mathrm{HI}}})$.

What we are hereby proposing is a representation framework that applies to faults effects. It could abstract different faults and fault tolerant mechanisms implemented as recovery functions or task extra-executions resulting into larger task execution times and worst-case execution times.

5   Case Study

Three case studies are presented for illustrating the flexibility and the effectiveness of the Worst-Case Set representation in modeling execution conditions, MC tasks and fault effects.

5.1   Test Case 1

As a first test case, it is the ns task from the Mälardalen WCET benchmark executed a multi-core platform, [20]; in this configuration, ns executes alone on one core while other tasks and the OS execute on different cores making interference through shared resources. The ns considered here has four scenarios $S = \{s^1, s^2, s^3, s^4\}$ depending only to the task inputs; $s^4 \equiv s^{worst}$ dominates all the other scenarios, s^3 dominates s^2 and s^1, and so on. The four traces of execution time measurements are $trace\_1$, $trace\_2$, $trace\_3$ and $trace\_4$, respectively for s^1, s^2, s^3 and s^4 and they embed all the known task behaviors, worst-case included. From $S = \{s^1, s^2, s^3, s^4\}$, it is possible to define four criticality levels $(\text{LO}, \text{MI} - 2, \text{MI} - 3, \text{HI})$ for the scenarios and their respective pWCETs, $S = \{s^1 \equiv s^{\text{LO}}, s^2 \equiv s^{\text{MI}-2}, s^3 \equiv s^{\text{MI}-3}, s^4 \equiv s^{\text{HI}}\}$. Figure 3 presents the four traces of measurements under s^{LO}, $s^{\text{MI}-1}$, $s^{\text{MI}-2}$ and s^{HI}, respectively $trace\_1$, $trace\_2$, $trace\_3$ and $trace\_4$. The criticality levels depend on the execution considered and the dominance between them. Figure 6 compares the pWCETs $(\mathcal{C}^{\text{LO}}, \mathcal{C}^{\text{MI}-2}, \mathcal{C}^{\text{MI}-3}, \mathcal{C}^{\text{HI}})$. s^{HI} is confirmed to be the worst-case scenario and $\mathcal{C}^{\text{HI}}$ is the HI-criticality task pWCET.

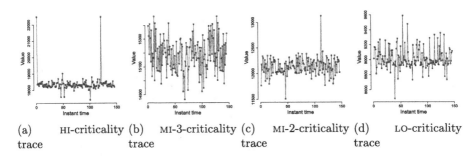

(a)   HI-criticality (b)   MI-3-criticality (c)   MI-2-criticality (d)   LO-criticality
trace   trace   trace   trace

Fig. 3. ns multi-scenario benchmark with four execution time measurements traces; the execution time in ordinate are in CPU cycles.

5.2   Test Case 2

The second test case is the *lms* task from the Mälardalen WCET benchmark. The task is executed in a multi-core platform concurrently with other interfering task and RTEMS OS within the same core as well as outside it, [20]. While executing, *lms* experience two scenarios $S = \{s^1, s^2\}$ from OS and environmental conditions possible at runtime; two traces of execution time exist, *trace2_1* and *trace2_2* and they cover all the conditions *lms* can experience. *trace2_2* dominates *trace2_1* in terms of measured execution times. The two scenarios define LO-criticality and HI-criticality conditions and pWCET models.

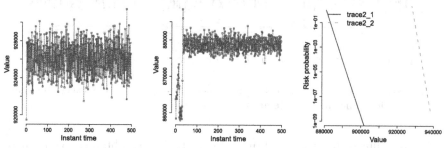

(a) *lms* HI-criticality trace; the execution time in ordinate are in CPU cycles

(b) *lms* LO-criticality trace; the execution time in ordinate are in CPU cycles

(c) icdf pWCET with execution time in abscissa are in CPU cycles; ordinate has log scale

Fig. 4. *lms* benchmark with two traces of execution time measurements *trace2_1* and *trace2_2*.

Figure 4 presents the two traces of measurements and a comparison between $\mathcal{C}^{\mathrm{LO}}$ and $\mathcal{C}^{\mathrm{HI}}$. In particular, Fig. 4(c) illustrates the partial ordering between the two criticality level with *trace2_2* dominating *trace2_2*.

5.3   Test Case 3

The last test case composes of an artificial task τ_1, $\tau_1 = ((\overline{\mathcal{C}}_1, \langle \overline{C}_1, 10^{-9} \rangle, (\hat{C}_1^{\mathrm{LO}}, \hat{C}_1^{\mathrm{MI}}, \hat{C}_1^{\mathrm{HI}})], 50, 50)$. Task period and deadline coincides and are equal to 50 CPU cycles.

τ_1 can execute under its normal functional behavior (no-fault present) s^{LO}, under fault condition s^{MI} or under the worst fault condition s^{HI}.

τ_1 is an artificial tasks since its normal execution time are considered to follow a Normal distribution and not measured from a benchmark. For s^{MI}, the penalties δ are extracted randomly from a uniform distribution with a defined MTTF applied with Eq. (8). Finally, for s^{HI}, the penalties δ are extracted the same uniform distribution but with a smaller MTTF and exhibiting more frequent faults (more critical).

The MC Worst-Case Set representation combines criticality levels (scenarios) and probabilities such that: $\langle C_{1,1}^{\text{LO}} = 8, 10^{-6} \rangle$, $\langle C_{1,2}^{\text{LO}} = 10, 10^{-9} \rangle$, $\langle C_{1,1}^{\text{MI}} = 12, 10^{-6} \rangle$ and $\langle C_{1,2}^{\text{MI}} = 14, 10^{-9} \rangle$, and $\langle C_{1,1}^{\text{HI}} = 16, 10^{-6} \rangle$ and $\langle C_{1,2}^{\text{HI}} = 17, 10^{-9} \rangle$, from the Normal and Uniform laws applied. $p = 10^{-6}$ and $p = 10^{-9}$ are chosen arbitrarily, but any exceeding probability applies.

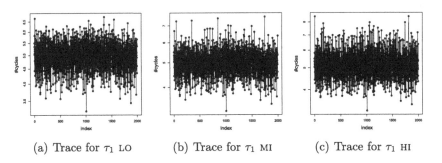

(a) Trace for τ_1 LO   (b) Trace for τ_1 MI   (c) Trace for τ_1 HI

Fig. 5. Measurements of τ_1 execution time under $S = \{s^{\text{LO}}, s^{\text{MI}}, s^{\text{HI}}\}$ execution conditions. Execution times in ordinate are in CPU cycles.

Figure 5 illustrates traces of measurements for τ_1 obtained as formerly described. With s^{HI} the faults are more frequent, making it the most-critical scenario, and that reflects in the pWCET threshold. Figure 7 details the pWCETs for $S = \{s^{\text{LO}}, s^{\text{MI}}, s^{\text{HI}}\}$; $\mathcal{C}^{\text{HI}}$ is named *task1_hi*, $\mathcal{C}^{\text{MI}}$ is named *task1_mi* and $\mathcal{C}^{\text{LO}}$ is named *task1_lo*. The dominance between scenarios and criticality levels is confirmed validating the MC task model for faults.

To note that between pWCETs in Fig. 7 there is not strict dominance, since curves overlaps for small values. The broader dominance as well as the validity

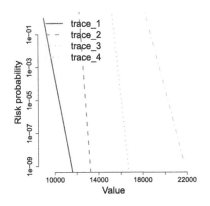

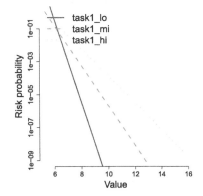

Fig. 6. *ns* pWCET comparison; pWCET represented as icdf and execution time in abscissa are in CPU cycles and ordinate has log scale.

Fig. 7. τ_1 pWCETs represented as icdf; execution times values are in CPU cycles and ordinate has log scale.

of the probabilistic representation is guaranteed at larger values and smaller probabilities, $p \leq 10^{-3}$.

6   Conclusion

The work proposed is a probabilistic representation framework for task execution behaviors named Worst-Case Set, which relies on multiple pWCETs for characterizing the diverse scenarios $s^j = f(I, Env, Map, \ldots)$ affecting task executions; probabilities and coverage of multiple execution conditions make the Worst-Case Set flexible and accurate for task models. The Worst-Case Set is applied for MC models (for different scenario, each with a criticality level associated) and for faults and fault effects on task executions (also related to criticality). A case study is presented to validate the probabilistic representation and illustrate its flexibility in modeling multiple task behaviors from diverse scenarios and criticalities.

Future work will apply the Worst-Case Set representations to probabilistic schedulability analysis as well as to develop schedulability strategies that will leverage probabilities and multiple criticality levels. It has been assumed that S was known; future work will be devoted to investigate scenarios complexity and scenarios completeness for safety and criticality guarantees to probabilistic task models.

References

1. Abella, J., Quiñones, E., Wartel, F., Vardanega, T., Cazorla, F.J.: Heart of gold: Making the improbable happen to increase confidence in MBPTA. In: 26th Euromicro Conference on Real-Time System (eCRTS) (2014)
2. Abeni, L., Buttazzo, G.: QoS guarantee using probabilistic deadlines. In: IEEE Euromicro Conference on Real-Time Systems (ECRTS99) (1999)
3. Altmeyer, S., Cucu-Grosjean, L., Davis, R.I.: Static probabilistic timing analysis for real-time systems using random replacement caches. Real-Time Syst. **51**(1), 77–123 (2015)
4. Burns, A., Davis, R.: Mixed-criticality systems: a review (2015). http://www-users. cs.york.ac.uk/~burns/review.pdf
5. Cucu-Grosjean, L.: Independence - a misunderstood property of and for (probabilistic) real-time systems. In: the 60th Anniversary of A. Burns, York (2013)
6. Cucu-Grosjean, L., Santinelli, L., Houston, M., Lo, C., Vardanega, T., Kosmidis, L., Abella, J., Mezzeti, E., Quinones, E., Cazorla, F.J.: Measurement-based probabilistic timing analysis for multi-path programs. In: 23rd Euromicro Conference on Real-Time Systems (ECRTS). IEEE (2012)
7. Cucu-Grosjean, L., Santinelli, L., Houston, M., Lo, C., Vardanega, T., Kosmidis, L., Abella, J., Mezzetti, E., Quiñones, E., Cazorla, F.J.: Measurement-based probabilistic timing analysis for multi-path programs. In: (ECRTS) (2012)
8. Davis, R.I., Santinelli, L., Altmeyer, S., Maiza, C., Cucu-Grosjean, L.: Analysis of probabilistic cache related pre-emption delays. In: Proceedings of the 25th IEEE Euromicro Conference on Real-Time Systems (ECRTS) (2013)

9. Díaz, J., Garcia, D., Kim, K., Lee, C., Bello, L., Lopez, J.M., Mirabella, O.: Stochastic analysis of periodic real-time systems. In: 23rd of the IEEE Real-Time Systems Symposium (RTSS) (2002)
10. Edgar, S., Burns, A.: Statistical analysis of WCET for scheduling. In: 22nd of the IEEE Real-Time Systems Symposium (2001)
11. Guet, F., Santinelli, L., Morio, J.: On the reliability of the probabilistic worst-case execution time estimates. In: 8th European Congress on Embedded Real Time Software and Systems (ERTS) (2016)
12. Guo, Z., Santinelli, L., Yang, K.: Edf schedulability analysis on mixed-criticality systems with permitted failure probability. In: 21st IEEE International Conference on Embedded and Real-Time Computing System and Applications (2015)
13. Hansen, J., Hissam, S., Moreno, G.: Statistical- based WCET estimation and validation. In: the 9th International Workshop on Worst-Case Execution Time (WCET) (2009)
14. Hardy, D., Puaut, I.: Static probabilistic worst case execution time estimation for architectures with faulty instruction caches. In: International Conference on Real-Time Networks and Systems (RTNS) (2013)
15. Lehoczky, J.: Real-time queueing theory. In: 10th of the IEEE Real-Time Systems Symposium (RTSS 1996) (1996)
16. Lima, G., Dias, D., Barros, E.: Extreme value theory for estimating task execution time bounds: a careful look. In: 28th Euromicro Conference on Real-Time System (eCRTS) (2016)
17. Milutinovic, S., Abella, J., Cazorla, F.J.: Modelling probabilistic cache representativeness in the presence of arbitrary access patterns. In: 19th IEEE International Symposium on Real-Time Distributed Computing, (ISORC) (2016)
18. Panerati, J., Abdi, S., Beltrame, G.: Balancing system availability and lifetime with dynamic hidden Markov models. In: NASA/ESA Conference on Adaptive Hardware and Systems (AHS) (2014)
19. Pathan, R.M.: Fault-tolerant and real-time scheduling for mixed-criticality systems. Real-Time Syst. **50**, 509–547 (2014)
20. Santinelli, L., Guet, F., Morio, J.: Revising measurement-based probabilistic timing analysis. In: Proceedings of the IEEE Real-Time and Embedded Technology and Applications Symposium (RTAS) (2017)
21. Santinelli, L., Guo, Z., George, L.: Fault-aware sensitivity analysis for probabilistic real-time systems. In: 2016 IEEE International Symposium on Defect and Fault Tolerance in VLSI and Nanotechnology Systems (DFT) (2016)
22. Tia, T., Deng, Z., Shankar, M., Storch, M., Sun, J., Wu, L., Liu, J.: Probabilistic performance guarantee for real-time tasks with varying computation times. In: IEEE Real-Time and Embedded Technology and Applications Symposium (1995)
23. Vestal, S.: Preemptive scheduling of multi-criticality systems with varying degrees of execution time assurance. In: Proceedings of the 28th IEEE International Real-Time Systems Symposium (RTSS). IEEE Computer Society (2007)
24. Wilhelm, R., Engblom, J., Ermedahl, A., Holsti, N., Thesing, S., Whalley, D.B., Bernat, G., Ferdinand, C., Heckmann, R., Mitra, T., Mueller, F., Puaut, I., Puschner, P.P., Staschulat, J., Stenström, P.: The worst-case execution-time problem - overview of methods and survey of tools. ACM Trans. Embedded Comput. Syst. **7**(3), 36:1–36:53 (2008)
25. Wong, H.C., Burns, A.: Schedulability analysis for the abort-and-restart (AR) model. In: Proceedings of the 22nd International Conference on Real-Time Networks and Systems (RTNS) (2014)

Timed and Hybrid Systems

Nested Timed Automata with Invariants

Yuwei Wang[1], Guoqiang Li[1(✉)], and Shoji Yuen[2]

[1] School of Software, Shanghai Jiao Tong University, Shanghai, China
{wangywgg,li.g}@sjtu.edu.cn
[2] Graduate School of Information Science, Nagoya University, Nagoya, Japan
yuen@is.nagoya-u.ac.jp

Abstract. Invariants are usually adopted into timed systems to constrain the time passage within each control location. It is well-known that a timed automaton with invariants can be encoded to an equivalent one without invariants. When recursions are taken into consideration, few results show whether invariants affect expressiveness. This paper investigates the effect of invariants to Nested Timed Automata (NeTAs), a typical real-timed recursive system. In particular, we study the reachability problem for NeTA-Is, which extend NeTAs with invariants. It is shown that the reachability problem is undecidable on NeTA-Is with a single global clock, while it is decidable when no invariants are given. Furthermore, we also show that the reachability is decidable if the NeTA-Is contains no global clocks by showing that a *good* stack content still satisfies well-formed constraints.

1 Introduction

From the past century, many research studies have been carried out on modeling and verification of real time systems. The pioneer work can be traced to *Timed Automata (TAs)* [1,2], which is one of the most successful models among them due to its simplicity, effectiveness and fruitful results. A TA is a finite automaton with a finite set of *clocks* that grow uniformly. Besides the constraints assigned on the transitions of TAs, they can also be assigned to each control location, named *invariants*, to constrain time passages of models. Invariants usually play a crucial role in the application modelling and verification [3], since in reality a system is not allowed to stay in one location for arbitrarily long time. It is well-known that TAs with and without invariants have the same expressive power [3]. However, little research has been conducted in investigating the impact of invariants on the reachability problem of timed systems with recursions.

This paper proposes an extension of *Nested Timed Automata* (NeTAs) [4,5], called NeTA-Is. A NeTA is a pushdown system whose stack contains TAs with global clocks passing information among different contexts. TAs in the stack can either be proceeding, in which clocks proceed as time elapses, or frozen, where clocks remain unchanged. NeTA-Is naturally extend NeTAs with invariants at each control location that must be fulfilled in all valid runs. Studies in [5] have shown that in NeTAs, (i) the reachability with a single global clock is decidable, and (ii) the reachability with multiple global clocks is undecidable. While in this paper, we show that (i) the reachability problem of a NeTA-I is undecidable even

© Springer International Publishing AG 2017
K.G. Larsen et al. (Eds.): SETTA 2017, LNCS 10606, pp. 77–93, 2017.
https://doi.org/10.1007/978-3-319-69483-2_5

with a single global clock by encoding Minsky machines to NeTA-Is, and (ii) it is decidable when the NeTA-I has no global clocks by showing that a *good* stack content still satisfies well-formed constraints [6].

Related Work. *Timed Automata* (TAs) [1,2] are the first model for real-timed systems. TAs are essentially finite automata extended with real-valued variables, called *clocks*. The reachability of TAs is shown to be decidable based on construction of regions and zones. It is also shown that invariants do not affect the decidability and thus only a syntactic sugar. Based on timed automata, lots of extensions are proposed and investigated especially for a recursive structure.

Dense Timed Pushdown Automata (DTPDAs) [7] combine timed automata and pushdown automata, where each stack frame containing not only a stack symbol but also a real-valued clock behaves as a basic unit of push/pop operations. The reachability of a DTPDA is shown to be decidable by encoding it to a PDA using the region technique. Another decidability proof is given in [6] through a general framework, *well-structured pushdown systems*. We adopt this framework in this paper to prove the decidability of reachability of *Constraint DTPDAs*, which extend DTPDAs with clock constraints on each location.

Recursive Timed Automata (RTAs) [8] contain finite components, each of which is a special timed automaton and can recursively invoke other components. Two mechanisms, *pass-by-value* and *pass-by-reference*, can be used to passing clocks among different components. A clock is *global* if it is always passed by reference, whereas it is *local* if it is always passed by value. Although the reachability problem of RTAs is undecidable, it is decidable if all clocks are global or all clocks are local.

Similarly, the reachability problems of both *Timed Recursive State Machines* (TRSMs), which combine *recursive state machines* (RSMs) and TAs, and *Extended Pushdown Timed Automata* (EPTAs), which augment *Pushdown Timed Automata* (PTAs) with an additional stack, are undecidable, while they are decidable in some restricted subclasses [9].

To the best of our knowledge, all these prior formal models focusing on timed systems with recursive structures lacks discussions of the impact of invariants, including DTPDAs, RTAs, TRSMs, EPTAs and NeTAs.

Paper Organization. The remainder of this paper is structured as follows: In Sect. 2 we introduce basic terminologies and notations. Section 3 defines syntax and the semantics of NeTA-Is. Section 4 shows that the reachability problem of NeTA-Is is Turing-complete. Section 5 introduces a model *Constraint DTPDAs* and shows its decidability. Section 6 is devoted to proofs of decidability results of NeTA-Is without global clocks by encoding it to a Constraint DTPDA. Section 7 concludes this paper with summarized results.

2   Preliminaries

For finite words $w = aw'$, we denote $a = head(w)$ and $w' = tail(w)$. The concatenation of two words w, v is denoted by $w.v$, and ϵ is the empty word.

Let $\mathbb{R}^{\geq 0}$ and $\mathbb{N}$ denote the sets of non-negative real numbers and natural numbers, respectively. Let ω denote the first limit ordinal. Let $\mathcal{I}$ denote the set of *intervals*. An interval is a set of numbers, written as (a, b'), $[a, b]$, $[a, b')$ or $(a, b]$, where $a, b \in \mathbb{N}$ and $b' \in \mathbb{N} \cup \{\omega\}$. For a number $r \in \mathbb{R}^{\geq 0}$ and an interval $I \in \mathcal{I}$, we use $r \in I$ to denote that r belongs to I.

Let $X = \{x_1, \ldots, x_n\}$ be a finite set of *clocks*. The set of *clock constraints*, $\Phi(X)$, over X is defined by $\phi ::= \top \mid x \in I \mid \phi \wedge \phi$ where $x \in X$ and $I \in \mathcal{I}$. An operation of *extracting constraint* $EC(\phi, x)$ is defined by induction over its argument ϕ.

$$
\begin{aligned}
EC(\top, x) &= [0, \omega) \\
EC(x \in I, x) &= I \\
EC(y \in I, x) &= [0, \omega) \ if \ x \neq y \\
EC(\phi_1 \wedge \phi_2, x) &= EC(\phi_1, x) \bigcap EC(\phi_2, x)
\end{aligned}
$$

A *clock valuation* $\nu : X \to \mathbb{R}^{\geq 0}$, assigns a value to each clock $x \in X$. ν_0 denotes the clock valuation assigning each clock in X to 0. For a clock valuation ν and a clock constraint ϕ, we write $\nu \models \phi$ to denote that ν satisfies the constraint ϕ. Given a clock valuation ν and a time $t \in \mathbb{R}^{\geq 0}$, $(\nu + t)(x) = \nu(x) + t$, for $x \in X$. A clock assignment function $\nu[y_1 \leftarrow b_1, \cdots, y_n \leftarrow b_n]$ is defined by $\nu[y_1 \leftarrow b_1, \cdots, y_n \leftarrow b_n](x) = b_i$ if $x = y_i$ for $1 \leq i \leq n$, and $\nu(x)$ otherwise. $\mathcal{V}al(X)$ is used to denote the set of clock valuation of X.

2.1 Timed Automata

A timed automaton is a finite automaton augmented with a finite set of clocks [1,2].

Definition 1 (Timed Automata). *A timed automaton (TA) is a tuple $\mathcal{A} = (Q, q_0, X, \mathbb{I}, \Delta) \in \mathscr{A}$, where*

- Q *is a finite set of control locations, with the initial location $q_0 \in Q$,*
- X *is a finite set of clocks,*
- $\mathbb{I} : Q \to \Phi(X)$ *is a function assigning each location with a clock constraint on X, called* invariants.
- $\Delta \subseteq Q \times \mathcal{O} \times Q$, *where $\mathcal{O}$ is a set of* operations. *A transition $\delta \in \Delta$ is a triplet (q_1, ϕ, q_2), written as $q_1 \xrightarrow{\phi} q_2$, in which ϕ is either of*
Local ϵ, *an empty operation,*
Test $x \in I$? *where $x \in X$ is a clock and $I \in \mathcal{I}$ is an interval,*
Reset $x \leftarrow 0$ *where $x \in X$, and*
Value passing $x \leftarrow x'$ *where $x, x' \in X$.*

Given a TA $\mathcal{A} \in \mathscr{A}$, we use $Q(\mathcal{A})$, $q_0(\mathcal{A})$, $X(\mathcal{A})$, $\mathbb{I}(\mathcal{A})$ and $\Delta(\mathcal{A})$ to represent its set of control locations, initial location, set of clocks, function of invariants and set of transitions, respectively. We will use similar notations for other models.

We call the four operations **Local**, **Test**, **Reset**, and **Value passing** as *internal* actions which will be used in Definition 3.

Definition 2 (Semantics of TAs). *Given a TA $(Q, q_0, X, \mathbb{I}, \Delta)$, a configuration is a pair (q, ν) of a control location $q \in Q$ and a clock valuation ν on X. The transition relation of the TA is represented as follows,*

– Progress transition: $(q, \nu) \xrightarrow{t}_{\mathscr{A}} (q, \nu + t)$, where $t \in \mathbb{R}^{\geq 0}$, $\nu \models \mathbb{I}(q)$ and $(\nu + t) \models \mathbb{I}(q)$.

– Discrete transition: $(q_1, \nu_1) \xrightarrow{\phi}_{\mathscr{A}} (q_2, \nu_2)$, if $q_1 \xrightarrow{\phi} q_2 \in \Delta$, $\nu_1 \models \mathbb{I}(q_1)$, $\nu_2 \models \mathbb{I}(q_2)$ and one of the following holds,

   • **Local** $\phi = \epsilon$, then $\nu_1 = \nu_2$.
   • **Test** $\phi = x \in I?$, $\nu_1 = \nu_2$ and $\nu_2(x) \in I$ holds. The transition can be performed only if the value of x belongs to I.
   • **Reset** $\phi = x \leftarrow 0$, $\nu_2 = \nu_1[x \leftarrow 0]$. This operation resets clock x to 0.
   • **Value passing** $\phi = x \leftarrow x'$, then $\nu_2 = \nu_1[x \leftarrow \nu_1(x')]$. The transition passes value of clock x' to clock x.

The initial configuration is (q_0, ν_0).

Remark 1. The TA definition in Definition 1 follows the style in [4] and is slightly different from the original definition in [1]. In [1], several test and reset operations could be performed in a single discrete transition. It can be shown that our definition of TA can soundly simulate the time traces in the original definition.

3   Nested Timed Automata with Invariants

A *nested timed automaton with invariants (NeTA-I)* extended from NeTAs[1] [5] is a pushdown system whose stack alphabet is timed automata. It can either behave like a TA (internal operations), push or fpush the current working TA to the stack, pop a TA from the stack or reference global clocks. Global clocks can be used to constrain the global behavior or passing value of local clocks among different TAs. The invariants can be classified into *global invariants*, which are constraints on global clocks, and *local invariants*, which are constraints on local clocks. In the executions of a NeTA-I, all invariants must be satisfied at all reachable configurations, including global invariants and local invariants. Note that because the stack contains only information belonging to TAs and does not contain the global clock valuation, there is no need to check global invariants in the stack.

Definition 3 (Nested Timed Automata with Invariants). *A nested timed automaton with invariants (NeTA-I) is a tuple* $\mathcal{N} = (T, \mathcal{A}_0, X, C, \mathbb{I}, \Delta)$, *where*

   – *T is a finite set of TAs* $\{\mathcal{A}_0, \mathcal{A}_1, \cdots, \mathcal{A}_n\}$, *with the initial TA* $\mathcal{A}_0 \in T$. *We assume the sets of control locations of* $\mathcal{A}_i$, *denoted by* $Q(\mathcal{A}_i)$, *are mutually disjoint, i.e.,* $Q(\mathcal{A}_i) \cap Q(\mathcal{A}_j) = \emptyset$ *for* $i \neq j$. *For simplicity, we assume that each* $\mathcal{A}_i$ *in T shares the same set of local clocks X.*
   – *C is a finite set of global clocks, and X is the finite set of k local clocks.*
   – $\mathbb{I} : Q \to \Phi(C)$ *is a function that assigns to each control location an invariant on global clocks. For clarity,* $\mathbb{I}(q)$ *denotes the global invariant in q, and* $\mathbb{I}(\mathcal{A}_i)(q)$ *denotes the local invariant in q where* $q \in \mathcal{A}_i$.
   – $\Delta \subseteq Q \times (Q \cup \{\varepsilon\}) \times Actions^+ \times Q \times (Q \cup \{\varepsilon\})$ *describes transition rules below, where* $Q = \cup_{\mathcal{A}_i \in T} Q(\mathcal{A}_i)$.

---

[1] The NeTAs here are called "NeTA-Fs" in [5].

A transition rule is described by a sequence of Actions $= \{internal, push,$
$fpush, pop, c \in I, c \leftarrow 0, x \leftarrow c, c \leftarrow x\}$ *where* $c \in C$ *and* $x \in X$.

Internal $(q, \varepsilon, internal, q', \varepsilon)$, *which describes an internal transition in the work-*
ing TA with $q, q' \in Q(\mathcal{A}_i)$.

Push $(q, \varepsilon, push, q_0(\mathcal{A}_{i'}), q)$, *which interrupts the currently working TA* $\mathcal{A}_i$ *at*
$q \in Q(\mathcal{A}_i)$ *and pushes it to the stack with all local clocks of* $\mathcal{A}_i$. *The local*
clocks in the stack generated by **Push** *operation are proceeding, i.e., still*
evolve as time elapses. Then, a TA $\mathcal{A}_{i'}$ *newly starts.*

Freeze-Push (F-Push) $(q, \varepsilon, fpush, q_0(\mathcal{A}_{i'}), q)$, *which is similar to* **Push**
except that all local clocks in the stack generated by **F-Push** *are frozen (i.e.*
stay the same as time elapses).

Pop $(q, q', pop, q', \varepsilon)$, *which restarts* $\mathcal{A}_{i'}$ *in the stack from* $q' \in Q(\mathcal{A}_{i'})$ *after* $\mathcal{A}_i$
has finished at $q \in Q(\mathcal{A}_i)$ *and all local clocks restart with values in the top*
stack frame.

Global-test $(q, \varepsilon, c \in I?, q', \varepsilon)$, *which tests whether the value of a global clock* c
is in I *with* $q, q' \in Q(\mathcal{A}_i)$.

Global-reset $(q, \varepsilon, c \leftarrow 0, q', \varepsilon)$ *with* $c \in C$, *which resets the global clock* c *to 0*
with $q, q' \in Q(\mathcal{A}_i)$.

Global-load $(q, \varepsilon, x \leftarrow c, q', \varepsilon)$, *which assigns the value of a global clock* c *to a*
local clock $x \in X$ *in the working TA with* $q, q' \in Q(\mathcal{A}_i)$.

Global-store $(q, \varepsilon, c \leftarrow x, q', \varepsilon)$, *which assigns the value of a local clock* $x \in X$
of the working TA to a global clock c *with* $q, q' \in Q(\mathcal{A}_i)$.

Definition 4 (Semantics of NeTA-Is). *Given a NeTA-I* $(T, \mathcal{A}_0, X, C, \mathbb{I}, \Delta)$,
let $Val_X = \{\nu : X \to \mathbb{R}^{\geq 0}\}$ *and* $Val_C = \{\mu : C \to \mathbb{R}^{\geq 0}\}$. *A configuration of a*
NeTA-I is an element $(\langle q, \nu, \mu \rangle, v)$ *with a control location* $q \in Q$, *a local clock*
valuation $\nu \in Val_X$, *a global clock valuation* $\mu \in Val_C$ *and a stack* $v \in (Q \times$
$\{0, 1\} \times Val_X)^*$. *We say a stack* v *is* good, *written as* $v^{\Uparrow}$, *if all local invariants*
are satisfied in v, *i.e., for each content* $\langle q_i, flag_i, \nu_i \rangle$ *in* v *with* $q_i \in Q(\mathcal{A}_j)$, $\nu_i \models$
$\mathbb{I}(\mathcal{A}_j)(q_i)$ *holds. We also denote* $v + t$ *by setting* $\nu_i := progress(\nu_i, t, flag_i)$ *of*

each $\langle q_i, flag_i, \nu_i \rangle$ *in the stack where* $progress(\nu, t, flag) = \begin{cases} \nu + t & \text{if } flag = 1 \\ \nu & \text{if } flag = 0 \end{cases}$

- Progress transition: $(\langle q, \nu, \mu \rangle, v) \xrightarrow{t} (\langle q, \nu + t, \mu + t \rangle, v + t)$ *for* $t \in \mathbb{R}^{\geq 0}$, *where*
  $q \in Q(\mathcal{A}_i)$, $\nu \models \mathbb{I}(\mathcal{A}_i)(q)$, $\mu \models \mathbb{I}(q)$, $(\nu + t) \models \mathbb{I}(\mathcal{A}_i)(q)$, $(\mu + t) \models \mathbb{I}(q)$, $v^{\Uparrow}$
  and $(v + t)^{\Uparrow}$.

- Discrete transition: $(\langle q, \nu, \mu \rangle, v) \xrightarrow{\varphi} (\langle q', \nu', \mu' \rangle, v')$, *where* $q \in Q(\mathcal{A}_i)$, $q' \in$
  $Q(\mathcal{A}'_i)$, $\nu \models \mathbb{I}(\mathcal{A}_i)(q)$, $\mu \models \mathbb{I}(q)$, $\nu' \models \mathbb{I}(\mathcal{A}'_i)(q')$, $\mu' \models \mathbb{I}(q')$, $v^{\Uparrow}$, $v'^{\Uparrow}$, *and*
  one of the following holds.

  • **Internal** $(\langle q, \nu, \mu \rangle, v) \xrightarrow{\varphi} (\langle q', \nu', \mu \rangle, v)$, *if* $(q, \varepsilon, internal, q', \varepsilon) \in \Delta$ *and*
    $\langle q, \nu \rangle \xrightarrow{\varphi} \langle q', \nu' \rangle$ *is in Definition 2.*

  • **Push** $(\langle q, \nu, \mu \rangle, v) \xrightarrow{push} (\langle q_0(\mathcal{A}_{i'}), \nu_0, \mu \rangle, \langle q, 1, \nu \rangle.v)$, *if* $(q, \varepsilon, push,$
    $q_0(\mathcal{A}_{i'}), q) \in \Delta$.

  • **F-Push** $(\langle q, \nu, \mu \rangle, v) \xrightarrow{f\text{-}push} (\langle q_0(\mathcal{A}_{i'}), \nu_0, \mu \rangle, \langle q, 0, \nu \rangle.v)$, *if* $(q, \varepsilon, fpush,$
    $q_0(\mathcal{A}_{i'}), q) \in \Delta$.

  • **Pop** $(\langle q, \nu, \mu \rangle, \langle q', flag, \nu' \rangle.w) \xrightarrow{pop} (\langle q', \nu', \mu \rangle, w)$, *if* $(q, q', pop, q', \varepsilon) \in \Delta$.

  • **Global-test** $(\langle q, \nu, \mu \rangle, v) \xrightarrow{c \in I?} (\langle q', \nu, \mu \rangle, v)$, *if* $(q, \varepsilon, c \in I?, q', \varepsilon) \in \Delta$ *and*
    $\mu(c) \in I$.

- **Global-reset** $(\langle q,\nu,\mu\rangle,v) \xrightarrow{c\leftarrow 0} (\langle q',\nu,\mu[c \leftarrow 0]\rangle,v)$, *if* $(q,\varepsilon,c \leftarrow 0,q',\varepsilon) \in \Delta$.
- **Global-load** $(\langle q,\nu,\mu\rangle,v) \xrightarrow{x\leftarrow c} (\langle q',\nu[x \leftarrow \mu(c)],\mu\rangle,v)$, *if* $(q,\varepsilon,x \leftarrow c,q',\varepsilon) \in \Delta$.
- **Global-store**$(\langle q,\nu,\mu\rangle,v) \xrightarrow{c\leftarrow x} (\langle q',\nu,\mu[c \leftarrow \nu(x)]\rangle,v)$, *if* $(q,\varepsilon,c \leftarrow x,q',\varepsilon) \in \Delta$.

The initial configuration of a NeTA-I is $(\langle q_0(\mathcal{A}_0),\nu_0,\mu_0\rangle,\varepsilon)$, *where* $\nu_0(x) = 0$ *for* $x \in X$ *and* $\mu_0(c) = 0$ *for* $c \in C$. *We use* $\longrightarrow$ *to range over these transitions, and* $\longrightarrow^*$ *is the reflexive and transitive closure of* $\longrightarrow$.

Intuitively, in a stack $v = (q_1, flag_1, \nu_1)\dots(q_n, flag_n, \nu_n)$, q_i is the control location of the pushed/fpushed TA, $flag_i \in \{0,1\}$ is a flag for whether the TA is pushed ($flag_i = 1$) or fpushed ($flag_i = 0$) and ν_i is a clock valuation for the local clocks of the pushed/fpushed TA.

4   Undecidability Results of NeTA-Is

In this section, we prove undecidability of NeTA-Is by encoding the halting problem of Minsky machines [10] to NeTA-Is with a single global clock.

Definition 5 (Minsky Machine). *A Minsky machine* $\mathcal{M}$ *is a tuple* (L, C, D) *where:*

- *L is a finite set of states, and* $l_f \in L$ *is the terminal state,*
- *C* = $\{ct_1, ct_2\}$ *is the set of two counters, and*
- *D is the finite set of transition rules of the following types,*
  - **increment counter** $d = inc(l, ct_i, l_k)$: *start from* l, $ct_i := ct_i + 1$, *goto* l_k,
  - **test-and-decrement counter** $d = dec(l, ct_i, l_k, l_m)$: *start from* l, *if* $(ct_i > 0)$ *then* $(ct_i := ct_i - 1$, *goto* $l_k)$ *else goto* l_m,
  where $ct_i \in C$, $d \in D$ *and* $l, l_k, l_m \in L$.

In this encoding, we use three TAs, $\mathcal{A}_0, \mathcal{A}_1$ and $\mathcal{A}_2$. Each TA has three local clocks x_0, x_1 and x_2. $\mathcal{A}_0$ is a special TA, as two local clocks of $\mathcal{A}_0$, x_1 and x_2 encode values of two counters as $x_i = 2^{-ct_i}$ for $i = 1, 2$. Decrementing and incrementing the counter ct_i are simulated by doubling and halving of the value of the local clock x_i in $\mathcal{A}_0$, respectively. In all TAs, x_0 is used to prevent time progress. In $\mathcal{A}_1$ and $\mathcal{A}_2$, x_1 and x_2 are used for temporarily storing value. We use only one global clock c to pass value among different TAs.

There are two types of locations in the encoding, q-locations and e-locations. All q-locations are assigned with invariants $x_0 \in [0,0]$. These invariants ensures that in all reachable configurations at q-locations, the value of x_0 must be 0. So time does not elapse at q-locations.

The idea of doubling or halving of x_i in $\mathcal{A}_0$ is as follows. First the value of x_i is stored to the global clock c. Then the current TA $\mathcal{A}_0$ is fpushed to the stack and through transitions in $\mathcal{A}_1$ and $\mathcal{A}_2$, the global clock c is doubled or halved. Later $\mathcal{A}_0$ is popped back and the value of c is loaded to x_i. Since all locations are q-locations in $\mathcal{A}_0$, time does not elapse in $\mathcal{A}_0$. This ensures that while doubling or halving a local clock, the other one is left unchanged.

The encoding is shown formally as follows.

A Minsky machine $\mathcal{M} = (L, C, D)$ can be encoded into a NeTA-I $\mathcal{N} = (T, \mathcal{A}_0, X, C', \mathbb{I}, \Delta)$, with $T = \{\mathcal{A}_0, \mathcal{A}_1, \mathcal{A}_2\}$ where

$$Q(\mathcal{A}_0) = \{q_l \mid l \in L\} \bigcup \{q_1^{inc,i,l_k} \mid inc(ct_i, l, l_k) \in D\}$$
$$\bigcup \{q_j^{dec,i,l_k} \mid dec(ct_i, l, l_k, l_m) \in D, 1 \leq j \leq 2\}$$
$$Q(\mathcal{A}_1) = \{q_j^{inc,i,l_k} \mid inc(ct_i, l, l_k) \in D, 2 \leq j \leq 8\}$$
$$\bigcup \{e_j^{inc,i,l_k} \mid inc(ct_i, l, l_k) \in D, j = 1, 2 \text{ or } 4\}$$
$$\bigcup \{q_j^{dec,i,l_k} \mid dec(ct_i, l, l_k, l_m) \in D, 3 \leq j \leq 7\}$$
$$\bigcup \{e_2^{dec,i,l_k} \mid dec(ct_i, l, l_k, l_m) \in D\}$$
$$Q(\mathcal{A}_2) = \{q_j^{inc,i,l_k} \mid inc(ct_i, l, l_k) \in D, 9 \leq j \leq 11\}$$
$$\bigcup \{e_3^{inc,i,l_k} \mid inc(ct_i, l, l_k) \in D\}$$
$$\bigcup \{q_j^{dec,i,l_k} \mid dec(ct_i, l, l_k, l_m) \in D, 8 \leq j \leq 10\}$$
$$\bigcup \{e_1^{dec,i,l_k} \mid dec(ct_i, l, l_k, l_m) \in D\}$$

- $X = \{x_0, x_1, x_2\}$ and $C' = \{c\}$.
- $\mathbb{I}(A_i)(q_-) = x_0 \in [0, 0]$ and $\mathbb{I}(A_i)(e_-) = \top$ where $0 \leq i \leq 2$ and $_-$ denotes any valid symbol. Here q_- denotes the q-location, which is labeled with q, and e_- denotes the e-location, which is labeled with e.
- Δ is shown implicitly in the following simulations due to limited space.

  • **increment counter** simulate $inc(l, ct_i, l_k)$. Initially $\nu(x_i) = d$ with $0 < d \leq 1$. In q_{l_k}, x_i will be halved. The value of x_i is stored to the global clock c and context is changed to $\mathcal{A}_1$. Then the value of c is halved. Although the timed elapsed in state e_2^{inc,i,l_k} and e_3^{inc,i,l_k} are nondeterministic, to reach the location q_{l_k}, the value of x_1 and c must coincide (i.e., they reach 1 together) at state e_4^{inc,i,l_k}. The readers can check that timed elapsed in e_1^{inc,i,l_k} must be $1 - d$, in e_2^{inc,i,l_k} and e_4^{inc,i,l_k} must be $d/2$, and in e_3^{inc,i,l_k} must be $1 - d/2$

  $$q_l \xrightarrow{c \leftarrow x_i} q_1^{inc,i,l_k} \xrightarrow{fpush} e_1^{inc,i,l_k} \xrightarrow{x_0 \leftarrow 0} q_2^{inc,i,l_k} \xrightarrow{c \in [1,1]?} q_3^{inc,i,l_k} \xrightarrow{c \leftarrow 0}$$
  $$e_2^{inc,i,l_k} \xrightarrow{x_0 \leftarrow 0} q_4^{inc,i,l_k} \xrightarrow{fpush} e_3^{inc,i,l_k} \xrightarrow{x_0 \leftarrow 0} q_9^{inc,i,l_k} \xrightarrow{c \in [1,1]?} q_{10}^{inc,i,l_k}$$
  $$\xrightarrow{c \leftarrow x_1} q_{11}^{inc,i,l_k} \xrightarrow{pop} q_4^{inc,i,l_k} \xrightarrow{x_2 \leftarrow 0} e_4^{inc,i,l_k} \xrightarrow{x_0 \leftarrow 0} q_5^{inc,i,l_k} \xrightarrow{c \in [1,1]?} q_6^{inc,i,l_k}$$
  $$\xrightarrow{x_1 \in [1,1]?} q_7^{inc,i,l_k} \xrightarrow{c \leftarrow x_2} q_8^{inc,i,l_k} \xrightarrow{pop} q_1^{inc,i,l_k} \xrightarrow{x_i \leftarrow c} q_{l_k}$$

  • **test-and-decrement counter** simulate $dec(l, ct_i, l_k, l_m)$. Initially $\nu(x_i) = d$ with $0 < d \leq 1$. At the beginning of the simulation, $x_i = 1$ is tested, which encodes the zero test of ct_i. In q_{l_k}, x_i will be doubled. The readers can also check that to reach the location q_{l_k}, timed elapsed in e_1^{dec,i,l_k} must be $1 - d$, and in e_2^{dec,i,l_k} must be d.

  $$q_l \xrightarrow{x_i \in [1,1]?} q_{l_m} \text{ and}$$
  $$q_l \xrightarrow{x_i \in (0,1)?} q_1^{dec,i,l_k} \xrightarrow{c \leftarrow x_i} q_2^{dec,i,l_k} \xrightarrow{fpush} q_3^{dec,i,l_k} \xrightarrow{x_1 \leftarrow c} q_4^{dec,i,l_k} \xrightarrow{fpush}$$
  $$e_1^{dec,i,l_k} \xrightarrow{x_0 \leftarrow 0} q_8^{dec,i,l_k} \xrightarrow{c \in [1,1]?} q_9^{dec,i,l_k} \xrightarrow{c \leftarrow x_1} q_{10}^{dec,i,l_k} \xrightarrow{pop} q_4^{dec,i,l_k} \xrightarrow{\epsilon}$$
  $$e_2^{dec,i,l_k} \xrightarrow{x_0 \leftarrow 0} q_5^{dec,i,l_k} \xrightarrow{c \in [1,1]?} q_6^{dec,i,l_k} \xrightarrow{c \leftarrow x_1} q_7^{dec,i,l_k} \xrightarrow{pop}$$
  $$q_2^{dec,i,l_k} \xrightarrow{x_i \leftarrow c} q_{l_k}$$

Theorem 1. *The reachability of a NeTA-I with a single global clock is undecidable.*

Remark 2. The invariants here are used to prevent time progress, and it can not be simulated by the traditional approach if pop rules are allowed, i.e., simply resetting x_0 to 0 first, and then using a test transition $x_0 \in [0,0]$? at the tail. For example, in the pop rule $q_{11}^{inc,i,l_k} \xrightarrow{pop} q_4^{inc,i,l_k}$, the state q_{11}^{inc,i,l_k} is the final state of $\mathcal{A}_2$, and there is no way using only test transition $x_0 \in [0,0]$? to promise time not elapsing in q_{11}^{inc,i,l_k}. Because after popping, we can not check values of the local clocks in the original TA $\mathcal{A}_2$, which has been already popped from the stack. Of course, if we introduce a fresh global clock, say c_0, the test transition $c_0 \in [0,0]$? can prevent time progress. Then it is actually an encoding from a Minsky machine to a NeTA with two global clocks and without invariants, which is consistent with results in [5].

5   Constraint Dense Timed Pushdown Automata

In this section, we first present syntax and semantics of Constraint Dense Timed Pushdown Automata. Later, we introduce digiwords and operations which are used for encoding from a Constraint Dense Timed Pushdown Automaton to a snapshot pushdown system. Finally, the decidability of reachability of a snapshot pushdown system is shown by observing that it is a growing WSPDS with a well-formed constraint [6].

Definition 6 (Constraint Dense Timed Pushdown Automata). *A constraint dense timed pushdown automaton (Constraint DTPDA) is a tuple $\mathcal{D} = \langle S, s_0, \Gamma, X, \mathbb{I}, \Delta \rangle \in \mathcal{D}$, where*

- *S is a finite set of states with the initial state $s_0 \in S$,*
- *Γ is a finite stack alphabet,*
- *X is a finite set of clocks (with $|X| = k$),*
- *$\mathbb{I} : S \to \Phi(X)$ is a function that assigns to each state an invariant, and*
- *$\Delta \subseteq S \times Action^+ \times S$ is a finite set of transitions.*

A (discrete) transition $\delta \in \Delta$ is a sequence of actions $(s_1, o_1, s_2), \cdots, (s_i, o_i, s_{i+1})$ written as $s_1 \xrightarrow{o_1; \cdots ; o_i} s_{i+1}$, in which o_j (for $1 \le j \le i$) is one of the followings,

- **Local** ϵ, *an empty operation,*
- **Test** ϕ, *where $\phi \in \Phi(X)$ is a clock constraint,*
- **Reset** $x \leftarrow 0$ *where $x \in X$,*
- **Value passing** $x \leftarrow x'$ *where $x, x' \in X$,*
- **Push** $push(\gamma)$, *where $\gamma \in \Gamma$ is a stack symbol,*
- **F-Push** $fpush(\gamma)$, *where $\gamma \in \Gamma$ is a stack symbol, and*
- **Pop** $pop(\gamma)$, *where $\gamma \in \Gamma$ is a stack symbol.*

Definition 7 (Semantics of Constraint DTPDAs). *For a Constraint DTPDA $\langle S, s_0, \Gamma, X, \mathbb{I}, \Delta \rangle$, a configuration is a triplet (s, w, ν) with a state $s \in S$, a stack $w \in (\Gamma \times (\mathbb{R}^{\ge 0})^k \times \{0, 1\} \times \Phi(X))^*$, and a clock valuation ν on X. Similarly, a stack w good, written as $w^{\Uparrow}$, if for each content $(\gamma_i, \bar{t}_i, flag_i, \phi_i)$ in w, we have $\nu[x_1 \leftarrow t_1, \cdots, x_k \leftarrow t_k] \models \phi_i$ where*

$\bar{t}_i = (t_1, \cdots, t_k)$. For $w = (\gamma_1, \bar{t}_1, flag_1, \phi_1). \cdots .(\gamma_n, \bar{t}_n, flag_n, \phi_n)$, a t-time passage on the stack, written as $w + t$, is $(\gamma_1, progress'(\bar{t}_1, t, flag_1), flag_1, \phi_1)$. $\cdots .(\gamma_n, progress'(\bar{t}_n, t, flag_n), flag_n, \phi_n)$ where

$$progress'(\bar{t}, t, flag) = \begin{cases} (t_1 + t, \cdots, t_k + t) & if\ flag = 1\ and\ \bar{t} = (t_1, \cdots, t_k) \\ \bar{t} & if\ flag = 0 \end{cases}$$

The transition relation of the Constraint DTPDA is defined as follows:

- Progress transition: $(s, w, \nu) \xrightarrow{t}_{\mathscr{D}} (s, w + t, \nu + t)$, where $t \in \mathbb{R}^{\geq 0}$, $w^{\Uparrow}$, $\nu \models \mathbb{I}(s)$, $(w + t)^{\Uparrow}$ and $(\nu + t) \models \mathbb{I}(s)$.
- Discrete transition: $(s_1, w_1, \nu_1) \xrightarrow{o}_{\mathscr{D}} (s_2, w_2, \nu_2)$, if $s_1 \xrightarrow{o} s_2$, $w_1^{\Uparrow}, \nu_1 \models \mathbb{I}(s_1)$, $w_2^{\Uparrow}, \nu_2 \models \mathbb{I}(s_2)$ and one of the following holds,
  - **Local** $o = \epsilon$, then $w_1 = w_2$, and $\nu_1 = \nu_2$.
  - **Test** $o = \phi$, then $w_1 = w_2$, $\nu_1 = \nu_2$ and $\nu_1 \models \phi$.
  - **Reset** $o = x \leftarrow 0$, then $w_1 = w_2$, $\nu_2 = \nu_1[x \leftarrow 0]$.
  - **Value passing** $o = x \leftarrow x'$, then $w_1 = w_2, \nu_2 = \nu_1[x \leftarrow \nu_1(x')]$.
  - **Push** $o = push(\gamma)$, then $\nu_2 = \nu_0$, $w_2 = (\gamma, (\nu_1(x_1), \cdots, \nu_1(x_k)), 1, \mathbb{I}(s_1)).w_1$ for $X = \{x_1, \cdots, x_k\}$.
  - **F-Push** $o = fpush(\gamma)$, then $\nu_2 = \nu_0$, $w_2 = (\gamma, (\nu_1(x_1), \cdots, \nu_1(x_k)), 0, \mathbb{I}(s_1)).w_1$ for $X = \{x_1, \cdots, x_k\}$.
  - **Pop** $o = pop(\gamma)$, then $\nu_2 = \nu_1[x_1 \leftarrow t_1, \cdots, x_k \leftarrow t_k]$, $w_1 = (\gamma, (t_1, \cdots, t_k), flag, \phi).w_2$.

The initial configuration $\kappa_0 = (s_0, \epsilon, \nu_0)$. We use $\longrightarrow_{\mathscr{D}}$ to range over these transitions, and $\longrightarrow_{\mathscr{D}}^*$ is the transitive closure of $\longrightarrow_{\mathscr{D}}$.

Intuitively, in a stack $w = (\gamma_1, \bar{t}_1, flag_1, \phi_1). \cdots .(\gamma_n, \bar{t}_n, flag_n, \phi_n)$, γ_i is a stack symbol, $\bar{t}_i$ is k-tuple of clocks values of $x_1, \cdots, x_k$ respectively, $flag_i = 1$ if the stack frame is pushed and $flag_i = 0$ if fpushed and ϕ_i is a clock constraint.

Example 1. Figure 1 shows transitions between configurations of a Constraint DTPDA with $S = \{s_1, s_2, s_3, \cdots\}$, $X = \{x_1, x_2\}$, $\Gamma = \{a, b, d\}$ and $\mathbb{I} = \{\mathbb{I}(s_1) = x_1 \in [0, 1) \wedge x_2 \in [3, 4), \mathbb{I}(s_2) = x_1 \in [0, 3), \mathbb{I}(s_3) = \top, \cdots\}$. Values changed from the last configuration are in bold. For simplicity, we omit some transitions and start from s_1. From s_1 to s_2, a discrete transition $fpush(d)$ pushes d to the stack with the values of x_1 and x_2, frozen. After pushing, value of x_1 and x_2 will be reset to zero. Then, at state s_2, a progress transition elapses 2.6 time units, and each value grows older for 2.6 except for frozen clocks in the top. From s_2 to s_3, the batched transition first pops symbol d from the stack and clock values are recovered from the poped clocks. Then, the value of x_1 is reset to 0. Note that the invariants are always satisfied in these reachable configurations.

In the following subsections, we denote the set of finite multisets over D by $\mathcal{MP}(D)$, and the union of two multisets M, M' by $M \uplus M'$. We regard a finite set as a multiset with the multiplicity 1, and a finite word as a multiset by ignoring the ordering. Let $frac(t) = t - floor(t)$ for $t \in \mathbb{R}^{\geq 0}$.

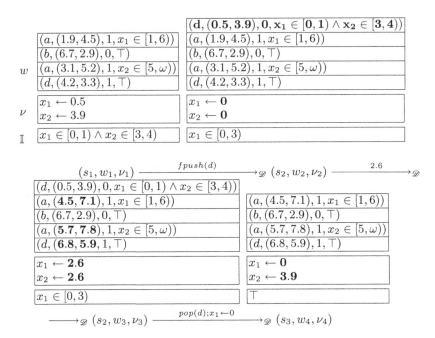

Fig. 1. An example of constraint DTPDAs

5.1 Digiword and Its Operations

Let $\langle S, s_0, \Gamma, X, \mathbb{I}, \Delta \rangle$ be a Constraint DTPDA, and let n be the largest integer (except for ω) appearing in $\mathbb{I}$ and Δ.

Definition 8 (Two Subsets of Intervals). *Let*

$$Intv(n) = \{\mathbf{r}_{2i} = [i,i] \mid 0 \leq i \leq n\} \cup \{\mathbf{r}_{2i+1} = (i, i+1) \mid 0 \leq i < n\} \cup \{\mathbf{r}_{2n+1} = (n, \omega)\}$$

Let $\mathcal{I}(n)$ denote a subset of intervals $\mathcal{I}$ such that all integers appearing in $\mathcal{I}(n)$ are less than or equal to n. For $v \in \mathbb{R}^{\geq 0}$, $proj(v) = \mathbf{r}_i$ if $v \in \mathbf{r}_i \in Intv(n)$.

Example 2. In Example 1, $n = 6$ and we have 13 intervals in $Intv(6)$,

$$0 \ \ \mathbf{r}_1 \ \ 1 \ \ \mathbf{r}_3 \quad 2 \ \ \mathbf{r}_5 \quad 3 \ \ \mathbf{r}_7 \quad 4 \ \ \mathbf{r}_9 \quad 5 \ \ \mathbf{r}_{11} \quad 6 \quad \mathbf{r}_{13}$$

$$\mathbf{r}_0 \qquad \mathbf{r}_2 \qquad \mathbf{r}_4 \qquad \mathbf{r}_6 \qquad \mathbf{r}_8 \qquad \mathbf{r}_{10} \qquad \mathbf{r}_{12}$$

$\mathcal{I}(6)$ contains intervals (a, b'), $[a, b]$, $[a, b')$ and $(a, b]$ where $a, b \in \{0, 1, \ldots, 6\}$ and $b' \in \{0, 1, \ldots, 6, \omega\}$.

$Intv(n)$ intend to contain digitizations of clocks, e.g., if a clock has value 1.9, then we say it is in $\mathbf{r}_3$. $\mathcal{I}(n)$ intend to contain intervals in invariants, e.g., an invariant $x \in [1, 2] \wedge y \in (3, 4)$ can be split into two intervals $[1, 2]$ and $(3, 4)$. Both $Intv(n)$ and $\mathcal{I}(n)$ are finite sets.

Definition 9 (Digitization). A digitization $\mathtt{digi} : \mathcal{MP}((X \cup \Gamma) \times \mathbb{R}^{\geq 0} \times \{0,1\} \times \mathcal{I}(n)) \rightarrow \mathcal{MP}((X \cup \Gamma) \times Intv(n) \times \{0,1\} \times \mathcal{I}(n))^*$ is defined as follows.

For $\bar{\mathcal{Y}} \in \mathcal{MP}((X \cup \Gamma) \times \mathbb{R}^{\geq 0} \times \{0,1\} \times \mathcal{I}(n))$, $\mathtt{digi}(\bar{\mathcal{Y}})$ is a word $Y_0 Y_1 \cdots Y_m$, where $Y_0, Y_1, \cdots, Y_m$ are multisets that collect $(x, proj(t), flag, I)$'s having the same $frac(t)$ for $(x, t, flag, I) \in \bar{\mathcal{Y}}$. Among them, Y_0 (which is possibly empty) is reserved for the collection of $(x, proj(t), flag, I)$ with $frac(t) = 0$ and $t \leq n$ (i.e., $proj(t) = \mathbf{r}_{2i}$ for $0 \leq i \leq n$). We assume that Y_i except for Y_0 is non-empty (i.e., $Y_i = \emptyset$ with $i > 0$ is omitted), and Y_i's are sorted by the increasing order of $frac(t)$ (i.e., $frac(t) < frac(t')$ for $(x, proj(t), flag, I) \in Y_i$ and $(x', proj(t'), flag', I') \in Y_{i+1}$).

For $Y_i \in \mathcal{MP}((X \cup \Gamma) \times Intv(n) \times \{0,1\} \times \mathcal{I}(n))$, we define the projections by $prc(Y_i) = \{(x, proj(t), 1, I) \in Y_i\}$ and $frz(Y_i) = \{(x, proj(t), 0, I) \in Y_i\}$. We overload the projections on $\bar{Y} = Y_0 Y_1 \cdots Y_m \in (\mathcal{MP}((X \cup \Gamma) \times Intv(n) \times \{0,1\} \times \mathcal{I}(n)))^*$ such that $frz(\bar{Y}) = frz(Y_0) frz(Y_1) \cdots frz(Y_m)$ and $prc(\bar{Y}) = prc(Y_0) prc(Y_1) \cdots prc(Y_m)$.

For a stack frame $v = (\gamma, (t_1, \cdots, t_k), flag, \phi)$ of a Constraint DTPDA, we denote a word $(\gamma, t_1, flag, EC(\phi, x_1)) \cdots (\gamma, t_k, flag, EC(\phi, x_k))$ by $dist(v)$. Given a state s and a clock valuation ν, we define a word $time(s, \nu) = (x_1, \nu(x_1), 1, EC(\mathbb{I}(s), x_1)) \ldots (x_k, \nu(x_k), 1, EC(\mathbb{I}(s), x_k))$ where $x_1 \ldots x_k \in X$.

Example 3. For the configuration $\varrho_1 = (s_1, v_4 \cdots v_1, \nu_1)$ in Example 1, let $\bar{\mathcal{Y}} = dist(v_4) \uplus \ldots \uplus dist(v_1) \uplus time(s_1, \nu_1)$, and $\bar{Y} = \mathtt{digi}(\bar{\mathcal{Y}})$, i.e.,

$$\bar{\mathcal{Y}} = \{(a, 1.9, 1, [1, 6)), (a, 4.5, 1, [0, \omega)), (b, 6.7, 0, [0, \omega)), (b, 2.9, 0, [0, \omega)),$$
$$(a, 3.1, 1, [0, \omega)), (a, 5.2, 1, [5, \omega)), (d, 4.2, 1, [0, \omega)),$$
$$(d, 3.3, 1, [0, \omega)), (x_1, 0.5, 1, [0, 1)), (x_2, 3.9, 1, [3, 4))\}$$
$$\bar{Y} = \{(a, \mathbf{r}_7, 1, [0, \omega))\}\{(a, \mathbf{r}_{11}, 1, [5, \omega)), (d, \mathbf{r}_9, 1, [0, \omega))\}\{(d, \mathbf{r}_7, 1, [0, \omega))\}$$
$$\{(x_1, \mathbf{r}_1, 1, [0, 1)), (a, \mathbf{r}_9, 1, [0, \omega))\}\{(b, \mathbf{r}_{13}, 0, [0, \omega))\}\{(x_2, \mathbf{r}_7, 1, [3, 4)),$$
$$(a, \mathbf{r}_3, 1, [1, 6)), (b, \mathbf{r}_5, 0, [0, \omega))\}$$
$$prc(\bar{Y}) = \{(a, \mathbf{r}_7, 1, [0, \omega))\}\{(a, \mathbf{r}_{11}, 1, [5, \omega)), (d, \mathbf{r}_9, 1, [0, \omega))\}\{(d, \mathbf{r}_7, 1, [0, \omega))\}$$
$$\{(x_1, \mathbf{r}_1, 1, [0, 1)), (a, \mathbf{r}_9, 1, [0, \omega))\}\{(x_2, \mathbf{r}_7, 1, [3, 4)), (a, \mathbf{r}_3, 1, [1, 6))\}$$
$$frz(\bar{Y}) = \{(b, \mathbf{r}_{13}, 0, [0, \omega))\}\{(b, \mathbf{r}_5, 0, [0, \omega))\}$$

Definition 10 (Digiwords and k-pointers). A word $\bar{Y} \in (\mathcal{MP}((X \cup \Gamma) \times Intv(n) \times \{0,1\} \times \mathcal{I}(n)))^*$ is called a digiword. We say a digiword $\bar{Y}$ is good, written as $\bar{Y}^{\Uparrow}$, if for all $(x, \mathbf{r}_i, flag, I)$ in $\bar{Y}$, $\mathbf{r}_i \subseteq I$. We denote $\bar{Y}|_\Lambda$ for $\Lambda \subseteq \Gamma \cup X$, by removing $(x, \mathbf{r}_i, flag, I)$ with $x \notin \Lambda$. A k-pointer $\bar{\rho}$ of $\bar{Y}$ is a tuple of k pointers to mutually different k elements in $\bar{Y}|_\Gamma$. We refer to the element pointed by the i-th pointer by $\bar{\rho}[i]$. From now on, we assume that a digiword has two pairs of k-pointers $(\bar{\rho}_1, \bar{\rho}_2)$ and $(\bar{\tau}_1, \bar{\tau}_2)$ that point to only proceeding and frozen clocks, respectively. We call $(\bar{\rho}_1, \bar{\rho}_2)$ proceeding k-pointers and $(\bar{\tau}_1, \bar{\tau}_2)$ frozen k-pointers. We also assume that they do not overlap each other, i.e., there are no i, j, such that $\bar{\rho}_1[i] = \bar{\rho}_2[j]$ or $\bar{\tau}_1[i] = \bar{\tau}_2[j]$.

$\bar{\rho}_1$ and $\bar{\rho}_2$ intend the store of values of the proceeding clocks at the last and one before the last **Push**, respectively. $\bar{\tau}_1$ and $\bar{\tau}_2$ intend similar for frozen clocks at **F-Push**.

Definition 11 (Embedding over Digiwords). *For digiwords* $\bar{Y} = Y_1 \cdots Y_m$ *and* $\bar{Z} = Z_1 \cdots Z_{m'}$ *with pairs of k-pointers* $(\bar{\rho}_1, \bar{\rho}_2), (\bar{\tau}_1, \bar{\tau}_2)$, *and* $(\bar{\rho}'_1, \bar{\rho}'_2), (\bar{\tau}'_1, \bar{\tau}'_2)$, *respectively, we define an embedding* $\bar{Y} \sqsubseteq \bar{Z}$, *if there exists a monotonic injection* $f : [1..m] \rightarrow [1..m']$ *such that* $Y_i \subseteq Z_{f(i)}$ *for each* $i \in [1..m]$, $f \circ \bar{\rho}_i = \bar{\rho}'_i$ *and* $f \circ \bar{\tau}_i = \bar{\tau}'_i$ *for* $i = 1, 2$.

The embedding $\sqsubseteq$ is a well-quasi-ordering which will be exploited in Sect. 5.3.

Definition 12 (Operations on Digiwords). *Let* $\bar{Y} = Y_0 \cdots Y_m, \bar{Y}' = Y'_0 \cdots Y'_{m'} \in (\mathcal{MP}((X \cup \Gamma) \times Intv(n) \times \{0,1\} \times \mathcal{I}(n)))^*$ *such that* $\bar{Y}$ *(resp.* $\bar{Y}'$) *has two pairs of proceeding and frozen k-pointers* $(\bar{\rho}_1, \bar{\rho}_2)$ *and* $(\bar{\tau}_1, \bar{\tau}_2)$ *(resp.* $(\bar{\rho}'_1, \bar{\rho}'_2)$ *and* $(\bar{\tau}'_1, \bar{\tau}'_2)$). *We define digiword operations as follows.*

- **Decomposition**: *Let* $Z \in \mathcal{MP}((X \cup \Gamma) \times Intv(n) \times \{0,1\} \times \mathcal{I}(n))$. *If* $Z \subseteq Y_j$, $decomp(\bar{Y}, Z) = (Y_0 \cdots Y_{j-1}, Y_j, Y_{j+1} \cdots Y_m)$.
- **Refresh** $refresh(\bar{Y}, s)$ *for* $s \in S$ *is obtained by updating all elements* $(x, \mathbf{r}_i, 1, I)$ *with* $(x, \mathbf{r}_i, 1, EC(\mathbb{I}(s), x))$ *for* $x \in X$.
- **Init** $init(\bar{Y})$ *is obtained by removing all elements* $(x, \mathbf{r}, 1, I)$ *from* $\bar{Y}$ *and inserting* $(x, \mathbf{r}_0, 1, [0, w])$ *to* Y_0 *for all* $x \in X$.
- **Insert**$_x$ $insert_x(\bar{Y}, x, y)$ *adds* $(x, \mathbf{r}_i, 1, I)$ *to* Y_j *for* $(y, \mathbf{r}_i, 1, I) \in Y_j$, $x, y \in X$.
- **Insert**$_I$: *Let* $Z \in \mathcal{MP}((X \cup \Gamma) \times Intv(n) \times \{0,1\} \times \mathcal{I}(n))$ *with* $(x, \mathbf{r}_i, flag, I) \in Z$ *for* $x \in X \cup \Gamma$. $insert_I(\bar{Y}, Z)$ *inserts* Z *to* $\bar{Y}$ *such that*

$$\begin{cases} either\ take\ the\ union\ of\ Z\ and\ Y_j\ for\ j > 0,\ or\ put\ Z\ at\ any\ place\ after\ Y_0 \\ \qquad\qquad if\ i\ is\ odd \\ take\ the\ union\ of\ Z\ and\ Y_0 \qquad if\ i\ is\ even \end{cases}$$

- **Delete.** $delete(\bar{Y}, x)$ *for* $x \subseteq X$ *is obtained from* $\bar{Y}$ *by deleting the element* $(x, \mathbf{r}, 1, I)$ *indexed by* x.
- **Permutation.** *Let* $\bar{V} = prc(\bar{Y}) = V_0 V_1 \cdots V_k$ *and* $\bar{U} = frz(\bar{Y}) = U_0 U_1 \cdots U_{k'}$. *A one-step permutation* $\bar{Y} \Rightarrow \bar{Y}'$ *is given by* $\Rightarrow = \Rightarrow_s \cup \Rightarrow_c$, *defined below. We denote* $inc(V_j)$ *for* V_j *in which each* $\mathbf{r}_i$ *is updated to* $\mathbf{r}_{i+1}$ *for* $i < 2n + 1$.

$(\Rightarrow_s)$ *Let*
$$\begin{cases} decomp(U_0 \,.\, inc(V_0) \,.\, tl(\bar{Y}), V_k) = (\bar{Y}^k_{\dashv}, \hat{Y}^k, \bar{Y}^k_{\vdash}) \\ decomp(insert_I((\hat{Y}^k \setminus V_k) \,.\, \bar{Y}^k_{\vdash}, V_k), V_k) = (\bar{Z}^k_{\dashv}, \hat{Z}^k, \bar{Z}^k_{\vdash}). \end{cases}$$
For j *with* $0 \leq j < k$, *we repeat to set*
$$\begin{cases} decomp(\bar{Y}^{j+1}_{\dashv} \,.\, \bar{Z}^{j+1}_{\dashv}, V_j) = (\bar{Y}^j_{\dashv}, \hat{Y}^j, \bar{Y}^j_{\vdash}) \\ decomp(insert_I((\hat{Y}^j \setminus V_j) \,.\, \bar{Y}^j_{\vdash}, V_j), V_j) = (\bar{Z}^j_{\dashv}, \hat{Z}^j, \bar{Z}^j_{\vdash}). \end{cases}$$
Then, $\bar{Y} \Rightarrow_s \bar{Y}' = \bar{Y}^0_{\dashv} \bar{Z}^0_{\dashv} \hat{Z}^0 \bar{Z}^1_{\dashv} \hat{Z}^1 \cdots \bar{Z}^k_{\dashv} \hat{Z}^k \bar{Z}^k_{\vdash}$.
$(\Rightarrow_c)$ *Let* $\bar{Y}^k_{\dashv} = U_0 \cup inc(V_k)$ *and* $\bar{Z}^k_{\dashv} = inc(V_0) Y_1 \cdots (Y_{i'} \setminus V_k) \cdots Y_m$.
For j *with* $0 \leq j < k$, *we repeat to set*
$$\begin{cases} decomp(\bar{Y}^{j+1}_{\dashv}.\bar{Z}^{j+1}_{\dashv}, V_j) = (\bar{Y}^j_{\dashv}, \hat{Y}^j, \bar{Y}^j_{\vdash}) \\ decomp(insert_I((\hat{Y}^j \setminus V_j).\bar{Y}^j_{\vdash}, V_j), V_j) = (\bar{Z}^j_{\dashv}, \hat{Z}^j, \bar{Z}^j_{\vdash}). \end{cases}$$
Then, $\bar{Y} \Rightarrow_c \bar{Y}' = \bar{Y}^0_{\dashv} \bar{Z}^0_{\dashv} \hat{Z}^0 \bar{Z}^1_{\dashv} \hat{Z}^1 \cdots \bar{Z}^{k-1}_{\dashv} \hat{Z}^{k-1} \bar{Z}^{k-1}_{\vdash}$.

$(\bar{\rho}_1, \bar{\rho}_2)$ *is updated to correspond to the permutation accordingly, and* $(\bar{\tau}_1, \bar{\tau}_2)$ *is kept unchanged.*

- **Rotate:** *For proceeding k-pointers* $(\bar{\rho}_1, \bar{\rho}_2)$ *of* $\bar{Y}$ *and* $\bar{\rho}$ *of* $\bar{Z}$, *let* $\bar{Y}|_\Gamma \Rightarrow^* \bar{Z}|_\Gamma$ *such that the permutation makes* $\bar{\rho}_1$ *match with* $\bar{\rho}$. *Then,* $rotate_{\bar{\rho}_1 \mapsto \bar{\rho}}(\bar{\rho}_2)$ *is the corresponding k-pointer of* $\bar{Z}$ *to* $\bar{\rho}_2$.
- **Map**$_{\rightarrow}^{flag}$ $map_{\rightarrow}^{fl}(\bar{Y}, \gamma)$ *for* $\gamma \in \Gamma$ *is obtained from* $\bar{Y}$ *by, for each* $x_i \in X$, *replacing* $(x_i, \mathbf{r}_j, 1, I)$ *with* $(\gamma, \mathbf{r}_j, fl, I)$. *Accordingly, if* $fl = 1$, $\bar{\rho}_1[i]$ *is updated to point to* $(\gamma, \mathbf{r}_j, 1, I)$, *and* $\bar{\rho}_2$ *is set to the original* $\bar{\rho}_1$. *If* $fl = 0$, $\bar{\tau}_1[i]$ *is updated to point to* $(\gamma, \mathbf{r}_j, 0, I)$, *and* $\bar{\tau}_2$ *is set to the original* $\bar{\tau}_1$.
- **Map**$_{\leftarrow}^{flag}$ $map_{\leftarrow}^{fl}(\bar{Y}, \bar{Y}', \gamma)$ *for* $\gamma \in \Gamma$ *is obtained,*
  - **(if** $fl = 1$**)** *by replacing each* $\bar{\rho}_1[i] = (\gamma, \mathbf{r}_j, 1, I)$ *in* $\bar{Y}|_\Gamma$ *with* $(x_i, \mathbf{r}_j, 1, I)$ *for* $x_i \in X$. *Accordingly, new* $\bar{\rho}_1$ *is set to the original* $\bar{\rho}_2$, *and new* $\bar{\rho}_2$ *is set to* $rotate_{\bar{\rho}'_1 \mapsto \bar{\rho}_2}(\bar{\rho}'_2)$. $\bar{\tau}_1$ *and* $\bar{\tau}_2$ *are kept unchanged.*
  - **(if** $fl = 0$**)** *by replacing each* $\bar{\tau}_1[i] = (\gamma, \mathbf{r}_j, 0, I)$ *in* $\bar{Y}|_\Gamma$ *with* $(x_i, \mathbf{r}_j, 1, I)$ *for* $x_i \in X$. *Accordingly, new* $\bar{\tau}_1$ *is set to the original* $\bar{\tau}_2$, *and new* $\bar{\tau}_2$ *is set to* $\bar{\tau}'_2$. $\bar{\rho}_1$ *and* $\bar{\rho}_2$ *are kept unchanged.*

We will use these operations on digiwords for encoding in the next subsection.

5.2   Snapshot Pushdown System

In this subsection, we show that a Constraint DTPDA is encoded into its digitization, called a *snapshot pushdown system* (snapshot PDS), which keeps the digitization of all clocks in the top stack frame, as a *digiword*. The keys of the encoding are, (1) when a pop occurs, the time progress recorded at the top stack symbol is propagated to the next stack symbol after finding a permutation by matching between proceeding k-pointers $\bar{\rho}_2$ and $\bar{\rho}'_1$, and (2) only invariants in the top stack frame need to be checked. Before showing the encoding, we first define the encoded configuration, called *snapshot configuration*.

Definition 13 (Snapshot Configuration). *Let* $\pi : \varrho_0 = (s_0, \epsilon, \nu_0) \longrightarrow_{\mathscr{D}}^*$ $\varrho = (s, w, \nu)$ *be a transition sequence of a Constraint DTPDA from the initial configuration. If* π *is not empty, we refer the last step as* $\lambda : \varrho' \longrightarrow_{\mathscr{D}} \varrho$, *and the preceding sequence by* $\pi' : \varrho_0 \longrightarrow_{\mathscr{D}}^* \varrho'$. *Let* $w = v_m \cdots v_1$. *A snapshot is* $snap(\pi) = (\bar{Y}, flag(v_m))$, *where* $\bar{Y} = \mathtt{digi}(\uplus_i dist(v_i) \uplus time(s, \nu))$. *Let a k-pointer* $\bar{\xi}(\pi)$ *be* $\bar{\xi}(\pi)[i] = (\gamma, proj(t_i), flag(v_m), I)$ *for* $(\gamma, t_i, flag(v_m), I) \in dist(v_m)$. *A snapshot configuration* $Snap(\pi)$ *is inductively defined from* $Snap(\pi')$.

$$
\begin{cases}
(s_0, snap(\epsilon)) & if\ \pi = \epsilon.(\bar{\rho}_1, \bar{\rho}_2)\ and\ (\bar{\tau}_1, \bar{\tau}_2)\ are\ \text{undefined}. \\
(s', snap(\pi)\ tail(Snap(\pi'))) & if\ \lambda\ is\ \textbf{Timeprogress}\ with\ \bar{Y}' \Rightarrow^* \bar{Y}. \\
\qquad Then, the\ permutation\ \bar{Y}' \Rightarrow^* \bar{Y}\ updates\ (\bar{\rho}'_1, \bar{\rho}'_2)\ to\ (\bar{\rho}_1, \bar{\rho}_2). \\
(s', snap(\pi)\ tail(Snap(\pi'))) & if\ \lambda\ is\ \textbf{Local, Test, Reset, Value} - \text{passing}. \\
(s, snap(\pi)\ Snap(\pi')) & if\ \lambda\ is\ \textbf{Push}.\ Then,\ (\bar{\rho}_1, \bar{\rho}_2) = (\bar{\xi}(\pi), \bar{\rho}'_1). \\
(s, snap(\pi)\ Snap(\pi')) & if\ \lambda\ is\ \textbf{F} - \textbf{Push}.\ Then,\ (\bar{\tau}_1, \bar{\tau}_2) = (\bar{\xi}(\pi), \bar{\tau}'_1). \\
(s, snap(\pi)\ tail(tail(Snap(\pi')))) & if\ \lambda\ is\ \textbf{Pop}. \\
\quad If\ flag = 1, (\bar{\rho}_1, \bar{\rho}_2) = (\bar{\rho}'_2, rotate_{\bar{\rho}''_1 \mapsto \bar{\rho}'_2}(\bar{\rho}''_2)); otherwise, (\bar{\tau}_1, \bar{\tau}_2) = (\bar{\tau}'_2, \bar{\tau}''_2).
\end{cases}
$$

We refer $head(Snap(\pi'))$ by $\bar{Y}'$, $head(tail(Snap(\pi')))$ by $\bar{Y}''$. Pairs of pointers of $\bar{Y}$, $\bar{Y}'$, and $\bar{Y}''$ are denoted by $(\bar{\rho}_1, \bar{\rho}_2)$, $(\bar{\rho}'_1, \bar{\rho}'_2)$, and $(\bar{\rho}''_1, \bar{\rho}''_2)$, respectively. If not mentioned, pointers are kept as is.

Definition 14 (Snapshot PDS). *For a Constraint DTPDA* $\langle S, s_0, \Gamma, X, \mathbb{I}, \nabla \rangle$, *a snapshot PDS* $\mathcal{S}$ *is a PDS (with possibly infinite stack alphabet)*

$$\langle S \cup \{s_{err}\}, s_0, (\mathcal{MP}((X \cup \Gamma) \times Intv(n) \times \{0,1\} \times \mathcal{I}(n)))^* \times \{0,1\}, \Delta_d \rangle.$$

with the initial configuration $\langle s_0, (\{(x, \mathbf{r}_0, 1, EC(\mathbb{I}(s_0), x)) \mid x \in X\}, 1) \rangle$. *For simplicity, we define* $s'' = \begin{cases} s' & \text{if } \bar{Y}'^{\Uparrow}, \\ s_{err} & \text{otherwise} \end{cases}$ *where* s_{err} *is a special error state that is used to indicate invariants are violated. Then* Δ_d *consists of:*

Progress $\langle s, (\bar{Y}, flag) \rangle \hookrightarrow_{\mathcal{S}} \langle s'', (\bar{Y}', flag) \rangle$ *for* $\bar{Y} \Rightarrow^* \bar{Y}'$, *where* $s' = s$.

Local $(s \xrightarrow{\epsilon} s' \in \Delta)$ $\langle s, (\bar{Y}, flag) \rangle \hookrightarrow_{\mathcal{S}} \langle s'', (\bar{Y}', flag) \rangle$, *where* $\bar{Y}' = refresh(\bar{Y}, s')$.

Test $(s \xrightarrow{\phi} s' \in \Delta)$ $\langle s, (\bar{Y}, flag) \rangle \hookrightarrow_{\mathcal{S}} \langle s'', (\bar{Y}', flag) \rangle$, *where* $\bar{Y}' = refresh(\bar{Y}, s')$, *if for every* $(x, \mathbf{r}_i, flag, I) \in \bar{Y}$ *with* $x \in X$, $\mathbf{r}_i \subseteq EC(\phi, x)$ *holds,*

Reset $(s \xrightarrow{x \leftarrow 0} s' \in \Delta$ *with* $\lambda \subseteq X)$ $\langle s, (\bar{Y}, flag) \rangle \hookrightarrow_{\mathcal{S}} \langle s'', (\bar{Y}', flag) \rangle$, *where* $\bar{Y}' = refresh(insert_I(delete(\bar{Y}, x), (x, \mathbf{r}_0, 1, [0, w))), s')$.

Value-passing $(s \xrightarrow{x \leftarrow y} s' \in \Delta$ *with* $x, y \in X)$ $\langle s, (\bar{Y}, flag) \rangle \hookrightarrow_{\mathcal{S}} \langle s'', (\bar{Y}', flag) \rangle$, *where* $\bar{Y}' = refresh(insert_x(delete(\bar{Y}, x), x, y), s')$.

Push $(s \xrightarrow{push(\gamma)} s' \in \Delta;\ fl = 1)$ *and* **F-Push** $(s \xrightarrow{fpush(\gamma)} s' \in \Delta;\ fl = 0)$ $\langle s, (\bar{Y}, flag) \rangle \hookrightarrow_{\mathcal{S}} \langle s'', (\bar{Y}', fl)(\bar{Y}, flag) \rangle$, *where* $\bar{Y}' = refresh(init(map_{\rightarrow}^{fl}(\bar{Y}, \gamma), s')$.

Pop $(s \xrightarrow{pop(\gamma)} s' \in \Delta)$ $\langle s, (\bar{Y}, flag)(\bar{Y}'', flag') \rangle \hookrightarrow_{\mathcal{S}} \langle s'', (\bar{Y}', flag') \rangle$, *where* $\bar{Y}' = refresh(map_{\leftarrow}^{flag}(\bar{Y}, \bar{Y}'', \gamma), s')$.

By induction on the number of steps of transitions, the encoding relation between a Constraint DTPDA and a snapshot PDS is observed.

Lemma 1. *Let us denote* ϱ_0 *and* ϱ *(resp.* $\langle s_0, \tilde{w}_0 \rangle$ *and* $\langle s, \tilde{w} \rangle$*) for the initial configuration and a configuration of a Constraint DTPDA (resp. its encoded snapshot PDS* $\mathcal{S}$*).*

(Preservation) *If* $\pi : \varrho_0 \longrightarrow_{\mathscr{D}}^* \varrho$, *there exists* $\langle s, \tilde{w} \rangle$ *such that* $\langle s_0, \tilde{w}_0 \rangle \hookrightarrow_{\mathcal{S}}^*$ $\langle s, \tilde{w} \rangle$ *and* $Snap(\pi) = \langle s, \tilde{w} \rangle$.

(Reflection) *If* $\langle s_0, \tilde{w}_0 \rangle \hookrightarrow_{\mathcal{S}}^* \langle s, \tilde{w} \rangle$,
  $s = s_{err}$ *is an error state, or*
  $s \neq s_{err}$ *and there exists* $\pi : \varrho_0 \longrightarrow_{\mathscr{D}}^* \varrho$ *with* $Snap(\pi) = \langle s, \tilde{w} \rangle$.

5.3 Well-Formed Constraint

A snapshot PDS is *a growing WSPDS* (Definition 6 in [6]) and $\Downarrow_T$ gives a *well-formed constraint* (Definition 8 in [6]). Let us recall the definitions.

Let P be a set of control locations and let Γ be a stack alphabet. Different from an ordinary definition of PDSs, we do not assume that P and Γ are finite, but associated with well-quasi-orderings (WQOs) $\preceq$ and $\leq$, respectively. Note that the embedding $\sqsubseteq$ over digiwords is a WQO by Higman's lemma.

For $w = \alpha_1\alpha_2 \cdots \alpha_n, v = \beta_1\beta_2 \cdots \beta_m \in \Gamma^*$, let $w \ll v$ if $m = n$ and $\forall i \in [1..n].\alpha_i \leq \beta_i$. We extend $\ll$ on configurations such that $(p, w) \ll (q, v)$ if $p \preceq q$ and $w \ll v$ for $p, q \in P$ and $w, v \in \Gamma^*$. A partial function $\psi \in \mathcal{P}Fun(X, Y)$ is *monotonic* if $\gamma \leq \gamma'$ with $\gamma \in dom(\psi)$ implies $\psi(\gamma) \ll \psi(\gamma')$ and $\gamma' \in dom(\psi)$.

A *a well-structured PDS* (WSPDS) is a triplet $\langle (P, \preceq), (\Gamma, \leq), \Delta \rangle$ of a set $(P, \preceq)$ of WQO states, a WQO stack alphabet $(\Gamma, \leq)$, and a finite set $\Delta \subseteq \mathcal{P}Fun(P \times \Gamma, P \times \Gamma^{\leq 2})$ of monotonic partial functions. A WSPDS is *growing* if, for each $\psi(p, \gamma) = (q, w)$ with $\psi \in \Delta$ and $(q', w') \geqslant (q, w)$, there exists (p', γ') with $(p', \gamma') \geqslant (p, \gamma)$ such that $\psi(p', \gamma') \geqslant (q', w')$.

A well-formed constraint describes a syntactical feature that is preserved under transitions. Theorem 5 in [6] ensures the reachability of a growing WSPDS when it has a well-formed constraint.

Definition 15 (Well-formed constraint). *Let a configuration $(s, \tilde{w})$ of a snapshot PDS S. An element in a stack frame of $\tilde{w}$ has a* parent *if it has a corresponding element in the next stack frame. The transitive closure of the parent relation is an* ancestor. *An element in $\tilde{w}$ is* marked, *if its ancestor is pointed by a pointer in some stack frame. We define a* projection $\Downarrow_\Upsilon (\tilde{w})$ *by removing unmarked elements in $\tilde{w}$. We say that $\tilde{w}$ is* well-formed *if $\Downarrow_\Upsilon (\tilde{w}) = \tilde{w}$.*

The idea of $\Downarrow_\Upsilon$ is to remove unnecessary elements (i.e., elements not related to previous actions) from the stack content. Note that a configuration reachable from the initial configuration by $\hookrightarrow^*_S$ is always well-formed. Since a snapshot PDS is a growing WSPDS with $\Downarrow_\Upsilon$, we conclude Theorem 2 from Lemma 1.

Theorem 2. *The reachability of a Constraint DTPDA is decidable.*

6 Decidability Results of NeTA-Is

In this section, we encode NeTA-I with no global clocks to constraint DTPDAs and thus show the decidability of the former model.

Given a NeTA-I $\mathcal{N} = (T, \mathcal{A}_0, X, C, \mathbb{I}, \Delta)$ with no global clocks ($C = \emptyset$), we define the target Constraint DTPDA $\mathcal{E}(\mathcal{N}) = \langle S, s_0, \Gamma, X, \mathbb{I}', \nabla \rangle$ such that

- $S = \Gamma = \bigcup_{\mathcal{A}_i \in T} Q(\mathcal{A}_i)$ is the set of all control locations of TAs in T.
- $s_0 = q_0(\mathcal{A}_0)$ is the initial control location of the initial TA $\mathcal{A}_0$.
- $X = \{x_1, ..., x_k\}$ is the set of k local clocks.
- $\mathbb{I}' : S \to \Phi(X)$ is a function such that $\mathbb{I}'(s) = \mathbb{I}(\mathcal{A}_i)(s)$ where $s \in Q(\mathcal{A}_i)$.
- ∇ is the union $\bigcup_{\mathcal{A}_i \in T} \Delta(\mathcal{A}_i) \bigcup \mathcal{H}(\mathcal{N})$ where
  $\begin{cases} \Delta(\mathcal{A}_i) = \{\textbf{Local}, \textbf{Test}, \textbf{Reset}, \textbf{Value-passing}\}, \\ \mathcal{H}(\mathcal{N}) \text{ consists of rules below.} \end{cases}$

Push	$q \xrightarrow{push(q)} q_0(\mathcal{A}_{i'})$	if $(q, \varepsilon, push, q_0(\mathcal{A}_{i'}), q) \in \Delta(\mathcal{N})$
F − Push	$q \xrightarrow{fpush(q)} q_0(\mathcal{A}_{i'})$	if $(q, \varepsilon, f\text{-}push, q_0(\mathcal{A}_{i'}), q) \in \Delta(\mathcal{N})$
Pop	$q \xrightarrow{pop(q')} q'$	if $(q, q', pop, q', \varepsilon) \in \Delta(\mathcal{N})$

Definition 16. *Let $\mathcal{N}$ be a NeTA-I $(T, \mathcal{A}_0, X, C, \mathbb{I}, \Delta)$ with no global clocks and let $\mathcal{E}(\mathcal{N})$ be the encoded constraint DTPDA $\langle S, s_0, \Gamma, X, \mathbb{I}', \nabla \rangle$. For a configuration $\kappa = (\langle q, \nu, \mu \rangle, v)$ of $\mathcal{N}$ such that $v = (q_1, flag_1, \nu_1) \ldots (q_n, flag_n, \nu_n)$, $[\![\kappa]\!]$ denotes a configuration $(q, \overline{w}(\kappa), \nu)$ of $\mathcal{E}(\mathcal{N})$ where $\overline{w}(\kappa) = w_1 \cdots w_n$ with $w_i = (q_i, \nu_i, flag_i, \mathbb{I}(q_i))$.*

We can prove that transitions are preserved and reflected by the encoding.

Lemma 2. *For a NeTA-I $\mathcal{N}$ with no global clocks, its encoded Constraint DTPDA $\mathcal{E}(\mathcal{N})$, and configurations κ, κ' of $\mathcal{N}$,*

(Preservation) *if $\kappa \longrightarrow \kappa'$, then $[\![\kappa]\!] \longrightarrow_{\mathscr{D}}^* [\![\kappa']\!]$, and*
(Reflection) *if $[\![\kappa]\!] \longrightarrow_{\mathscr{D}}^* \varrho$, there exists κ' with $\varrho \longrightarrow_{\mathscr{D}}^* [\![\kappa']\!]$ and $\kappa \longrightarrow^* \kappa'$.*

Theorem 3. *The reachability of a NeTA-I with no global clocks is decidable.*

7   Conclusion

This paper proposes a model NeTA-Is by extending NeTAs with invariants assigned to each control location. We have shown that the reachability problem of a NeTA-I with a single global clock is undecidable, while that of a NeTA-I without global clocks is decidable. Compared to the different result of NeTA [5], it is revealed that unlike that of timed automata, invariants affect the expressiveness of timed recursive systems. Hence, when adopting timed recursive systems to model and verify complex real-time systems, one should carefully consider the introduction of invariants.

Acknowledgements. This work is supported by National Natural Science Foundation of China with grant Nos. 61472240, 61672340, 61472238, and the NSFC-JSPS bilateral joint research project with grant No. 61511140100.

References

1. Alur, R., Dill, D.L.: A theory of timed automata. Theoret. Comput. Sci. **126**, 183–235 (1994)
2. Henzinger, T.A., Nicollin, X., Sifakis, J., Yovine, S.: Symbolic model checking for real-time systems. Inf. Comput. **111**, 193–244 (1994)
3. Bengtsson, J., Yi, W.: Timed automata: semantics, algorithms and tools. In: Desel, J., Reisig, W., Rozenberg, G. (eds.) ACPN 2003. LNCS, vol. 3098, pp. 87–124. Springer, Heidelberg (2004). doi:10.1007/978-3-540-27755-2_3
4. Li, G., Cai, X., Ogawa, M., Yuen, S.: Nested timed automata. In: Braberman, V., Fribourg, L. (eds.) FORMATS 2013. LNCS, vol. 8053, pp. 168–182. Springer, Heidelberg (2013). doi:10.1007/978-3-642-40229-6_12
5. Li, G., Ogawa, M., Yuen, S.: Nested timed automata with frozen clocks. In: Sankaranarayanan, S., Vicario, E. (eds.) FORMATS 2015. LNCS, vol. 9268, pp. 189–205. Springer, Cham (2015). doi:10.1007/978-3-319-22975-1_13
6. Cai, X., Ogawa, M.: Well-structured pushdown system: case of dense timed pushdown automata. In: Codish, M., Sumii, E. (eds.) FLOPS 2014. LNCS, vol. 8475, pp. 336–352. Springer, Cham (2014). doi:10.1007/978-3-319-07151-0_21

7. Abdulla, P.A., Atig, M.F., Stenman, J.: Dense-timed pushdown automata. In: Proceedings of the LICS 2012, pp. 35–44. IEEE Computer Society (2012)
8. Trivedi, A., Wojtczak, D.: Recursive timed automata. In: Bouajjani, A., Chin, W.-N. (eds.) ATVA 2010. LNCS, vol. 6252, pp. 306–324. Springer, Heidelberg (2010). doi:10.1007/978-3-642-15643-4_23
9. Benerecetti, M., Minopoli, S., Peron, A.: Analysis of timed recursive state machines. In: Proceedings of the TIME 2010, pp. 61–68. IEEE Computer Society (2010)
10. Minsky, M.: Computation: Finite and Infinite Machines. Prentice-Hall, Upper Saddle River (1967)

Multi-core Cyclic Executives
for Safety-Critical Systems

Calvin Deutschbein[1], Tom Fleming[2], Alan Burns[2], and Sanjoy Baruah[3(✉)]

[1] The University of North Carolina at Chapel Hill, Chapel Hill, USA
[2] The University of York, York, UK
[3] Washington University in St. Louis, St. Louis, USA
baruah@wustl.edu

Abstract. In a cyclic executive, a series of pre-determined frames are executed in sequence; once the series is complete the sequence is repeated. Within each frame individual units of computation are executed, again in a pre-specified sequence. The implementation of cyclic executives upon multi-core platforms is considered. A Linear Programming (LP) based formulation is presented of the problem of constructing cyclic execu-tives upon multiprocessors for a particular kind of recurrent real-time workload – collections of implicit-deadline periodic tasks. Techniques are described for solving the LP formulation under different kinds of restric-tions in order to obtain preemptive and non-preemptive cyclic executives.

1   Introduction and Motivation

Real-time scheduling theory has made great advances over the past several decades. Despite these advances, interactions with industrial collaborators in highly safety-critical application domains, particularly those that are subject to stringent certification requirements, reveal that the use of the very simple *cyclic executive* approach [1] remains surprisingly wide-spread for scheduling safety-critical systems. A cyclic executive (CE) is a simple deterministic scheme that consists, for a single processor, of the repeated execution of a series of *frames*, each comprising a sequence of *jobs* that execute in their defining sequence and must complete by the end of the frame. Although there are a number of draw-backs to using cyclic executives (some are discussed in Sect. 2), this approach offers two significant advantages, *predictability* and *low run-time overhead*, that are responsible for their continued widespread use in scheduling highly safety-critical systems.

Highly safety-critical real-time systems have traditionally been implemented upon custom-built single-core processors that are designed to guarantee pre-dictable timing behavior during run-time. As safety-critical software has become more computation-intensive, however, it has proved too expensive to custom-build hardware powerful enough to accommodate the computational require-ments of such software; hence, there is an increasing trend towards implementing safety-critical systems upon commercial off-the-shelf (COTS) platforms. Most COTS processors today tend to be multi-core ones; this motivates our research

© Springer International Publishing AG 2017
K.G. Larsen et al. (Eds.): SETTA 2017, LNCS 10606, pp. 94–109, 2017.
https://doi.org/10.1007/978-3-319-69483-2_6

described here into the construction of CEs that are suitable for execution upon multi-core processors.

This research. We derive several approaches to constructing cyclic executives for implicit-deadline periodic task systems upon identical multiprocessors. These approaches share the commonality that they are all based upon formulating the schedule construction problem as a linear program (LP). Cyclic executives in which jobs may be preempted can be derived from solutions to such LPs; since efficient polynomial-time algorithms are known for solving LPs, this approach enables us to design algorithms for constructing preemptive CEs that have running time polynomial in the size of the CE.

In order to construct non-preemptive CEs from a solution to the LP, the LP must be further constrained to require that some variables may only take on integer values: this is an *integer* linear program, or ILP. Solving an ILP is known to be NP-hard [8], and hence unlikely to be solvable exactly in polynomial time. However, the optimization community has recently been devoting immense effort to devise extremely efficient implementations of ILP solvers, and highly optimized libraries with such efficient implementations are widely available today. Since CEs are constructed prior to run-time, we believe that it is reasonable to attempt to solve ILPs exactly rather than only approximately, and seek to obtain ILP formulations *that we will seek to solve exactly* to construct non-preemptive multiprocessor CEs for implicit-deadline periodic task systems. However if this is not practical for particular problem instances, we devise an approximation algorithm with polynomial running time for constructing non-preemptive CEs, and evaluate the performance of this approximation algorithm vis-a-vis the exact one both via the theoretical metric of speedup factor, and via simulation experiments on synthetically generated workloads. We additionally show that for a particular kind of workload that is quite common in practice – systems of *harmonic* tasks – even better results are obtainable.

2   Cyclic Executives

In this section we provide a brief introduction to the cyclic executive approach to hard-real-time scheduling. This is by no means comprehensive or complete; for a textbook description, please consult [11, Chap. 5.2–5.4].

In the cyclic executive approach, a schedule called a *major schedule* is determined prior to run-time, which describes the sequence of actions (i.e., computations) to be performed during some fixed period of time called the *major cycle*. The actions of a major schedule are executed cyclically, going back to the beginning at the start of each major cycle.[1] The major schedule is further divided into

---

[1] Multiple major schedules may be defined for a single system, specifying the desired system behavior for different *modes* of system operation; switching between modes is accomplished by swapping the major schedule used. If a major cycle is of not too large a duration, then switches between modes may be restricted to only occur at the end of major cycles.

one or more *frames* (also known as minor schedules or minor cycles). Each frame is allocated a fixed length of time during which the computations assigned to that frame must be executed. Timing correctness is monitored at frame boundaries via hardware interrupts generated by a timer circuit: if the computations assigned to a frame are discovered to have not completed by the end of the frame then a *frame overrun* error is flagged and control transferred to an error-handling routine.

The chief benefits of the cyclic executive approach to scheduling are its implementation simplicity and efficiency, and the timing predictability it offers: if we have a reliable upper bound on the execution duration of each computation then an application's schedulability is determined by construction (i.e., if we are successful in building the CE then we can be assured that all deadlines are met).

The chief challenge lies in constructing the schedules. This problem is rendered particularly challenging by the requirement that for implementation efficiency considerations, timing monitoring is performed only at frame boundaries — as stated above, a timer is set at the start of a frame to go off at the end of the frame, at which point in time it is verified that all actions assigned to that frame have indeed completed execution (if not, corrective action must be taken via a call to error-handling routines). CE's are typically used for periodic workloads. Hence the schedule-generation approach proposed in [1] requires that at least one frame lie within the interval formed by the instants that each action — "job" — become available for execution, and the instant that it has a deadline. For efficiency considerations, it is usually required that all tasks have a period that is a multiple of the minor cycle, and a deadline that is no smaller than the minor cycle duration. Schedule construction is in general highly intractable for many interesting models of periodic processes [1]; however, heuristics have been developed that permit system developers to construct such schedules for reasonably complex systems (as Baker & Shaw have observed [1], "if we do not insist on optimality, practical cases can be scheduled using heuristics").

In this paper, we model our periodic workload as a task system of implicit-deadline periodic tasks. Some of our results additionally require that the tasks have harmonic periods: for any pair of tasks τ_i and τ_j, it is the case that T_i divides T_j exactly or T_j divides T_i exactly. Although this does constitute a significant restriction on the periodic task model, many safety-critical systems appear to respect this restriction.

3   Workload Model

Throughout this paper we assume that we are given a task system $\tau = \{\tau_i = (C_i, T_i)\}_{i=1}^{N}$ of N implicit-deadline periodic[2] tasks that are to be scheduled upon an m-processor identical multiprocessor platform. The worst-case execution time (WCET) of τ_i is C_i, and its period is T_i. Let P denote the least common multiple (lcm) of the periods of all the tasks in τ (P is often called the *hyper-period* of τ),

---

[2] We highlight that these are periodic, not sporadic, tasks: τ_i generates jobs at time-instants $k \times T_i$, for all $k \in \mathbb{N}$.

N and m	Number of **tasks** and **processors**
$\tau_i = (C_i, T_i)$	The i'th task has worst-case execution time C_i and period T_i
P	$\text{lcm}_{i=1}^{N}\{T_i\}$ – the *hyperperiod*. Selected as major cycle duration
F	$\text{gcd}_{i=1}^{N}\{T_i\}$. Selected as minor cycle (frame) duration
f	The amount of execution that a single processor can accommodate in one frame. Upon unit-speed processors, $f = F$
Φ_k	The k'th frame, for $k \in \{1, 2, \ldots, P/F\}$
n	The total number of *jobs* in one hyperperiod. $n = \sum_{i=1}^{N}(P/T_i)$
$j_i = (a_i, c_i, d_i)$	The i'th job, $1 \le i \le n$. Its arrival time, WCET, and absolute deadline
$\mathcal{J}$	The collection of these n jobs
x_{ijk}	LP variable: the fraction of the i'th job assigned to the j'th processor during the k'th frame

Fig. 1. Some of the notation used in this paper

and let F denote the greatest common divisor (gcd) of the periods of all the tasks in τ. P is selected as the duration of the major cycle, and F the duration of the minor cycle, of the CE's we will construct.

Some further notation and terminology: Let $\mathcal{J} = \{j_1, j_2, \ldots, j_n\}$ denote all the jobs generated by τ that have their arrival times and deadlines within the interval $[0, P)$, and let a_i, c_i and d_i denote the arrival time, WCET, and (absolute) deadline respectively of job j_i. (We will often represent a job j_i by an ordered 3-tuple of its parameters: $j_i \overset{\text{def}}{=} (a_i, c_i, d_i)$. We refer to the interval $[a_i, d_i)$ as the *scheduling window* of this job j_i.) Note that the number of jobs n may in general take on a value that is exponential in the number of tasks N. Since we are seeking to explicitly construct a schedule for the n jobs, we believe that it is reasonable to evaluate the efficiency of algorithms for constructing these schedules in terms of the number of jobs n to be scheduled rather than in terms of the number of periodic tasks N.

Without loss of generality, we assume that the tasks are indexed according to non-decreasing periods: $T_i \le T_{i+1}$ for all i, $1 \le i < N$. For *harmonic* task systems τ, the tasks have harmonic periods: T_i divides T_{i+1} exactly for all i, $1 \le i < N$.

Example 1. Consider a system τ comprising three tasks τ_1, τ_2, and τ_3, with periods $T_1 = 4$, $T_2 = 6$, and $T_3 = 12$. $P = \text{lcm}(4, 6, 12) = 12$; $F = \text{gcd}(4, 6, 12) = 2$. (Therefore, minor cycle duration is 2, and major cycle duration is 12.) For this τ, $\mathcal{J}$ comprises the six jobs j_1–j_6 depicted in Fig. 2. There are $(12/2) =$ six frames or minor cycles within the major cycle – these are labeled in the figure as $\Phi_1, \Phi_2, \ldots, \Phi_6$ with Φ_k spanning the interval $[2(k-1), 2k]$.

4   Representing Cyclic Executives as Linear Programs

In this section we represent the problem of constructing a cyclic executive as a linear program. We start out with a brief review of some well-known facts concerning linear programs that we will use in later sections of the paper.

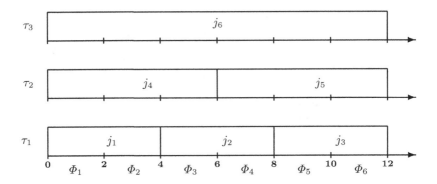

Fig. 2. The jobs generated by the task system of Example 1.

4.1 Some Linear Programming Background

In an integer linear program (ILP), one is given a set of v variables, some or all of which are restricted to take on integer values only, a collection of "constraints" that are expressed as *linear* inequalities over these v variables, and an "objective function," also expressed as a linear inequality of these variables. The set of all points in v-dimensional space over which all the constraints hold is called the *feasible region* for the integer linear program. The goal is to find the extremal (maximum or minimum, as specified) value of the objective function over the feasible region.

A linear program (LP) is like an ILP, without the constraint that some of the variables are restricted to take on integer values only. That is, in an LP over a given set of v variables, one is given a collection of constraints that are expressed as linear inequalities over these v variables, and an objective function, also expressed as a linear inequality of these variables. The region in v-dimensional space over which all the constraints hold is again called the feasible region for the linear program, and the goal is to find the extremal value of the objective function over the feasible region. A region is said to be *convex* if, for any two points $\mathbf{p_1}$ and $\mathbf{p_2}$ in the region and any scalar $\lambda, 0 \leq \lambda \leq 1$, the point $(\lambda \cdot \mathbf{p_1} + (1 - \lambda) \cdot \mathbf{p_2})$ is also in the region. A *vertex* of a convex region is a point $\mathbf{p}$ in the region such that there are no distinct points $\mathbf{p_1}$ and $\mathbf{p_2}$ in the region, and a scalar $\lambda, 0 < \lambda < 1$, such that $[\mathbf{p} \equiv \lambda \cdot \mathbf{p_1} + (1 - \lambda) \cdot \mathbf{p_2}]$.

It is known that an LP can be solved in polynomial time by the ellipsoid algorithm [9] or the interior point algorithm [7].

We now state without proof some basic facts concerning linear programming optimization problems.

Fact 1. *The feasible region for a LP problem is convex, and the objective function reaches its optimal value at a vertex point of the feasible region.*

An optimal solution to an LP problem that is a vertex point of the feasible region is called a **basic solution** to the LP problem.

Fact 2. *A basic solution to an LP can be found in polynomial time.*

Fact 3. *Consider a linear program on v variables with each variable subject to the constraint that it be ≥ 0 (such constraints are called non-negativity constraints). Suppose that in addition to these non-negativity constraints there are c other linear constraints. If $c < v$, then at most v of the variables have non-zero values at each vertex of the feasible region (including at all basic solutions).*

4.2   An LP Representation of CEs

Given a periodic task system comprising N tasks for which an m-processor cyclic executive is to be obtained, we now describe the construction of a linear program with $\left(N \times m \times (P/F)\right)$ variables, each of which is subject to a non-negativity constraint (i.e., each may only take on a value ≥ 0), and $\left(n + (m+N) \times (P/F)\right)$ additional linear constraints.

4.2.1   Variables

We will have a variable x_{ijk} denote the fraction of job j_i that is scheduled upon the j'th processor during the k'th frame. The index i takes on each integer value in the range $[1, n]$ (recall that n denotes the total number of jobs generated by all the periodic tasks over the hyper-period). For each i,

– The index j takes on each integer value in the range $[1, m]$.
– Note that job j_i may only execute within those frames that are contained in the scheduling window – the interval $[a_i, d_i]$ – of job j_i. The index k, therefore, only takes on values over the range of frame-numbers of those frames contained within $[a_i, d_i]$.

The total number of x_{ijk} variables is equal to $\left(N \times m \times (P/F)\right)$, where N denotes the number of periodic tasks, m denotes the number of processors, and P/F represents the number of minor cycles.

4.2.2   An Objective Function

Let f denote the amount of computing that can be accomplished by a processor executing for the duration F of an entire frame; for unit-speed processors, $f = F$. We will define the following objective function for our LP:

$$\text{minimize } f \tag{1}$$

The value of f obtained by solving the LP represents the minimum amount of computation needed to be completed by an individual processor within a duration F; if the available processors can indeed accommodate this amount of computation, then the solution is a feasible one.

4.2.3   Constraints

Since the x_{ijk} variables represent fractions of jobs, they must all be assigned values that are ≥ 0; hence, they are all subject to non-negativity constraints. In addition, these variables are used to construct a linear program representation of a CE, via the following constraints:

1. We represent the requirement that each job must receive the required amount of execution by having the constraints

$$\sum_{\text{all } j,k} x_{ijk} = 1 \text{ for each } i, \ 1 \leq i \leq n \tag{2}$$

   There are n such constraints, one per job.
2. We represent the requirement that each processor may be assigned no more than f units of execution during each minor cycle by having the constraints

$$\sum_{\text{all } i} x_{ijk} \cdot c_i \leq f \text{ for each } j, \ 1 \leq j \leq m \text{ and } k, \ 1 \leq k \leq P/F \tag{3}$$

   There are $m \times (P/F)$ such constraints.
3. We represent the requirement that each job may be assigned no more than f units of execution during each minor cycle by having the constraints

$$\sum_{\text{all } j} x_{ijk} \cdot c_i \leq f \text{ for each } i, \ 1 \leq i \leq n \text{ and } k, \ 1 \leq k \leq P/F \tag{4}$$

   There are $N \times (P/F)$ such constraints.

The total number of constraints is thus equal to $[n + (m + N) \times (P/F)]$.

4.2.4   Solving the LP

With regards to the LP constructed above, observe that

1. Given an assignment of integer values (i.e., either 0 or 1) to each of the x_{ijk} variables that satisfy the constraints of the LP, we may construct a non-preemptive cyclic executive in the following manner: for each x_{ijk} that is assigned the value 1, schedule the execution of job j_i on the j'th processor during the k'th frame.
2. Given an assignment of non-negative values to the x_{ijk} variables that satisfy the constraints of the LP, we may construct a global preemptive cyclic executive in the following manner. For each x_{ijk} that is assigned a non-zero value, schedule job j_i for a duration $x_{ijk} \times c_i$ on the j'th processor during the k'th frame. (Of course, care must be taken to ensure that during each frame no job executes concurrently upon two different processors – we will see in Sect. 5 below how this is ensured.)

That is, an integer solution to the ILP yields a non-preemptive cyclic executive while a fractional solution yields a global preemptive cyclic executive. We discuss the problem of obtaining such solutions, and thereby obtaining preemptive and non-preemptive cyclic executives respectively, in Sects. 5 and 6 respectively.

5   Preemptive Cyclic Executives

In this section we discuss the problem of constructing preemptive cyclic executives for implicit-deadline periodic task systems by obtaining solutions to the linear program described above.

Let us suppose that we have solved the linear program, and have thus obtained an assignment of non-negative values to the x_{ijk} variables that satisfy the constraints of the LP. We now describe the manner in which we construct a preemptive cyclic executive for the k_o'th frame Φ_{k_o}; the entire cyclic executive is obtained by repeating this procedure for each k_o, $1 \leq k_o \leq (P/F)$.

For each job j_{i_o} observe that

$$\chi_{i_o} \overset{\text{def}}{=} \sum_{j=1}^{m} x_{i_o j k_o}$$

represents the total amount of execution assigned to job j_{i_o} during frame Φ_{k_o} in the solution to the LP. By Constraint 4 of the LP, it follows that $\chi_{i_o} \leq f$ for each job j_{i_o}; i.e. no job is assigned more than f units of execution over the frame. Additionally, it follows from summing Constraint 3 of the LP over all m processors (i.e., for all values of the variable j in Constraint 3) that

$$\left(\sum_{i_o=1}^{n} \chi_{i_o} \right) \leq m \times f.$$

We have thus shown that (i) no individual job is scheduled during the frame for more than the computing capacity of a single processor during one frame, and (ii) the total amount of execution scheduled over the interval does not exceed the cumulative computing capacity of the frame (across all m processors). We may therefore construct a schedule within the frame using McNaughton's wrap-around rule [12] in the following manner:

1. We order the jobs that receive any execution within frame Φ_{k_o} arbitrarily.
2. Then we begin placing jobs on the processors in order, filling the j'th processor entirely before starting the $(j + 1)$'th processor. Thus, a job j_{i_o} may be split across processors, assigned to the last t time units of the frame on the j'th processor and the first $(\chi_{i_o} - t)$ time units of the frame on the $(j + 1)$'th processor; since $\chi_{i_o} \leq f$, these assignments will not overlap in time.

It is evident that this can all be accomplished efficiently within run-time polynomial in the representation of the task system.

Implementation. In Sect. 6.2 below, we describe experiments that we have conducted comparing ILP-based exact and LP-based approximate algorithms for constructing *non*-preemptive CEs. These experiments required us to solve LPs, similar to the kind described here, using the Gurobi Optimization tool [6]; performance of the Gurobi Optimization tool scaled very well with the size of the task system in these experiments.

6   Non-preemptive Cyclic Executives

We now discuss the process of obtaining 0/1 integer solutions to the linear program defined in Sect. 4.2; as discussed there, such a solution can be used to construct non-preemptive cyclic executives for the periodic task system represented using the linear program.

Let us start out observing that in order for a non-preemptive cyclic executive to exist, it is necessary that any job fits into an individual frame; i.e., that

$$\max_{i=1}^{N}\{C_i\} \le f \tag{5}$$

Any task system for which this condition does not hold cannot be scheduled non-preemptively.

Let us now take a closer look at the LP that was constructed in Sect. 4.2. Consider any 0/1 integer solution to this LP. Each x_{ijk} variable will take on value either zero or one in such a 0/1 integer solution; hence in the LP, the *Constraints 2 render the Constraints 4 redundant*. To see why this should be so, consider any job (say, j_{i_o}), and any frame (say, Φ_{k_o}). From Constraints 2 and the fact that each x_{ijk} variable is assigned a value of zero or one, it follows that in any 0/1 integer solution to the linear program we will have

$$\left(\sum_j x_{i_o j k_o} = 0\right) \text{ or } \left(\sum_j x_{i_o j k_o} = 1\right),$$

depending upon whether job j_{i_o} is scheduled (on any processor) within the Frame Φ_{k_o} or not. We thus see that at most one of the $x_{i_o j k_o}$'s can equal 1, from which it follows that Constraint 4 necessarily holds for job j_{i_o} within Frame Φ_{k_o}. We may therefore omit the Constraints 4 in the linear program. Hence for non-preemptive schedules, we have a somewhat simpler ILP that needs to be solved, comprising

$$\left(N \times m \times \frac{P}{F}\right) \text{ variables but only } \left(n + m \times \frac{P}{F}\right) \text{ constraints.}$$

6.1   An Approximation Algorithm

The problem of finding a 0/1 solution to a Linear Program is NP-hard in the strong sense; all algorithms known for obtaining such solutions have running time that is exponential in the number of variables and constraints. As we had mentioned earlier, this intractability of Integer Linear Programming does not necessarily rule out the ILP-based approach to constructing cyclic executives that we have described above, since excellent solvers have been implemented that are able to solve very large ILPs in reasonable amounts of time.

However, the fact of the matter is that not all ILPs can be solved efficiently. We now describe an approximation algorithm for constructing Cyclic Executives, that does not require us to solve ILPs exactly. The algorithm is approximate in the sense that it may fail to construct Cyclic Executives for some input instances for which CE's do exist (and could have been constructed using the exponential-time ILP-based method discussed above). In Theorem 1 we quantify the non-optimality of our approximation algorithm.

Our algorithm starts out constructing the linear program as described in Sect. 4.2, but without the Constraints 4 (as discussed above, the Constraints 2 render these redundant). However, rather than seeking to solve the NP-hard problem of obtaining a 0/1 integer solution to this problem, we instead replace the 0/1 integer constraints with the requirement that each x_{ijk} variable be a non-negative real number no larger than one (i.e., that $0 \leq x_{ijk} \leq 1$ for all variables x_{ijk}), and then obtain a basic solution[3] to the resulting linear program (without the constraint that variables take on integer values). As stated in Fact 2 of Sect. 4.1, such a basic solution can be found efficiently in polynomial time.

Recall that our LP has $\left(N \times m \times \frac{P}{F}\right)$ variables but only $\left(n + m \times \frac{P}{F}\right)$ constraints. By Fact 3 of Sect. 4.1, at most $\left(n + m \times \frac{P}{F}\right)$ of the variables will take on non-zero values at the basic solution. Some of these non-zero values will be equal to one – each such value determines the frame and processor upon which a job is to be scheduled in the cyclic executive. I.e., *for each x_{ijk} that is assigned a value equal to one in the basic solution, we assign job j_i to the j'th processor during frame Φ_k*.

It remains to schedule the jobs which were not assigned as above — these are the jobs for which Constraint 2 was satisfied in the LP solution by having multiple non-zero terms on the LHS. This is done according to the following procedure; the correctness of this procedure is proved in [10].

1. Consider all the variables $X \overset{\text{def}}{=} x_{ijk}$ that have been assigned non-zero values strictly less than one in the basic solution. That is,

$$X \overset{\text{def}}{=} \left\{ x_{ijk} \text{ such that } 0 < x_{ijk} < 1 \text{ in the basic solution} \right\}$$

2. Construct a bipartite graph with
   (a) A vertex for each job j_{i_o} such that there is some (one or more) $x_{i_o jk} \in X$. Let V_1 denote the set of all such vertices that are added.
   (b) A vertex for each ordered pair $[j_o, k_o]$ such that there is some (one or more) $x_{ij_o k_o} \in X$. Let V_2 denote the set of all such vertices that are added.
   (c) For each $x_{i_o j_o k_o} \in X$ add an edge in this bipartite graph from the vertex in V_1 corresponding to job j_{i_o}, to the vertex in V_2 corresponding to ordered pair $[j_o, k_o]$.
3. It has been shown in [10] that there is a matching in this bipartite graph that includes all the vertices in V_1. Such bipartite matchings can be found in polynomial time using standard network-flow algorithms.

---

[3] Recall from Sect. 4.1 above that a *basic solution* to an LP is an optimal solution that is a vertex point of the feasible region defined by the constraints of the LP.

4. Once such a bipartite matching is obtained, each job corresponding to a vertex in V_1 is assigned to the processor and frame corresponding to the vertex in V_2 to which it has been matched. In this manner, each processor in each frame is guaranteed to be assigned at most one job during this process of assigning the jobs that were not already assigned in the basic solution.

6.2 Evaluating the Approximation Algorithm

We now compare the effectiveness of the polynomial-time approximation algorithm of Sect. 6.1 with that of the ILP-based exact algorithm (solving which takes exponential time in the worst case). We start out with theoretical evaluation: Corollary 1 quantifies the worst-case performance of the approximation algorithm via te speedup factor metric. We have also conducted some simulation experiments on randomly-generated workloads, to get a feel for typical (rather than worst-case) effectiveness – these are discussed in Sect. 6.2 below.

Theorem 1. *Let f_{opt} denote the minimum amount of computation that must be accommodated on an individual processor within each frame in any feasible m-processor CE for a given implicit-deadline periodic task system τ. Let $C_{\max}$ denote the largest WCET of any task in τ: $C_{\max} \stackrel{def}{=} \max_{\tau_i \in \tau}\{C_i\}$. The polynomial-time approximation algorithm of Sect. 6.1 above will successfully construct a CE for τ upon m processors, with each processor needing to accommodate no more than $(f_{opt} + C_{\max})$ amount of execution during any frame.*

Proof: Since (as we had argued in Sect. 4) an integer solution to the ILP represents an optimal CE, observe that the minimum value of f computed in an integer solution to an ILP would be equal to f_{opt}. And since the ILP is more constrained than the Linear Program, the minimum value for f computed in the (not necessary integral) solution to the LP obtained by the polynomial-time algorithm of Sect. 6.1 is $\le f_{opt}$. Let f_{LP} denote this minimum value of f computed as a solution to the LP; we thus have that $f_{LP} \le f_{opt}$.

In constructing the CE above, the polynomial-time algorithm of Sect. 6.1 schedules each job according to one of two rules:

1. If variable $x_{i_o j_o k_o}$ is assigned a value one in the solution to the LP, then job j_{i_o} is scheduled upon the j_o'th processor during frame Φ_{k_o}.
2. Any job j_{i_o} not scheduled as above is scheduled upon the processor-frame pair to which it gets matched in the bipartite matching.

Clearly, the jobs assigned according to the first rule would fit upon the processors if each had a computing capacity of f_{LP} within each frame. Now, observe that the matching in the bipartite graph assigns at most one job to each processor during any given frame; therefore, the *additional* execution assigned to any processor during any frame is $< C_{\max}$. Hence each processor could accommodate all the execution assigned it within each frame provided it had a computing capacity of at least $f_{LP} + C_{\max}$, which is $< (f_{opt} + C_{\max})$.   □

The *speedup factor* of an algorithm A is defined to be smallest positive real number x such that any task system that is successfully scheduled upon a particular platform by an optimal algorithm is successfully scheduled by algorithm A upon a platform in which the speed or computing capacity of all processors are scaled up by a factor $(1 + x)$.

Corollary 1: *The polynomial-time approximation algorithm of Sect. 6.1 has a speedup bound no larger than 2.*

Proof: By Theorem 1 above, If a CE can be constructed for task system τ by an optimal algorithm upon m speed-f_{opt} processors, it can be scheduled by the polynomial-time algorithm of Sect. 6.1 upon m speed-$\left(f_{\text{opt}} + C_{\text{max}} \right)$ processors. The corollary follows from the observation that C_{max} is necessarily $\leq f_{\text{opt}}$; hence $\left(f_{\text{opt}} + C_{\text{max}} \right) / f_{\text{opt}}$ is $\leq 2 f_{\text{opt}} / f_{\text{opt}} \leq 2$. □

Experimental Evaluation. We saw above (Corollary 1) that the polynomial-time approximation algorithm of Sect. 6.1 has a speedup factor no worse than 2. We have conducted some experiments on randomly-generated synthetic workloads to further compare the performance of the approximation algorithm with the exact approach of solving the ILP.

Workload generation. The task system parameters for each experiment were randomly generated using a variant of the methods used in prior research such as [3,5], in the following manner:

- Task utilizations (U_i) were generated using the UUniFast algorithm [2].
- Task periods were set to be at one of $F \times \{1, 2, 3, 4\}$ (the frame size F was set equal to 25ms in these experiments, in accordance with prior recent work on cyclic executives such as [3,5]). Periods were assigned randomly and uniformly over these four values. (Since we are restricting attention in this paper to implicit-deadline systems, job deadlines were set equal to their periods.)
- Task WCETs were determined as the product of utilization and period.
- All task systems in which one or more tasks had a WCET greater than minor cycle duration F, were discarded (since such systems are guaranteed to have no feasible non-preemptive schedules).
  (For some of our experiments, we needed task systems in which the largest WCET of any task (the parameter C_{max} of Theorem 1 was bounded at one-half of three-quarters the frame size. In generating task systems for these experiments, we discarded all task systems in which some task had WCET greater than the bound.)
- All the experiments assumed a four-processor platform ($m \leftarrow 4$).

Experiments conducted, and observations made. We conducted two sets of experiments; in each experiment within each set,

1. A task system was generated using the procedure detailed above, with a specified number of tasks, a specified total utilization, and for some experiments, a specified bound on $C_{\max}$.
   Each task system so generated was scheduled in two different ways.
2. First, it was scheduled non-preemptively by generating a linear program as described in Sect. 4.2, and then solved as an ILP using the Gurobi [6] optimization tool (instrumented to time out after two seconds of execution, earlier experiments indicating that for systems of 20 tasks on 4 processors, longer runs never improved upon the value obtained within the first two seconds).
3. Second, it was scheduled preemptively by solving the linear program obtained above as an LP (i.e., without any integrality constraints) using Gurobi, and then applying the technique described in Sect. 6.1 to obtain a non-preemptive cyclic executive. The maximum amount of computation assigned to any processor within an individual frame in this schedule was determined, and designated as $f_{\max}$.
4. The speedup factor needed by the polynomial-time approximation algorithm for this particular task system was then computed as

$$\max\left(1, \frac{f_{\max}}{F}\right)$$

(Recall that F denotes the frame size, chosen to equal 25 ms in our experiments.)

We now describe the two sets of experiments separately.

6.2.1 Variation of Speedup Factor with System Utilization

As explained above, the speedup bound of 2 identified in Corollary 1 above is a worst-case one. In this set of experiments, we set out to determine how the speedup factor of a randomly-generated system tends to depend upon the cumulative utilization of the task system. We therefore generated 400 task systems, each comprising 20 tasks, to have cumulative system utilization equal to U, for each value of U between 0 and 4 in steps of 0.05. The observed speedup factor needed by the approximation algorithm to schedule each task system was determined as described above, and the average and standard deviations computed. These values, plotted in Fig. 3, show a clear increasing trend: as overall utilization increases, so does the speedup factor needed to construct a non-preemptive schedule using the approximation algorithm.

6.2.2 Variation of Speedup Factor with $C_{\max}$

Theorem 1 reveals that the speedup factor depends upon the value of $C_{\max}$, the largest WCET of any individual task. To investigate this relationship, we generated 100 task systems with overall utilization U for each value of U between 2 and 4 in steps of 0.05, in which the value of $C_{\max}$ was bounded from above at half the frame size, three quarters the frame size, and the full frame size. The observed speedup factor needed by the approximation algorithm to schedule each

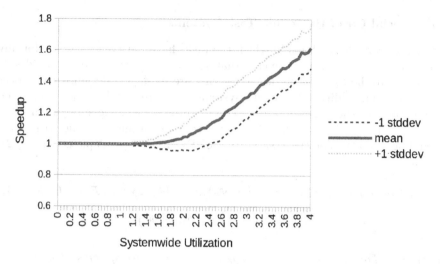

Fig. 3. Investigating how speedup factor changes with overall system utilization. The mean observed speedup factor over 400 task systems at each utilization is depicted, as is the range within one standard deviation from the mean.

task system was determined as described above, and the average over the 100 individual task systems at each data point computed. These values, plotted in Fig. 4, show a clear increasing trend within each system utilization: the larger the bound on $C_{\max}$, the greater the observed speedup factor.

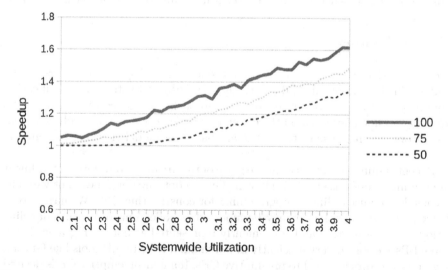

Fig. 4. Investigating how observed speedup factor depends upon $C_{\max}$, the largest WCET of any task. The mean observed speedup factor over 100 task systems is plotted, for $C_{\max}$ bounded at $\frac{1}{2}$, $\frac{3}{4}$, and 1 times the frame size.

6.3   Special Case: Harmonic Task Systems

Let us now consider systems in which the tasks have harmonic periods: for any pair of tasks τ_i and τ_j, it is the case that T_i divides T_j exactly or T_j divides T_i exactly. Many highly safety-critical systems are explicitly designed to respect this restriction; additionally, many systems that are not harmonic are often representable as the union of a few – two or three – harmonic sub-systems.

For any job j_i, let us define $\mathcal{F}_i$ to be the set of frames that lie within j_i's scheduling window. For the task system of Example 1 (as depicted in Fig. 2), e.g., we have

$$\mathcal{F}_1 = \{\Phi_1, \Phi_2\}, \mathcal{F}_2 = \{\Phi_3, \Phi_4\}, \mathcal{F}_3 = \{\Phi_5, \Phi_6\}, \mathcal{F}_4 = \{\Phi_1, \Phi_2, \Phi_3\}, \mathcal{F}_5 = \{\Phi_4, \Phi_5, \Phi_6\},$$

$$\text{and } \mathcal{F}_6 = \{\Phi_1, \Phi_2, \Phi_3, \Phi_4, \Phi_5, \Phi_6\}.$$

Lemma 1: *For any two jobs j_i and j_ℓ in harmonic task systems, it is the case that*

$$\left(\mathcal{F}_i \subseteq \mathcal{F}_j\right) \text{ or } \left(\mathcal{F}_j \subseteq \mathcal{F}_i\right) \text{ or } \left(\mathcal{F}_i \bigcap \mathcal{F}_j \text{ is empty}\right) \qquad \square$$

A polynomial-time approximation scheme (PTAS) was derived in [4] for the problem of *scheduling on restricted identical machines with nested processing set restrictions*; this PTAS can be directly applied to our problem of constructing non-preemptive cyclic executives for implicit-deadline periodic task systems with harmonic periods. This allows us to conclude that for the special case of harmonic task systems, polynomial-time approximation algorithms may be devised for constructing cyclic schedules that are accurate to any desired degree of accuracy.

7   Conclusions

Cyclic executives (CEs) are widely used in safety-critical systems industries, particularly in those application domains that are subject to statutory certification requirements. In our experience, current approaches to the construction of CEs are either ad hoc and based on the expertise and experience of individual system integrators, or make use of tools that are based on model checking or heuristic search.

Recent significant advances in the state of the art in the development of linear programming tools, as epitomized in the Gurobi optimizer [6], have motivated us to consider the use of linear programming for constructing CEs. We have shown that CEs for workloads that may be modeled as collections of implicit-deadline periodic tasks are easily and conveniently represented as linear programs (LPs). These LPs are solved very efficiently in polynomial time by LP tools like Gurobi; such solutions directly lead to preemptive CEs. If a non-preemptive CE is desired then one must solve an *integer* LP (ILP), which is a somewhat less tractable problem than solving LPs. However, our experiments indicate that Gurobi is able to solve most ILP problems representing non-preemptive CEs for collections of implicit-deadline periodic tasks quite effectively in a reasonable amount of time. We have also developed an approximation algorithm for constructing non-preemptive CEs that runs in polynomial time, and performs quite favorably in comparison to the exact algorithm in terms of both a worst-case quantitative metric (speedup factor) and in experiments on randomly-generated synthetic workloads.

Acknowledgements. This research is supported by NSF grants CNS 1409175 and CPS 1446631, AFOSR grant FA9550-14-1-0161, and ARO grant W911NF-14-1-0499.

References

1. Baker, T.P., Shaw, A.: The cyclic executive model and Ada. In: Proceedings of the IEEE Real-Time Systems Symposium, pp. 120–129 (1988)
2. Bini, E., Buttazzo, G.: Measuring the performance of schedulability tests. Real-Time Syst. **30**(1–2), 129–154 (2005)
3. Burns, A., Fleming, T., Baruah, S.: Cyclic executives, multi-core platforms and mixed criticality applications. In: Proceedings of the 2015 27th EuroMicro Conference on Real-Time Systems, ECRTS 2015. IEEE Computer Society Press, Lund (Sweden) (2015)
4. Epstein, L., Levin, A.: Scheduling with processing set restrictions: PTAS results for several variants. Int. J. Prod. Econ. **133**(2), 586–595 (2011)
5. Fleming, T., Burns, A.: Extending mixed criticality scheduling. In: Proceedings of the International Workshop on Mixed Criticality Systems (WMC), December 2013
6. Gurobi Optimization Inc: Gurobi Optimizer Reference Manual (2016). http://www.gurobi.com
7. Karmakar, N.: A new polynomial-time algorithm for linear programming. Combinatorica **4**, 373–395 (1984)
8. Karp, R.: Reducibility among combinatorial problems. In: Miller, R., Thatcher, J. (eds.) Complexity of Computer Computations, pp. 85–103. Plenum Press, New York (1972)
9. Khachiyan, L.: A polynomial algorithm in linear programming. Dokklady Akademiia Nauk SSSR **244**, 1093–1096 (1979)
10. Lenstra, J.K., Shmoys, D., Tardos, E.: Approximation algorithms for scheduling unrelated parallel machines. Math. Program. **46**, 259–271 (1990)
11. Liu, J.W.S.: Real-Time Systems. Prentice-Hall Inc., Upper Saddle River (2000)
12. McNaughton, R.: Scheduling with deadlines and loss functions. Manag. Sci. **6**, 1–12 (1959)

Compositional Hoare-Style Reasoning About Hybrid CSP in the Duration Calculus

Dimitar P. Guelev[1]([⊠]), Shuling Wang[2], and Naijun Zhan[2,3]

[1] Institute of Mathematics and Informatics,
Bulgarian Academy of Sciences, Sofia, Bulgaria
`gelevdp@math.bas.bg`
[2] State Key Laboratory of Computer Science, Institute of Software,
Chinese Academy of Sciences, Beijing, China
[3] University of Chinese Academy of Sciences, Beijing, China

Abstract. Deductive methods for the verification of hybrid systems vary on the format of statements in correctness proofs. Building on the example of Hoare triple-based reasoning, we have investigated several such methods for systems described in Hybrid CSP, each based on a different assertion language, notation for time, and notation for proofs, and each having its pros and cons with respect to expressive power, compositionality and practical convenience. In this paper we propose a new approach based on weakly monotonic time as the semantics for interleaving, the Duration Calculus (DC) with infinite intervals and general fixpoints as the logic language, and a new meaning for Hoare-like triples which unifies assertions and temporal conditions. We include a proof system for reasoning about the properties of systems written in the new form of triples that is complete relative to validity in DC.

1 Introduction

Hybrid systems exhibit combinations of discrete and continuous evolution, the typical example being a continuous plant with discrete control. A number of abstract models and requirement specification languages have been proposed for the verification of hybrid systems, the commonest model being *hybrid automata* [3,20,24]. Hybrid CSP (HCSP) [19,39] is a process algebra which extends CSP by constructs for continuous evolution described in terms of ordinary differential equations, with domain boundary- and communication-triggered interruptions. The mechanism of synchronization is message passing. Because of its compositionality, HCSP can be used to handle complex and open systems. Here follows an example of a simple generic HCSP description of a continuously evolving plant with discrete control:

(**while** $\top$ **do** $\langle F(\dot{x}, x, u) = 0 \rangle \unrhd sensor!x \to actuator?u)$ ||
(**while** $\top$ **do** (**wait** d; $sensor?s$; $actuator!C(s)$))

The plant evolves according to some continuous law F that depends on a control parameter u. The controller samples the state of the plant and updates the control parameter once every d time units.

© Springer International Publishing AG 2017
K.G. Larsen et al. (Eds.): SETTA 2017, LNCS 10606, pp. 110–127, 2017.
https://doi.org/10.1007/978-3-319-69483-2_7

In this paper we propose a Hoare-style proof system for reasoning about hybrid systems which are modelled in HCSP. The features of HCSP which are handled by the logic include communication, timing constraints, interrupts and continuous evolution governed by differential equations. Our proof system is based on the Duration Calculus (DC, [4,5]), which is a first-order real-time temporal logic and therefore enables the verification of HCSP systems for temporal properties. DC is an interval-based temporal logic. The form of the satisfaction relation in DC is $I, \sigma \models \varphi$, where φ is a temporal formula, I is an interpretation of the respective vocabulary over time, and σ is a *reference interval* of real time, unlike *point-based* TLs, where a reference time point is used. The advantages of intervals stem from the possibility to accommodate a complete execution of a process and have reference to termination time points of processes as well as the starting points. Pioneering work on interval-based reasoning includes Allen's interval algebra and Halpern and Shoham's logic [2,15]. ITLs have been studied in depth with respect to the various models of time by a number of authors, cf. e.g. [9]. Since an interval can be described as the pair of its endpoints, interval-based logics are also viewed as *two-dimensional* modal logics [35,36]. Interval Temporal Logic (ITL) was first proposed and developed by Moszkowski [7,26,27] for discrete time, as a reasoning tool for digital circuits. DC can be viewed as a theory in *real time* ITL. We use the infinite interval variant of DC which was proposed in [6], which allows intervals whose right end is ∞ for modelling non-terminating behaviour. We include an operator for Kleene star in order to model iterative behaviour, to facilitate the handling of liveness properties. Axioms and proof rules about infinite time and Kleene star in DC can be found in [11,13].

Hoare-style proof systems are about proving *triples* of the form $\{P\}\, \mathbf{code}\, \{Q\}$, which stand for the *partial correctness* property $P(x) \wedge \mathbf{code}(x, x') \rightarrow Q(x')$. The meaning of triples generalizes to the setting of reactive systems in various ways, the common feature of them all being that P and/or Q are *temporal* properties. In our system **code** is a HCSP term, P and Q are written in DC. The intended meaning is

> Given an infinite run which satisfies P at some initial subinterval, **code** causes it to satisfy also Q at the initial subinterval representing the (1) execution of **code**.

The initial subinterval which is supposed to satisfy P, can as well be a degenerate (0-length) one. Then P boils down to an assertion on the initial state. This interval can also be to the entire infinite run in question. In this case P can describe conditions provided by the environment throughout runs. Q is supposed to hold at an interval which exactly matches that of the execution of **code**. In case **code** does not terminate, this would be the whole infinite run too. Using our DC semantics $[\![.]\!]$ for HCSP terms, the validity of $\{P\}\, \mathbf{code}\, \{Q\}$ is defined as the validity of

$$P^\frown \top \Rightarrow \neg([\![\mathbf{code}]\!] \wedge \neg(Q^\frown \top))$$

at infinite intervals, which is equivalent to (1).

We exploit the form of triples to obtain a *compositional* proof system, with each rule corresponding to a basic HCSP construct. This forces proofs to follow the structure of the given HCSP term. Triples in this form facilitate assume-guarantee reasoning too. For instance,

$$\frac{\{A\}\, \mathbf{code}_1 \{B\} \qquad \{B\}\, \mathbf{code}_2 \{C\}}{\{A\}\, \mathbf{code}_1 \| \, \mathbf{code}_2 \{((B^\frown \top) \wedge C) \vee (B \wedge (C^\frown \top)\}}$$

where $\|$ denotes parallel composition, is an admissible rule in our proof system, despite not being among the basic rules. A detailed study of assume-guarantee reasoning about discrete-time reactive systems in terms of triples of a similar form with point-based temporal logic conditions can be found in [1].

The main result about our proof system in the paper is its completeness relative to validity in DC.

Structure of the paper. After brief preliminaries on HCSP, we propose a weakly-monotonic time semantics for it in terms of DC formulas and prove its equivalence to an appropriate operational semantics. Next we give our proof system and use the DC-based semantics to demonstrate its relative completeness. Finally we summarize a generalization of the approach where arbitrary fixpoints can be used instead of HCSP's tail recursion and the use of $\|$ and the respective rather involved proof rule can be eliminated. That turns out to require both the general fixpoint operator of DC [29] and the right-neighbourhood modality (cf. [4]) to handle the meaning of Ps in the presence of properly recursive calls. We conclude by discussing related work and make some remarks.

2   Preliminaries

Syntax and informal semantics of Hybrid CSP. Process terms have the syntax

$P, Q ::=$	**skip** $\|$	do nothing;
	$x_1, \ldots, x_n := e_1, \ldots, e_n \|$	simultaneous assignment;
	wait d $\|$ **await** b $\|$	fixed time delay; wait until b becomes true;
	$ch?x \| ch!e \| IO \|$	input and output; communication-guarded choice;
	$\langle F(\dot{x}, x) = 0 \wedge b \rangle \|$	x evolves according to F as long as b holds;
	$\langle F(\dot{x}, x) = 0 \wedge b \rangle \unrhd IO \|$	evolve by F until $\neg b$ or IO becomes ready; terminate, if $\neg b$ is reached first; otherwise execute IO;
	$P; Q \| P \| Q \|$	sequential composition; parallel composition
	if b **then** P **else** Q $\| P \sqcup Q \|$	conditional; internal non-deterministic choice;
	$\mu X.G$	recursion.

In the above BNF, IO has the following form:

$$ch_1?x_1 \to P_1 [] \ldots [] ch_k?x_k \to P_k [] ch_{k+1}!e_{k+1} \to P_{k+1} [] \ldots [] ch_n!e_n \to P_n \quad (2)$$

for some arbitrary k, n, $x_1, \ldots, x_k$, $e_{k+1}, \ldots, e_n$ and some distinct $ch_1, \ldots, ch_n$. IO engages in one of the indicated communications as soon as a partner process becomes ready, and then proceeds as the respective P_i. In $\mu X.G$, G has the syntax

$$G ::= H \mid \overrightarrow{x} := \overrightarrow{e}; P \mid \langle F(\dot{x}, x) = 0 \land b \rangle; P \mid \text{if } b \text{ then } G \text{ else } G$$
$$\mid G \sqcup G \mid G; P \mid \mu Y.G \mid H \| H$$

where H stands for arbitrary X-free terms, $Y \neq X$ and P can be any process term. This restricts X of $\mu X.G$ to be *guarded* in G and rules out occurrences of X of $\mu X.G$ in the scope of $\|$ in G. The communication primitives $ch?x$ and $ch!e$ are not mentioned in the syntax for G as they are treated as derived in this paper. They can be assigned the role of guards which $\overrightarrow{x} := \overrightarrow{e}$ has in (1). Obviously X is guarded in the $P_1, \ldots, P_n$ of IO as in (2) too. Below we focus on the commonest instance of μ

while b do $P \rightleftharpoons \mu X.\text{if } b \text{ then } (P; X) \text{ else skip}$ (3)

CSP's Kleene star $P^* \rightleftharpoons \mu X.(\text{skip} \sqcup (P; X))$, which stands for some unspecified number of successive executions of P, is handled similarly. We explain how our setting ports to general fixpoints, and some technical benefits from that, other than the obvious gain in expressive power, in a dedicated section.

The Duration Calculus. We use DC with infinite intervals as in [6,13] and Kleene star. The reader is referred to [4] for a comprehensive introduction. DC is a classical first-order predicate modal logic with one normal binary modality called *chop* and written $\frown$. The time domain is $\mathbb{R}^\infty = \mathbb{R} \cup \{\infty\}$. Satisfaction has the form $I, \sigma \models \varphi$ where I is an interpretation of the non-logical symbols, $\sigma \in \mathbb{I}(\mathbb{R}^\infty)$, $\mathbb{I}(\mathbb{R}^\infty) = \{[t_1, t_2]; t_1 \in \mathbb{R}, t_2 \in \mathbb{R}^\infty, t_1 \leq t_2\}$. *Flexible* non-logical symbols depend on the reference intervals for their meaning. *Chop* is defined by the clause

$$I, \sigma \models (\varphi \frown \psi) \text{ iff either there exists a } t \in \sigma \setminus \{\infty\} \text{ such that } I, [\min \sigma, t] \models \varphi$$
$$\text{and } I, [t, \max \sigma] \models \psi, \text{ or } \max \sigma = \infty \text{ and } I, \sigma \models \varphi.$$

Along with the usual first-order non-logical symbols, DC features boolean valued *state variables*, which form boolean combinations called *state expressions*. The value $I_t(S)$ of a state expression S is supposed to change between 0 and 1 only finitely many times in every bounded interval of time. *Duration terms* $\int S$ take a state expression S as the operand. The value of $\int S$ at interval σ is $\int_{\min \sigma}^{\max \sigma} I(S)(t)dt$, which is the combined length of the parts of σ which satisfy S. ℓ is used for $\int (0 \Rightarrow 0)$ and always evaluates to the length of the reference interval. Other common defined constructs include

$$\lceil S \rceil^0 \rightleftharpoons \int S = \ell, \quad \lceil S \rceil \rightleftharpoons \lceil S \rceil^0 \land \ell \neq 0, \quad \Diamond \varphi \rightleftharpoons (\top; \varphi; \top), \quad \Box \varphi \rightleftharpoons \neg \Diamond \neg \varphi.$$

In this paper we additionally use the following abbreviations:

$$\lceil S \rceil_{\text{fin}} \rightleftharpoons \lceil S \rceil \land \ell < \infty, \quad \lceil S \rceil^0_{\text{fin}} \rightleftharpoons \lceil S \rceil^0 \land \ell < \infty$$
$$\boxdot^\circ \varphi \rightleftharpoons \neg(\neg \varphi; \ell \neq 0), \quad \boxdot^\circ \varphi \rightleftharpoons \neg(\top; \ell \neq 0 \land \neg \varphi)$$

In Sect. 6 we use the *converse neighbourhood modality* $\diamondsuit_l^c$ of *Neighbourhood Logic* (and the corresponding system of DC), which is defined by the clause:

$$I, \sigma \models \diamondsuit_l^c \varphi \text{ iff } I, [\min \sigma, t] \models \varphi \text{ for some } t \in \mathbb{R} \cup \{\infty\}, t \geq \min \sigma.$$

We also use the least fixpoint operator μ. In formulas $\mu X.\varphi$, where X is a dedicated type of variable, φ can only have positive occurrences of X. The meaning of $\mu X.\varphi$, is defined by means of the operator of type $\mathcal{P}(\mathbb{I}(\mathbb{R}^\infty)) \to \mathcal{P}(\mathbb{I}(\mathbb{R}^\infty))$, which maps $A \subseteq \mathbb{I}(\mathbb{R}^\infty)$ to $\{\sigma \in \mathbb{I}(\mathbb{R}^\infty) : I_X^A, \sigma \models \varphi\}$. $I, \sigma \models \mu X.\varphi$ iff σ appears in the least fixpoint of this operator, which happens to be monotonic by virtue of the syntactic condition on φ. Using μ, Kleene star φ^* is defined as $\mu X.(\ell = 0 \vee \varphi \frown X)$.

3   Operational Semantics of HCSP

Ownership of variables. We write $\text{Var}(P)$ for the set of the program variables which occur in P. Expressions of the form $\dot{x}$ in continuous evolution process terms are, syntactically, just program variables, and are restricted not to appear in arithmetical expressions e outside the $F(\dot{x}, x)$ of continuous evolution terms, or on the left hand side of $:=$. As it becomes clear below, the dependency between x and $\dot{x}$ as functions of time is spelled out as part of the semantics of continuous evolution. We write $\text{Var}_{:=}(P)$ for all the variables in $\text{Var}(P)$ which occur on the left hand side of $:=$, in $ch?$ statements, and the x-es or $\dot{x}$-es in any of the forms of continuous evolution within P. Parallel composition $P\|Q$ is well-formed only if $\text{Var}_{:=}(P) \cap \text{Var}_{:=}(Q) = \emptyset$.

Modelling input and output. We treat $ch?x$ and $ch!e$ as derived constructs as they can be defined in terms of dedicated shared variables $ch?$, $ch!$ and ch after [28]:[1]

$$ch!e \rightleftharpoons ch := e; ch! := \top; \textbf{await } ch?; \textbf{await } \neg ch?; ch! := \bot; \textbf{await } \neg ch!$$
$$ch?x \rightleftharpoons ch? := \top; \textbf{await } ch!; x := ch; ch? := \bot; \textbf{await } \neg ch!$$

We assume that $ch!, ch \in \text{Var}_{:=}(ch!e)$ and $ch? \in \text{Var}_{:=}(ch?x)$. Communication-guarded external choice IO can be defined similarly. We omit the definition as it is lengthy and otherwise uninsightful. The other derived constructs are defined as follows:

$$\langle F(\dot{x}, x) = 0 \wedge b \rangle \trianglerighteq_d Q \rightleftharpoons t := 0; \langle F(\dot{x}, x) = 0 \wedge \dot{t} = 1 \wedge b \wedge t \leq d \rangle;$$
$$\text{if } \neg\overline{(\neg b)} \text{ then } Q \text{ else skip}$$

wait d $\rightleftharpoons \langle 0 = 0 \wedge \top \rangle \trianglerighteq_d \textbf{skip}$

$$\langle F(\dot{x}, x) = 0 \wedge b \rangle \trianglerighteq IO \rightleftharpoons \langle F(\dot{x}, x) = 0 \wedge b \wedge \bigwedge_{i \in I} \neg ch_i^* \rangle; \text{if } \bigwedge_{i \in I} \neg ch_i^* \text{ then skip else } IO$$

---

[1] Hoare style proof rules for a system with $ch?x$ and $ch!e$ appearing as primitive constructs were proposed by Zhou Chaochen *et al.* in [12]. That work features a different type of triples and follows the convention that process variables are not observable across threads thus ruling out a shared-variable emulation of communication.

Here ch_i^* stands for $ch_i?$, resp. $ch_i!$, depending on whether the respective action in IO is input or output. To account of the impossibility to mechanically (and computationally) tell apart $x < c$ from $x \leq c$ about time-dependent quantities x, in $\neg(\neg b)$ we use $\bar{a}$ for a condition that defines the topological closure of $\{\mathbf{x} : a(\mathbf{x})\}$. It is assumed that $\bar{b}$ admits a syntactical definition. E.g., $\overline{x < c}$ is $x \leq c$ for x being a continuous function of time.

Reduction of HCSP process terms. Next we define a *reduction relation* $P \xrightarrow{A,V} Q$ where V is a set of process variables and

$$A : \sigma \to (V' \cup \{r, n\} \to \mathbb{R}^\infty \cup \{0, 1\}), \tag{4}$$

where V' is a set of process variables, r and n are boolean variables outside V' and $\sigma \in \mathbb{I}$. In $\mathbb{R}^\infty \cup \{0, 1\}$ we emphasize the additional use of $0, 1 \in \mathbb{R}$ as truth values. We consider $P \xrightarrow{A,V} Q$ only for V such that $\mathrm{Var}_{:=}(P) \subseteq V \subseteq V'$. For HCSP terms P in the scope of a Q which on its turn is an operand of a $\|$, with no other $\|$s between this one and P, the semantics of P must specify the behaviour of all the variables from $\mathrm{Var}_{:=}(Q)$, which are *controlled* by the *enveloping thread Q* of P. $V' \setminus V$ is meant to include the variables which are not controlled by the enveloping thread of P but still may be accessed in it. In the sequel we write $\mathrm{dom} A$ for σ and $\mathrm{Var}(A)$ for V' from (4).

If $V \subseteq \mathrm{Var}(A)$, then $A|_V$ stands for the restriction of A to the variables from V. I.e., given A as in (4),

$$A|_V : \sigma \to (V \cup \{r, n\} \to \mathbb{R}^\infty \cup \{0, 1\}).$$

Given an arithmetic or boolean expression e such that $V(e) \subseteq V(A)$, we write $A_t(e)$ for the value of e under A at time $t \in \mathrm{dom} A$. Given A and B such that $\max \mathrm{dom} A = \min \mathrm{dom} B$, $\mathrm{Var}(A) = \mathrm{Var}(B)$ and $A_{\max \mathrm{dom} A}(x) = B_{\min \mathrm{dom} B}(x)$ for all $x \in \mathrm{Var}(A) \cup \{r, n\}$, $A; B$ is determined by the conditions $\mathrm{dom} A; B = \mathrm{dom} A \cup \mathrm{dom} B$, $(A; B)_t(x) = A_t(x)$ for $t \in \mathrm{dom} A$ and $(A; B)_t(x) = B_t(x)$ for $t \in \mathrm{dom} B$ for all $x \in \mathrm{Var}(A) \cup \{r, n\}$. A complete and possibly infinite behaviour of P can be defined as $A_1; A_2; \ldots$ where $P_{i-1} \xrightarrow{A_i, V} P_i$, and $P_0 = P$.

The auxiliary variables r and n. To handle the causal ordering of computation steps without having to account of the *negligibly* small time delays they contribute, we allow stretches of time in which continuous evolution is 'frozen', and which are meant to just keep apart time points with different successive variable values that have been obtained by computation. Intervals of negligible time are marked by the boolean variable r. P (or any of its descendant processes) claims exclusive control over the process variables during such intervals, thus achieving atomicity of assignment. Time used for computation steps by *any* process which runs in parallel with P or P itself is marked by n. Hence $A_t(r) \leq A_t(n)$, $t \in \mathrm{dom} A$ always holds in the mappings (4). As it becomes clear below, each operand P_i of a $P_1\|P_2$ has its own r, and no two such variables evaluate to 1 at the same time, which facilitates encoding the meaning of $\|$ by conjunction. In processes with loops and no other recursive calls, the rs can be enumerated globally. More involved form of recursive calls require the rs to be quantified away.

This approach is known as the *true synchrony hypothesis*. It was introduced to the setting of DC in [30] and was developed in [10,11] where properties φ of the overall behaviour of a process in terms of the relevant continuously evolving quantities are written $(\varphi/\neg N)$, *the projection of φ onto state $\neg N$*, which holds iff φ holds at the interval obtained by gluing the $\neg N$-parts of the reference one. The approach is alternative to the use of *super-dense chop* [16].

The reduction rules. To abbreviate conditions on A in the rules which generate the valid instances of $P \xrightarrow{A,V} Q$ below, given an $X \subseteq \mathrm{Var}(A)$ and a boolean or arithmetical expression e, we put:

$$\mathrm{const}(X, A) \ \rightleftharpoons \ \bigwedge_{x \in X} (\forall t \in \mathrm{dom}\,A)(A_t(x) = A_{\min \mathrm{dom}\,A}(x))$$
$$\mathrm{const}^{\circ}(X, A) \ \rightleftharpoons \ \bigwedge_{x \in X} (\forall t \in \mathrm{dom}\,A \setminus \{\max \mathrm{dom}\,A\})(A_t(x) = A_{\min \mathrm{dom}\,A}(x))$$
$$\mathrm{const}^{\circ}(e, A, a) \rightleftharpoons (\forall t \in \mathrm{dom}\,A \setminus \{\max \mathrm{dom}\,A\})(A_t(e) = a)$$

Below we omit the mirror images of rules about commutative connectives $\sqcup$ and $\|$.

$$\frac{}{\mathbf{skip} \xrightarrow{A,V} \checkmark} \quad \mathrm{max\,dom}\,A = \mathrm{min\,dom}\,A$$

$$\frac{\begin{array}{l} \mathrm{const}(V \setminus \{x_1, \ldots, x_n\}, A) \\ \mathrm{const}^{\circ}(\{x_1, \ldots, x_n\}, A) \\ \mathrm{const}^{\circ}(r \wedge n, A, 1) \\ A_{\max \mathrm{dom}\,A}(x_i) = A_{\min \mathrm{dom}\,A}(e_i), \ i = 1, \ldots, n \\ \mathrm{max\,dom}\,A < \infty \end{array}}{x_1, \ldots, x_n := e_1, \ldots, e_n \xrightarrow{A,V} \checkmark}$$

$$\frac{P \xrightarrow{A,V} P' \quad P' \neq \checkmark}{P;Q \xrightarrow{A,V} P';Q} \qquad \frac{P \xrightarrow{A,V} \checkmark \quad \mathrm{max\,dom}\,A < \infty}{P;Q \xrightarrow{A,V} Q} \qquad \frac{P \xrightarrow{A,V} P' \quad \mathrm{max\,dom}\,A = \infty}{P;Q \xrightarrow{A,V} P'}$$

$$\frac{\begin{array}{l} \mathrm{const}^{\circ}(F(\dot{x}, x), A, 0) \\ \mathrm{const}^{\circ}(A_t(\dot{x}) - \frac{d}{dt}A_t(x), A, 0) \\ \mathrm{const}^{\circ}(\neg r \wedge \neg n \wedge b, A, 1) \\ \mathrm{const}(V \setminus \{\dot{x}, x\}, A) \end{array}}{\langle F(\dot{x}, x) = 0 \wedge b \rangle \xrightarrow{A,V} \langle F(\dot{x}, x) = 0 \wedge b \rangle} \qquad \frac{\begin{array}{l} A_{\min \mathrm{dom}\,A}(b) = 0 \\ \mathrm{max\,dom}\,A = \mathrm{min\,dom}\,A \end{array}}{\langle F(\dot{x}, x) = 0 \wedge b \rangle \xrightarrow{A,V} \checkmark}$$

$$\frac{P \xrightarrow{A,V} R \quad A_{\min \mathrm{dom}\,A}(b) = 1}{\mathbf{if}\ b\ \mathbf{then}\ P\ \mathbf{else}\ Q \xrightarrow{A,V} R} \qquad \frac{Q \xrightarrow{A,V} R \quad A_{\min \mathrm{dom}\,A}(b) = 0}{\mathbf{if}\ b\ \mathbf{then}\ P\ \mathbf{else}\ Q \xrightarrow{A,V} R} \qquad \frac{P \xrightarrow{A,V} P'}{P \sqcup Q \xrightarrow{A,V} P'}$$

$$\frac{[\mu X.P/X]P \xrightarrow{A,V} Q}{\mu X.P \xrightarrow{A,V} Q}$$

$$\frac{\begin{array}{c}\text{Var}(A) \cup \text{Var}(B) \subseteq \text{Var}(C) \\ C|_{\text{Var}(A)\cup\{r,n\}} = A \\ C|_{\text{Var}(B)\cup\{r,n\}} = B \\ \text{const}^\circ(\neg r \wedge \neg n, C, 1) \\ V_1 \cap V_2 = \emptyset \\ P \xrightarrow{A,V_1} \checkmark \quad Q \xrightarrow{B,V_2} \checkmark\end{array}}{P\|Q \xrightarrow{C,V_1\cup V_2} \checkmark}$$

$$\frac{\begin{array}{c}\text{Var}(A) \cup \text{Var}(B) \subseteq \text{Var}(C) \\ C|_{\text{Var}(A)\cup\{r,n\}} = A \\ C|_{\text{Var}(B)\cup\{r,n\}} = B \\ \text{const}^\circ(\neg r \wedge \neg n, C, 1) \\ V_1 \cap V_2 = \emptyset \\ P \xrightarrow{A,V_1} P' \quad Q \xrightarrow{B,V_2} Q' \\ P' \neq \checkmark, \; Q' \neq \checkmark\end{array}}{P\|Q \xrightarrow{C,V_1\cup V_2} P'\|Q'}$$

$$\frac{\begin{array}{c}\text{Var}(A) \cup \text{Var}(B) \subseteq \text{Var}(C) \\ C|_{\text{Var}(A)\cup\{r,n\}} = A \\ C|_{\text{Var}(B)\cup\{r,n\}} = B \\ \text{const}^\circ(\neg r \wedge \neg n, C, 1) \\ V_1 \cap V_2 = \emptyset \\ P \xrightarrow{A,V_1} P' \quad Q \xrightarrow{B,V_2} \checkmark \\ P' \neq \checkmark\end{array}}{P\|Q \xrightarrow{C,V_1\cup V_2} P'}$$

$$\frac{\begin{array}{c}V \subseteq V' \\ \text{const}(V' \setminus V, B) \\ V' \cup \text{Var}(A) \subseteq \text{Var}(B) \\ B|_{\text{Var}(A)\cup\{r,n\}} = A \\ \text{const}^\circ(r \wedge n, A, 1) \\ P \xrightarrow{A,V} P' \quad P' \neq \checkmark\end{array}}{P\|Q \xrightarrow{B,V'} P'\|Q}$$

$$\frac{\begin{array}{c}V \subseteq V' \\ \text{const}(V' \setminus V, B) \\ V' \cup \text{Var}(A) \subseteq \text{Var}(B) \\ B|_{\text{Var}(A)\cup\{r,n\}} = A \\ \text{const}^\circ(r \wedge n, A, 1) \\ P \xrightarrow{A,V} \checkmark\end{array}}{P\|Q \xrightarrow{B,V'} Q}$$

4   A DC Semantics of Hybrid Communicating Sequential Processes

Given a process P, $[\![P]\!]$, with some subscripts to be specified below, is a DC formula which defines the class of DC interpretations that represent runs of P.

Process variables and their corresponding DC temporal variables. Real-valued process variables x are modelled by pairs of DC temporal variables x and x', which are meant to store the value of x at the beginning and at the end of the reference interval, respectively. The axiom

$$\Box \forall z \neg (x' = z ^\frown x \neq z).$$

entails that the values of x and x' are determined by the beginning and the end point of the reference interval, respectively. It can be shown that

$$\models_{DC} \Box \forall z \neg (x' = z ^\frown x \neq z) \Rightarrow$$
$$\Box(x = c \Rightarrow \neg(x \neq c ^\frown \top)) \wedge \Box(x' = c \Rightarrow \neg(\top ^\frown x' \neq c))$$

This is known as the locality principle in ITL about x. About primed variables x', the locality principle holds wrt the endpoints of reference intervals. Boolean process variables are similarly modelled by propositional temporal letters. For the sake of brevity we put

$$\mathrm{loc}(X) \rightleftharpoons \bigwedge_{\substack{x \in X \\ x \text{ is real}}} \Box \forall z \neg (x' = z \,\widehat{}\, x \neq z) \wedge \bigwedge_{\substack{x \in X \\ x \text{ is boolean}}} \Box \neg ((x' \,\widehat{}\, \neg x) \vee (\neg x' \,\widehat{}\, x))$$

In the sequel, given a DC term e or formula φ written using only unprimed variables, e' and φ' stand for the result of replacing all these variables by their respective primed counterparts.

Time derivatives of process variables. As mentioned above, terms of the form $\dot{x}$ where x is a process variable are treated as distinct process variables and are modelled by their respective temporal variables $\dot{x}$ and $\dot{x}'$. The requirement on $\dot{x}$ to be interpreted as the time derivative of x is incorporated in the semantics of continuous evolution statements.

Computation time and the parameters $[\![.]\!]$. As explained in Sect. 3, we allow stretches of time that are dedicated to computation steps and are marked by the auxiliary boolean process variable r. Such stretches of time are conveniently excluded when calculating the duration of process execution. To this end, in DC formulas, we use a state variable R which indicates the time taken by computation steps by the reference process. Similarly, a state variable N indicates time for computation steps by which any process that runs in parallel with the reference one, including the reference one. R and N match the auxiliary variables r and n from the operational semantics and, just like r and n, are supposed to satisfy the condition $R \Rightarrow N$. We assume that all continuous evolution becomes temporarily suspended during intervals in which computation is performed, with the relevant real quantities and their derivatives remaining frozen. To guarantee the atomicity of assignment, computation intervals of different processes are not allowed to overlap. As it becomes clear in the DC semantics of $\|$ below, P_i of $P_1\|P_2$ are each given its own variable R_i, $i = 1, 2$, to mark computation time, and R_1 and R_2 are required to satisfy the constraints $\neg(R_1 \wedge R_2)$ and $R_1 \vee R_2 \Leftrightarrow R$ where R is the variable which marks computation times for the whole of $P_1\|P_2$.

The semantics $[\![P]\!]_{R,N,V}$ of a HCSP term P is given in terms of the DC temporal variables which correspond to the process variables occurring in P, the state variables R and N, and the set of variables V which are controlled by P's immediately enveloping $\|$-operand.

Assignment. To express that the process variables from $X \subseteq V$ may change at the end of the reference interval only, and those from $V \setminus X$ remain unchanged, we write

$$\mathrm{const}(V, X) \rightleftharpoons \bigwedge_{\substack{x \in V \setminus X \\ x \text{ is real}}} \Box(x' = x) \wedge \bigwedge_{\substack{x \in V \setminus X \\ x \text{ is boolean}}} \Box(x \Leftrightarrow x') \wedge$$
$$\bigwedge_{\substack{x \in X \\ x \text{ is real}}} \Box^{\circ}(x' = x) \wedge \bigwedge_{\substack{x \in X \\ x \text{ is boolean}}} \Box^{\circ}(x' \Leftrightarrow x).$$

The meaning of simultaneous assignment is as follows:

$$[\![x_1,\ldots,x_n := e_1,\ldots,e_n]\!]_{R,N,V} \rightleftharpoons \lceil R \rceil_{\text{fin}} \wedge \text{const}(V,\{x_1,\ldots,x_n\}) \wedge$$
$$\bigwedge_{\substack{i=1,\ldots,n \\ x_i \text{ is real}}} x_i' = e_i \wedge \bigwedge_{\substack{i=1,\ldots,n \\ x_i \text{ is boolean}}} x_i' \Leftrightarrow e_i.$$

Parallel composition. Consider processes P_1 and P_2 and $V \supseteq \text{Var}_{:=}(P_1\|P_2)$. Let $\overline{1} = 2$, $\overline{2} = 1$. Let

$$\exists_\|(R,R_1,R_2,V,P_1,P_2)\varphi \rightleftharpoons \exists R_1 \exists R_2 \left(\begin{array}{c} \lceil (R_1 \vee R_2 \Leftrightarrow R) \wedge \neg(R_1 \wedge R_2) \rceil^0 \wedge \\ \bigwedge_{i=1}^2 \Box(\lceil R_i \rceil \Rightarrow \text{const}(V \setminus \text{Var}_{:=}(P_i))) \wedge \varphi \end{array} \right).$$

$\exists_\|(R,R_1,R_2,V_1,V_2)$ means that

- the R-subintervals for the computation steps of $P_1\|P_2$ can be divided into R_1- and R_2-subintervals to mark the computation steps of some sub-processes P_1 and P_2 of P which run in parallel;
- the variables which are not controlled by P_i remain unchanged during P_i's computation steps, $i = 1, 2$, and, finally,
- some property φ, which can involve R_1 and R_2, holds.

The universal dual $\forall_\|$ of $\exists_\|$ is defined in the usual way. Let V_i abbreviate $\text{Var}_{:=}(P_i)$. Now we can define $[\![P_1\|P_2]\!]_{R,N,V}$ as

$$\exists_\|(R,R_1,R_2,V,P_1,P_2) \bigvee_{i=1}^2 \left(\begin{array}{c} [\![P_i]\!]_{R_i,N,V_i} \wedge (\lceil N \wedge \neg R_{\overline{i}} \rceil^0_{\text{fin}} \widehat{} [\![P_{\overline{i}}]\!]_{R_{\overline{i}},N,V_{\overline{i}}} \widehat{} \lceil \neg R_{\overline{i}} \rceil^0) \vee \\ (\lceil N \wedge \neg R_i \rceil^0_{\text{fin}} \widehat{} [\![P_i]\!]_{R_i,N,V_i}) \wedge ([\![P_{\overline{i}}]\!]_{R_{\overline{i}},N,V_{\overline{i}}} \widehat{} \lceil \neg R_{\overline{i}} \rceil^0) \end{array} \right)$$
$$\tag{5}$$

To understand the four disjunctive members of Φ above, note that $P_1\|P_2$ always starts with some action on behalf of either P_1, or P_2, or both, in the case of continuous evolution. Hence (at most) one of P_i, $i = 1, 2$, needs to allow negligible time for $P_{\overline{i}}$'s first step. This is expressed by a $\lceil N \wedge \neg R_i \rceil^0_{\text{fin}}$ before $[\![P_i]\!]_{R,N,V_i}$. The amount of time allowed is finite and may be 0 in case both P_1 and P_2 start with continuous evolution in parallel. This makes it necessary to consider two cases, depending on which process starts first. If $P_{\overline{i}}$ terminates before P_i, then a $\lceil N \wedge \neg R_i \rceil^0$ interval follows $[\![P_i]\!]_{R_{\overline{i}},N,V_{\overline{i}}}$. This generates two more cases to consider, depending on the value of i.

The definitions of $[\![x_1,\ldots,x_n := e_1,\ldots,e_n]\!]_{R,N,V}$ and $[\![P_1\|P_2]\!]_{R,N,V}$ already appear in (3) and (5). Here follow the definitions for the rest of the basic constructs:

$$[\text{await } b]_{R,N,V} \quad \rightleftharpoons \text{const}(V) \wedge (\lceil \neg R \rceil \vee \ell = 0) \wedge \Box^\circ \neg \overline{b'} \wedge (\overline{b'} \vee \ell = \infty)$$

$$[\langle F(\dot{x}, x) = 0 \wedge b \rangle]_{R,N,V} \quad \rightleftharpoons \quad \Box \begin{pmatrix} \text{const}(V \setminus \{\dot{x}, x\}) \wedge \lceil \neg R \rceil \wedge \\ \begin{pmatrix} \lceil N \rceil \Rightarrow \text{const}(\{\dot{x}, x\}) \wedge \\ \forall ub\Box(\dot{x} \leq ub) \Rightarrow x' \leq x + ub \int \neg N \wedge \\ \forall lb\Box(\dot{x} \geq lb) \Rightarrow x' \geq x + lb \int \neg N \wedge \\ F(\dot{x}, x) = 0 \end{pmatrix} \wedge \boxdot^\circ b \end{pmatrix}$$
$$\frown (\neg b \wedge \ell = 0)$$

$$[P; Q]_{R,N,V} \quad \rightleftharpoons ([P]_{R,N,V} \frown \lceil N \wedge \neg R \rceil^0_{\text{fin}} \frown [Q]_{R,N,V})$$
$$[P \sqcup Q]_{R,N,V} \quad \rightleftharpoons [P]_{R,N,V} \vee [Q]_{R,N,V}$$

$$[\text{if } b \text{ then } P \text{ else } Q]_{R,N,V} \rightleftharpoons (b \wedge [P]_{R,N,V}) \vee (\neg b \wedge [Q]_{R,N,V})$$

$$[\text{while } b \text{ do } P]_{R,N,V} \quad \rightleftharpoons (b \wedge [P]_{R,N,V} \frown \lceil \neg R \rceil^0)^* \frown (\neg b \wedge \ell = 0)$$

To understand $[\langle F(\dot{x}, x) = 0 \wedge b \rangle]_{R,N,V}$, observe that, assuming I to be the DC interpretation in question and $\lambda t.I_t(\dot{x})$ to be continuous, the two inequalities in $[\langle F(\dot{x}, x) = 0 \wedge b \rangle]_{R,N,V}$ express that

$$I_{t_2}(x) - I_{t_1}(x) = \int_{t_1}^{t_2} I_t(\dot{x})(1 - I_t(N))dt$$

at all finite subintervals $[t_1, t_2]$ of the reference intervals. This means that both $\dot{x}$ and x are constant in N-subintervals, and $I_{t_2}(x) - I_{t_1}(x) = \int_{t_1}^{t_2} I_t(\dot{x})dt$ holds at $\neg N$-subintervals.

Completeness of $[.]$. Given a process term P, every DC interpretation I for the vocabulary of $[P]_N, N, V$ represents a valid behaviour of P with N being true in the subintervals which P uses for computation steps. To realize this, consider HCSP terms P, Q of the syntax

$$P, Q ::= \text{skip} \mid A; R \mid []_i A_i; R_i \mid \text{if } b \text{ then } P \text{ else } Q \mid P \sqcup Q$$
$$A \quad ::= x := e \mid \text{await } b \mid \langle F(\dot{x}, x) = 0 \wedge b \rangle \tag{6}$$

where R and R_i stand for a process term with no restrictions on its syntax (e.g., occurrences of **while**-terms are allowed). (6) is the *guarded normal form (GNF)* for HCSP terms, with the guards being the primitive terms of the form A, and can be established by induction on the construction of terms, with suitable equivalences for each combination of guarded operands that $\|$ may happen to have. E.g.,

$$\langle F_1(\dot{x}, x) = 0 \wedge b_1 \rangle; P_1 \| \langle F_2(\dot{x}, x) = 0 \wedge b_2 \rangle; P_2 \tag{7}$$

is equivalent to

$$\langle F_1(\dot{x}, x) = 0 \wedge F_2(\dot{x}, x) = 0 \wedge b_1 \wedge b_2 \rangle;$$
$$\text{if } b_1 \text{ then } \langle F_1(\dot{x}, x) = 0 \wedge b_1 \rangle; P_1 \| P_2$$
$$\text{else if } b_2 \ P_1 \| \langle F_2(\dot{x}, x) = 0 \wedge b_2 \rangle; P_2 \tag{8}$$
$$\text{else } P_1 \| P_2$$

Note that $[]_i A_i; R_i$ is a modest generalization of IO as defined in (2). Some combinations of operands of $\|$ require external choice to be extended this way, with the intended meaning being that if none of the A_is which have the forms $ch?x$ and $ch!e$ is ready to execute, then some other option can be pursued immediately. For example, driving $\|$ inwards may require using that

$$(ch_1?x \to P_1 []ch_2!e \to P_2 []ch_3?y \to P_3)\|ch_1!f; Q_1\|ch_2?z; Q_2 \equiv$$
$$((x := f; (P_1\|Q_1)\|ch_2?z; Q_2) \sqcup (z := e; (P_2\|Q_2)\|ch_1!f; Q_1))[]$$
$$ch_3?y; P_3\|ch_1!f; Q_1\|ch_2?z; Q_2.$$

On the RHS of $\equiv$ above, one of the assignments and the respective subsequent process are bound to take place immediately in case the environment is not ready to communicate over ch_3.

The GNF renders the correspondence between the semantics of guards and the As which appear in the rules for $\xrightarrow{A,V}$ explicit, thus making obvious that any finite prefix of a valid behaviour satisfies some *chop*-sequence of guards that can be generated by repeatedly applying the GNF a corresponding number of times and then using the distributivity of *chop* over disjunction, starting with the given process term, and then proceeding to transform the R-parts of guarded normal forms that appear in the process. The converse holds too. This entails that the denotational semantics is equivalent to the operational one.

5   Reasoning About Hybrid Communicating Sequential Processes with DC Hoare Triples

We propose reasoning in terms of triples of the form

$$\{A\}P\{G\}_V \tag{9}$$

where A and G are DC formulas, P is a HCSP term, and V is a set of program variables. V is supposed to denote the variables whose evolution needs to be specified in the semantics of P, e.g., an assignment $x := e$ in P is supposed to leave the values of the variables $y \neq x$ unchanged. This enables deriving, e.g., $\{y = 0\}x := 1\{y = 0\}_{\{x,y\}}$, which would be untrue for a y that belongs to a process that runs in parallel with the considered one and is therefore not a member of V. Triple (9) is valid, if

$$\models loc(V) \wedge \lceil R \Rightarrow N \rceil^0 \wedge (A^\frown \top) \Rightarrow \neg(\llbracket P \rrbracket_{R,N,V} \wedge \neg G^\frown \top) \tag{10}$$

Since R and N typically occur in $\llbracket P \rrbracket_{R,N,V}$, triples (9) can have occurrences of R and N in A and G too, with their intended meanings.

Next we propose axioms and rules for deriving triples about processes with each of the HCSP constructs as the main one in them. For P of one of the forms **skip**, $x_1, \ldots, x_n := e_1, \ldots, e_n$, and $\langle \mathcal{F}(\dot{x}, x) = 0 \wedge b \rangle$, we introduce the axioms

$$\{\top\}P\{\llbracket P \rrbracket_{R,N,V}\}_V.$$

where V can be any set of process variables such that $V \supseteq \mathrm{Var}_{:=}(P)$. Here follow the rules for reasoning about processes which are built using each of the remaining basic constructs:

$$(\mathbf{seq}) \quad \frac{\{A\}P\{G\}_V \qquad \{B\}Q\{H\}_V}{\mathrm{loc}(V) \wedge \lceil R \Rightarrow N\rceil^0 \wedge (A\,^\frown\top) \Rightarrow \neg(G\,^\frown\neg(\lceil N \wedge \neg R\rceil^0\,^\frown B\,^\frown\top))}{\{A\}P;Q\{(G\,^\frown\lceil N \wedge \neg R\rceil^0\,^\frown H)\}_V}$$

$$(\sqcup) \quad \frac{\{A\}P\{G\}_V \qquad \{B\}Q\{H\}_V}{\{A \wedge B\}P \sqcup Q\{G \vee H\}_V}$$

$$(\mathbf{if}) \quad \frac{\{A \wedge b\}P\{G\}_V \qquad \{A \wedge \neg b\}Q\{G\}_V}{\{A\}\mathbf{if}\ b\ \mathbf{then}\ P\ \mathbf{else}\ Q\{G\}_V}$$

$$(\mathbf{while}) \quad \frac{\{A\}P\{G\}_V}{\mathrm{loc}(V) \wedge \lceil R \Rightarrow N\rceil^0 \wedge (A\,^\frown\top) \Rightarrow \neg(G \wedge \ell < \infty\,^\frown\neg\Diamond_l^c(\lceil\neg R\rceil^0\,^\frown A))}{\{A\}\mathbf{while}\ b\ \mathbf{do}\ P\{((b \wedge G\,^\frown\lceil\neg R\rceil^0)^*\,^\frown\neg b \wedge \ell = 0)\}_V}$$

Parallel composition. The established pattern suggests the following proof rule ($\|$):

$$\frac{\{A_1\}P_1\{G_1\}_{\mathrm{Var}_{:=}(P_1)} \qquad \{A_2\}P_2\{G_2\}_{\mathrm{Var}_{:=}(P_2)}}{\begin{cases} \forall_\|(R, R_1, R_2, V, P_1, P_2) \left(\bigvee_{i=1}^2 \neg(\lceil N \wedge \neg R_i\rceil^0_{\mathrm{fin}}\,^\frown\neg[R_i/R]A_i\,^\frown\top) \wedge ([R_{\bar{i}}/R]A_{\bar{i}}\,^\frown\top) \right) \\ P_1 \| P_2 \\ \exists_\|(R, R_1, R_2, V, P_1, P_2) \left(\bigvee_{i=1}^2 \begin{array}{l} G_i \wedge (\lceil N \wedge \neg R_{\bar{i}}\rceil^0_{\mathrm{fin}}\,^\frown G_{\bar{i}}\,^\frown\lceil\neg R_{\bar{i}}\rceil^0) \\ \vee \\ (\lceil N \wedge \neg R_i\rceil^0_{\mathrm{fin}}\,^\frown G_i) \wedge (G_{\bar{i}}\,^\frown\lceil\neg R_{\bar{i}}\rceil^0) \end{array} \right) \end{cases}_V}$$

This rule can be shown to be complete as it straightforwardly encodes the semantics of $\|$. However, it is not convenient for proof search as it only derives triples with a special form of the condition on the righthand side and actually induces the use of $[\![P_i]\!]_{R_i, N, \mathrm{Var}_{:=}(P_i)}$ as G_i, which typically give excess detail. We discuss a way around this inconvenience below, together with the modifications of the setting which are needed in order to handle general HCSP fixpoints $\mu X.P$.

General rules. Along with the process-specific rules, we also need the rules

$$(N) \quad \frac{\mathrm{loc}(V) \wedge \lceil R \Rightarrow N\rceil^0 \wedge \Diamond_l^c A \Rightarrow G}{\{A\}P\{G\}_V} \qquad \mathrm{Var}_{:=}(P) \subset V$$

$$(\mathbf{K}) \quad \frac{\{A\}P\{G \Rightarrow H\}_V \qquad \{B\}P\{G\}_V \qquad \mathrm{loc}(V) \wedge \lceil R \Rightarrow N\rceil^0 \wedge \Diamond_l^c A \Rightarrow \Diamond_l^c B}{\{A\}P\{H\}_V}$$

These rules are analogous to the modal form N of Gödel's generalization rule (also known as the necessitation rule) and the modal axiom **K**.

Soundness and Completeness. The soundness of the proof rules is established by a straightforward induction on the construction of proofs with the definition of $[\![.]\!]_{R,N,V}$. The system is also complete relative to validity in DC.

This effectively means that we allow all valid DC formulas as axioms in proofs, or, equivalently, given some sufficiently powerful set of proof rules and axioms for the inference of valid *DC formulas* in DC with infinite intervals, our proof rules about triples suffice for deriving all the *valid triples*. Such systems are beyond the scope of our work. A Hilbert-style proof system for ITL with infinite intervals was proposed and shown to be complete with respect to an abstractly defined class of time models (linearly ordered commutative groups) in [13], building on similar work about finite intervals from [8].

The deductive completeness of our proof system boils down to the possibility to infer triples of the form $\{\top\}P\{G\}_V$ for any given term P and a certain strongest corresponding G, which turns out to be the DC formula $[\![P]\!]_{R,N,V}$ that we call the semantics of P. Then the validity of $\{\top\}P\{[\![P]\!]_{R,N,V}\}_V$ is used to derive any valid triple about P by a simple use of the proof rules **K** and N. The completeness now follows from the fact that $[\![P]\!]_{R,N,V}$ defines the class of all valid behaviours of P.

Proposition 1. *The triple*

$$\{\top\}P\{[\![P]\!]_{R,N,V}\}_V \tag{11}$$

is derivable for all process terms P and all V such that $\mathrm{Var}_{:=}(P) \supseteq V$.

Corollary 1 (relative completeness of the Hoare-style proof system). *Let A, G and P be such that (10) is valid. Then (9) is derivable in the extension of the given proof system by all DC theorems.*

6   General Fixpoints and Bottom-Up Proof Search in HCSP

To avoid the constraints on the form of the conclusion of rule ($\|$), we propose a set of rules which correspond to the various possible forms of the operands of the designated $\|$ in the considered HCSP term. These rules enable bottom-up proof search much like when using the rules for (just) CSP constructs, which is key to the applicability of classical Hoare-style proof. We propose separate rules for each combination of main connectives in the operands of $\|$, except $\|$ itself and the fixpoint construct. For instance, the equivalence between (7) and (8) suggests the following rule for this particular combination of $\|$ with the other connectives:

$$\frac{\begin{array}{l}\{P\}\langle F_1(\dot{x},x) = 0 \wedge F_2(\dot{x},x) = 0 \wedge b_1 \wedge b_2\rangle\{R\} \\ \{R \wedge b_1 \wedge \neg b_2\}\langle F_1(\dot{x},x) = 0 \wedge b_1\rangle; P_1 \| P_2\{Q\} \\ \{R \wedge b_2 \wedge \neg b_1\}P_1 \| \langle F_2(\dot{x},x) = 0 \wedge b_2\rangle; P_2\{Q\} \\ \{R \wedge \neg b_1 \wedge \neg b_2\}P_1 \| P_2\{Q\}\end{array}}{\{P\}\langle F_1(\dot{x},x) = 0 \wedge b_1\rangle; P_1 \| \langle F_2(\dot{x},x) = 0 \wedge b_2\rangle; P_2\{Q\}}$$

Rules like the above one use the possibility to drive $\|$ inwards by equivalences like that between (7) and (8), which can be derived for all combinations of the main connectives of $\|$'s operands, except for loops, and indeed can be used to eliminate $\|$. For **while**-loops, the GNF contains a copy of the loop on the RHS of *chop*:

$$\textbf{while } b \textbf{ do } P \equiv \textbf{if } b \textbf{ then } (P; \textbf{while } b \textbf{ do } P) \textbf{ else skip}. \tag{12}$$

Tail-recursive instances of $\mu X.G$ come handy in completing the elimination of $\|$ in such cases by standard means, namely, by treating equivalences such as (12) as the equations leading to definitions such as (3).

To handle general recursion in our setting, we need to take care of the special way in which we treat A from $\{A\}P\{G\}$ in (10). In the rule for $\{A\}\textbf{while } b \textbf{ do } P\{G\}$ clipping of initial G-subintervals of an A-interval are supposed to leave us with suffix subintervals which satisfy $(A^\frown\top)$, to provide for successive executions of P. With X allowed on the LHS of *chop* in the P of $\mu X.P$, special care needs to be taken for this to be guaranteed. To this end, instead of $(A^\frown\top)$, $\diamondsuit_l^c A$ is used to state that A holds at an interval that starts at the same time point as the reference one, and is not necessarily its subinterval. This is needed for reasoning from the viewpoint of intervals which accommodate nested recursive executions. The meaning of triples (9) becomes

$$\models \mathrm{loc}(V) \wedge \lceil R \Rightarrow N\rceil^0 \wedge \diamondsuit_l^c A \wedge \llbracket P\rrbracket_{R,N,V} \Rightarrow G. \tag{13}$$

In this setting, $\mu X.P$ admits the proof rule, where X does not occur in A:

$$(\mu) \quad \frac{\mathrm{loc}(V) \wedge \lceil R \Rightarrow N\rceil^0 \wedge \diamondsuit_l^c A \wedge G \Rightarrow [\diamondsuit_l^c A \wedge X/X]G \qquad \{A\}P\{G\}_V}{\{A\}\mu X.P\{\mu X.G\}_V}$$

This rule subsumes the one for **while** $-$ **do**, but only as part of a suitably revised variant of the whole proof system wrt (13). E.g., the rule for sequential composition becomes

$$\begin{array}{l} \{A\}P\{G\}_V \\ \{B\}Q\{H\}_V \\ \mathrm{loc}(V) \wedge \lceil R \Rightarrow N\rceil^0 \wedge \diamondsuit_l^c A \Rightarrow \neg(G \wedge \ell < \infty^\frown \neg\diamondsuit_l^c(\lceil N \wedge \neg R\rceil^0 {}^\frown B)) \\ \hline \{A\}P; Q\{(G^\frown\lceil N \wedge \neg R\rceil^0 {}^\frown H)\}_V \end{array}$$

7   Related Work

Our work was influenced by the studies on DC-based reasoning about process-algebraic specification languages in [14,17,18,38]. In a previous paper we proposed a calculus for HCSP [22], which was based on DC in a limited way and lacked compositionality. In [12,37] we gave other variants of compositional and sound calculi for HCSP with different assertion and temporal condition formats. Completeness was not considered. The approach in [12] and in this paper is largely drawn from [10] where computation time was proposed to be treated as negligible in order to simplify delay calculation, and the operator of projection was proposed to facilitate writing requirements with negligible time ignored.

Hoare-style reasoning about real-time systems was also studied in the literature with explicit time logical languages [21]. However, our view is that using temporal logic languages is preferable. Dedicated temporal constructs both lead to more readable specifications, and facilitate the identification of classes of requirements that can be subjected to automated analysis. Another approach to the verification of hybrid systems is Platzer's Differential Dynamic Logic [32]. However, the hybrid programs considered there have limited functionality. Communication, parallelism and interrupts are not handled. For logic compositionality, assume-guarantee reasoning has been studied for communication-based concurrency in CSP without timing in [25, 31].

Both in our work and in alternative approaches such as [25], the treatment of continuous evolution is somewhat separated from the analysis of the other basic process-algebraic constructs. Indeed we make a small step forward here by fully expressing the meaning of differential law-governed evolution in DC, which is theoretically sufficient to carry out all the relevant reasoning in the logic. Of course, the feasibility of such an approach is nowhere close to the state of art in the classical theory of ordinary differential equations. Indeed it would typically lead to formalized accounts of classical reasoning. Techniques for reasoning about the ODE-related requirements are the topic of separate studies, see, e.g., [23, 33, 34].

8   Concluding Remarks

We have presented a weakly monotonic time-based semantics and a corresponding Hoare style proof system for HCSP with both the semantics and the temporal conditions in triples being in first-order DC with infinite intervals and extreme fixpoints. The proof system is compositional but the proof rule for parallel composition introduces complications because of the special form of the triples that it derives. However, we have shown that HCSP equivalences that can serve as elimination rules for $\parallel$ can also be used to derive proof rules for $\parallel$ which do not bring the above difficulty and indeed are perfectly compatible with standard bottom-up proof search. Interestingly, the informal reading of the derived rules for $\parallel$ together with the ones which are inherited from CSP, does not require the mention of weakly monotonic time technicalities. This means that the use of this special semantics can be restricted to establishing the soundness of practically relevant proof systems and awareness of its intricacies is not essential for applying the system. The meaning of triples we propose subsumes classical pre-/postcondition Hoare triples and triples linking (hybrid) temporal conditions in a streamlined way. This is a corollary of the choice to use assumptions which hold at an arbitrary initial subintervals, which is also compatible with reasoning about invariants A in terms of statements of the form $(A^\frown \top) \Rightarrow \neg(\llbracket P \rrbracket \wedge \neg(A^\frown \top))$.

Acknowledgements. Dimitar Guelev was partially supported through Bulgarian NSF Contract DN02/15/19.12.2016. Shuling Wang and Naijun Zhan were partially supported by NSFC under grants 61625206, by "973 Program" under grant No. 2014CB340701, by CDZ project CAP (GZ 1023), and by the CAS/SAFEA International Partnership Program for Creative Research Teams.

References

1. Abadi, M., Lamport, L.: Composing specifications. ACM Trans. Program. Lang. Syst. **15**(1), 73–132 (1993)
2. Allen, J.F.: Maintaining knowledge about temporal intervals. Commun. ACM **26**(11), 832–843 (1983)
3. Alur, R., Courcoubetis, C., Henzinger, T.A., Ho, P.-H.: Hybrid automata: an algorithmic approach to the specification and verification of hybrid systems. In: Grossman, R.L., Nerode, A., Ravn, A.P., Rischel, H. (eds.) HS 1991-1992. LNCS, vol. 736, pp. 209–229. Springer, Heidelberg (1993). doi:10.1007/3-540-57318-6_30
4. Zhou, C., Hansen, M.R.: Duration Calculus: A Formal Approach to Real-Time Systems. EATCS. Springer, Heidelberg (2004). doi:10.1007/978-3-662-06784-0
5. Zhou, C., Hoare, C.A.R., Ravn, A.P.: A calculus of durations. Inf. Process. Lett. **40**(5), 269–276 (1991)
6. Zhou, C., Dang, V.H., Li, X.: A duration calculus with infinite intervals. In: Reichel, H. (ed.) FCT 1995. LNCS, vol. 965, pp. 16–41. Springer, Heidelberg (1995). doi:10. 1007/3-540-60249-6_39
7. Cau, A., Moszkowski, B., Zedan, H.: ITL web pages. http://www.antonio-cau.co. uk/ITL/
8. Dutertre, B.: On First-order Interval Temporal Logic. Report CSD-TR-94-3, Department of Computer Science, Royal Holloway, University of London (1995)
9. Goranko, V., Montanari, A., Sciavicco, G.: A road map of interval temporal logics and duration calculi. J. Appl. Non Classical Logics **14**(1–2), 9–54 (2004)
10. Guelev, D.P., Hung, D.V.: Prefix and projection onto state in duration calculus. In: Proceedings of TPTS 2002, ENTCS, vol. 65, no. 6. Elsevier Science (2002)
11. Guelev, D.P., Van Hung, D.: A relatively complete axiomatisation of projection onto state in the duration calculus. J. Appl. Non Class. Logics **14**(1–2), 151–182 (2004). Special Issue on Interval Temporal Logics and Duration Calculi
12. Guelev, D.P., Wang, S., Zhan, N., Zhou, C.: Super-dense computation in verification of hybrid CSP processes. In: Fiadeiro, J.L., Liu, Z., Xue, J. (eds.) FACS 2013. LNCS, vol. 8348, pp. 13–22. Springer, Cham (2014). doi:10.1007/ 978-3-319-07602-7_3
13. Wang, H., Xu, Q.: Completeness of temporal logics over infinite intervals. Discr. Appl. Math. **136**(1), 87–103 (2004)
14. Zhu, H., He, J.: A *DC*-based semantics for verilog. Technical report 183, UNU/IIST, P.O. Box 3058, Macau (2000)
15. Halpern, J.Y., Shoham, Y.: A propositional logic of time intervals. In: Proceedings of LICS 1986, pp. 279–292. IEEE Computer Society Press (1986)
16. Hansen, M.R., Zhou, C.: Chopping a point. In: BCS-FACS 7th Refinement Workshop, Electronic Workshops in Computing. Springer (1996)
17. Haxthausen, A.E., Yong, X.: Linking DC together with TRSL. In: Grieskamp, W., Santen, T., Stoddart, B. (eds.) IFM 2000. LNCS, vol. 1945, pp. 25–44. Springer, Heidelberg (2000). doi:10.1007/3-540-40911-4_3
18. He, J., Xu, Q.: Advanced features of duration calculus and their applications in sequential hybrid programs. Formal Asp. Comput. **15**(1), 84–99 (2003)
19. He, J.: From CSP to hybrid systems. In: Roscoe, A.W. (ed.) A Classical Mind, pp. 171–189. Prentice Hall International (UK) Ltd., Hertfordshire (1994)

20. Henzinger, T.A.: The theory of hybrid automata. In: Proceedings of LICS 1996, pp. 278–292. IEEE Computer Society Press (1996)
21. Hooman, J.: Extending Hoare logic to real-time. Formal Asp. Comput. **6**(6A), 801–826 (1994)
22. Liu, J., Lv, J., Quan, Z., Zhan, N., Zhao, H., Zhou, C., Zou, L.: A calculus for hybrid CSP. In: Ueda, K. (ed.) APLAS 2010. LNCS, vol. 6461, pp. 1–15. Springer, Heidelberg (2010). doi:10.1007/978-3-642-17164-2_1
23. Liu, J., Zhan, N., Zhao, H.: Computing semi-algebraic invariants for polynomial dynamical systems. In: Proceedings of EMSOFT 2011, pp. 97–106. ACM (2011)
24. Manna, Z., Pnueli, A.: Verifying hybrid systems. In: Grossman, R.L., Nerode, A., Ravn, A.P., Rischel, H. (eds.) HS 1991-1992. LNCS, vol. 736, pp. 4–35. Springer, Heidelberg (1993). doi:10.1007/3-540-57318-6_22
25. Misra, J., Chandy, K.M.: Proofs of networks of processes. IEEE Trans. Software Eng. **7**(4), 417–426 (1981)
26. Moszkowski, B.: Temporal logic for multilevel reasoning about hardware. IEEE Comput. **18**(2), 10–19 (1985)
27. Moszkowski, B.: Executing Temporal Logic Programs. Cambridge University Press, Cambridge (1986). http://www.cse.dmu.ac.uk/~cau/papers/tempura-book.pdf
28. Olderog, E.-R., Hoare, C.A.R.: Specification-oriented semantics for communicating processes. In: Diaz, J. (ed.) ICALP 1983. LNCS, vol. 154, pp. 561–572. Springer, Heidelberg (1983). doi:10.1007/BFb0036937
29. Pandya, P.K.: Some extensions to propositional mean-value calculus: expressiveness and decidability. In: Kleine Büning, H. (ed.) CSL 1995. LNCS, vol. 1092, pp. 434–451. Springer, Heidelberg (1996). doi:10.1007/3-540-61377-3_52
30. Pandya, P.K., Hung, D.: Duration calculus of weakly monotonic time. In: Ravn, A.P., Rischel, H. (eds.) FTRTFT 1998. LNCS, vol. 1486, pp. 55–64. Springer, Heidelberg (1998). doi:10.1007/BFb0055336
31. Pandya, P.K., Joseph, M.: P - a logic - a compositional proof system for distributed programs. Distrib. Comput. **5**, 37–54 (1991)
32. Platzer, A.: Differential dynamic logic for hybrid systems. J. Autom. Reasoning **41**(2), 143–189 (2008)
33. Prajna, S., Jadbabaie, A.: Safety verification of hybrid systems using barrier certificates. In: Alur, R., Pappas, G.J. (eds.) HSCC 2004. LNCS, vol. 2993, pp. 477–492. Springer, Heidelberg (2004). doi:10.1007/978-3-540-24743-2_32
34. Sankaranarayanan, S., Sipma, H.B., Manna, Z.: Constructing invariants for hybrid systems. In: Alur, R., Pappas, G.J. (eds.) HSCC 2004. LNCS, vol. 2993, pp. 539–554. Springer, Heidelberg (2004). doi:10.1007/978-3-540-24743-2_36
35. Venema, Y.: A modal logic for chopping intervals. J. Logic Comput. **1**(4), 453–476 (1991)
36. Venema, Y.: Many-dimensional modal logics. Ph.D. thesis, University of Amsterdam (1991)
37. Wang, S., Zhan, N., Guelev, D.: An assume/guarantee based compositional calculus for hybrid CSP. In: Agrawal, M., Cooper, S.B., Li, A. (eds.) TAMC 2012. LNCS, vol. 7287, pp. 72–83. Springer, Heidelberg (2012). doi:10.1007/978-3-642-29952-0_13
38. Yong, X., George, C.: An operational semantics for timed RAISE. In: Wing, J.M., Woodcock, J., Davies, J. (eds.) FM 1999. LNCS, vol. 1709, pp. 1008–1027. Springer, Heidelberg (1999). doi:10.1007/3-540-48118-4_4
39. Zhou, C., Wang, J., Ravn, A.P.: A formal description of hybrid systems. In: Alur, R., Henzinger, T.A., Sontag, E.D. (eds.) HS 1995. LNCS, vol. 1066, pp. 511–530. Springer, Heidelberg (1996). doi:10.1007/BFb0020972

Program Analysis

Termination of Semi-algebraic Loop Programs

Yi Li[⊠]

Chongqing Key Laboratory of Automated Reasoning and Cognition,
CIGIT, CAS, Chongqing, China
zm_liyi@163.com

Abstract. Program termination is a fundamental research topic in program analysis. In this paper, we investigate the termination of a class of semi-algebraic loop programs. We relate the termination of such a loop program to a certain semi-algebraic system. Also, we show that under some conditions, such a loop program does not terminate over the reals if and only if its corresponding semi-algebraic system has a real solution.

1 Introduction

Termination analysis of loop programs is very important for software correctness. A standard technique to prove the termination of a loop is to find a ranking function, which maps a program state into an element of some well-founded ordered set, such that the value descends whenever the loop completes an iteration. Several methods for synthesizing polynomial ranking functions have been presented in [1,3–6,9–13,15–17,20]. Additionally, the complexity of the linear ranking function problem for linear-constraint loops is discussed in [2,4,5].

It is well known that the termination of loop programs is undecidable, even for the class of linear programs. In [21], Tiwari proved that the termination of a class of single-path loops with linear guards and assignments is decidable over the reals. The termination of this kind of linear loop program was reconsidered in [18,23]. Braverman [8] generalized the work of Tiwari, and showed that termination of a simple of class linear loops over the integers is decidable. Xia et al. [22] gave the Non-Zero Minimum condition under which the termination problem of loops with linear updates and nonlinear polynomial loop conditions is decidable over the reals. In addition, there are some other methods for determining termination problem of loop programs. For instance, in [7] Bradley et al. applied finite difference trees to prove termination of multipath loops with polynomial guards and assignments. In [19], Zhan et al. analyzed the termination problems for multi-path polynomial programs with equational loop guards and established sufficient conditions for termination and nontermination.

In this paper, we investigate the termination of loop programs of the form

$$P: \textbf{while } C(\mathbf{x}) > 0 \textbf{ do}$$
$$\{F(\mathbf{x}, \mathbf{x}') \geq 0\} \tag{1}$$
$$\textbf{endwhile}$$

© Springer International Publishing AG 2017
K.G. Larsen et al. (Eds.): SETTA 2017, LNCS 10606, pp. 131–146, 2017.
https://doi.org/10.1007/978-3-319-69483-2_8

where $\mathbf{x}, \mathbf{x}' \in \mathbb{R}^n$, $C(\mathbf{x}) = (c_1(\mathbf{x}), ..., c_s(\mathbf{x})) \in (\mathbb{R}[\mathbf{x}])^s$ and $F(\mathbf{x}, \mathbf{x}') = (f_1(\mathbf{x}, \mathbf{x}'),$ $..., f_m(\mathbf{x}, \mathbf{x}')) \in (\mathbb{R}[\mathbf{x}])^m$. And c_i's and f_j's are all homogeneous polynomials. Given a polynomial $h(\mathbf{x})$ in $\mathbf{x}$, we say $h(\mathbf{x})$ is a homogeneous polynomial, if nonzero terms of $h(\mathbf{x})$ all have the same degree. Let $E(\mathbf{x}, \mathbf{x}') = (C(\mathbf{x})^T, F(\mathbf{x}, \mathbf{x}')^T)$. Let $S = \{(\mathbf{x}, \mathbf{x}') \in \mathbb{R}^{2n} : E(\mathbf{x}, \mathbf{x}') \rhd 0\}$ where $\rhd = \{\underbrace{>, \ldots, >}_{s \text{ times}}, \underbrace{\geq, \ldots, \geq}_{m \text{ times}}\}$.

Different from a program whose loop body is an assignment not an inequality, the program defined as above has nondeterministic update statements. Such inequalities in loop body of Program P may arise due to abstraction. We take an example from [20] to illustrate this. Consider the loop **while**$(i-j \geq 1)$**do** $(i, j) :=$ $(i - Nat, j + Pos)$**od**, where i, j are integer variables, Nat and Pos stand for any nonnegative and positive integer number respectively. It is easy to see that the update statements $i := i - Nat, j := j + Pos$ can be expressed by the inequalities $i' \leq i, j' \geq j + 1$. Thus, the above loop can be abstracted as **while**$(i - j \geq 1)$**do** $\{i' \leq i, j' \geq j + 1\}$**od**. More examples about programs with nondeterministic update statements can be found in [1,4–6,13].

For convenience, we say that Program P is defined by $E(\mathbf{x}, \mathbf{x}') \rhd 0$. And let $P \triangleq P(E(\mathbf{x}, \mathbf{x}'))$. Moreover, since $E(\mathbf{x}, \mathbf{x}') \rhd 0$ is a semi-algebraic system and all algebraic expressions of it are homogeneous, Program P is called homogeneous semi-algebraic loop program. The notion of semi-algebraic programs is not new, which first appears in [13]. Especially, if all the expressions in $E(\mathbf{x}, \mathbf{x}') \rhd 0$ are linear, then we call $E(\mathbf{x}, \mathbf{x}') \rhd 0$ the homogeneous linear semi-algebraic system, denoted $E(\mathbf{x}, \mathbf{x}')_{Lin} \rhd 0$, and the program defined by $E(\mathbf{x}, \mathbf{x}')_{Lin} \rhd 0$ is called homogeneous linear semi-algebraic program. At the same time, let $\widetilde{P}$ be the nonhomogeneous semi-algebraic program, i.e., $\widetilde{P}$ contains at least one nonhomogeneous expression. For *nonhomogeneous* linear semi-algebraic program $\widetilde{P}$, its termination has been widely studied in [1,4–6,11,17] by synthesizing linear polynomial ranking functions. For nonhomogeneous nonlinear semi-algebraic program $\widetilde{P}$, [13] presented an algorithm to computing non-linear polynomial ranking functions of $\widetilde{P}$ by semi-definite programming solving. In the above works, the synthesized ranking functions are all polynomial ranking functions. The reason is that polynomials can be dealt with more efficiently and more conveniently in many computer algebra systems. It is well known that the existence of ranking functions of a program exactly implies the program must terminate. However, for a given program, even if it has ranking functions, its ranking functions are not necessarily polynomials. It is easy to construct an example of a program which terminates but has no polynomial ranking functions. For example, consider the loop **while**$(x > 0)$**do** $\{x' = -x; \}$**od**. It is easy to see that the loop is terminating, but has no polynomial ranking functions. (Suppose that there is a polynomial ranking function $\rho(x) = \sum_{i=0}^{m} a_i x^i$ for the loop. By the ranking conditions of ranking functions in Definition 3, we have $\forall x, x > 0 \Rightarrow \rho(x) - \rho(x') = \rho(x) - \rho(-x) = \sum_{i=0}^{m} a_i(x^i - x'^i) = \sum_{i=0}^{m} a_i(x - x')(\sum_{j=0}^{i-1} x^j x'^{i-1-j}) = \sum_{i=0}^{m} a_i(2x)(\sum_{j=0}^{i-1} x^j(-x)^{i-1-j}) \geq 1$. However, the above formula cannot hold, since as $x \to 0$, we have $\rho(x) - \rho(x') \to 0 \ngeq 1$.

That is a contradiction. Therefore, the loop has no polynomial ranking functions.) Therefore, for the homogeneous semi-algebraic program P, the above methods based on synthesizing polynomial ranking functions may fail to check if P terminates. In this paper, we mainly consider the termination of P. This is because, Program $\widetilde{P}$ always can be equivalently converted to a homogeneous program such as P by introducing two additional variables z and z'. This will be further analyzed in Sect. 4.

The rest of the paper is organized as follows. In Sect. 2, we introduce some basic notions regarding on semi-algebraic systems, ranking functions. In Sect. 3, we establish some conditions for Program P such that under such conditions, checking if Program P does not terminate is equivalent to checking if a certain semi-algebraic system has a real solution. In other words, if such semi-algebraic system has a real solution, then Program P is non-terminating over the reals. Otherwise, a ranking function, which indicates Program P is terminating over the reals, can be constructed. In Sect. 4, we show that the termination of a nonhomogeneous semi-algebraic program can always be equivalently reduced to that of a homogeneous one obtained by introducing additional program variables. Section 5 concludes the paper.

2 Preliminaries

In the section, some basic definitions on ranking functions, semi-algebraic systems will be introduced.

Definition 1 *(Nontermination). Given the homogeneous semi-algebraic program P defined as before, we say that Program P is non-terminating over the reals, if there exists an infinite sequence $\{\mathbf{x}_i\}_{i=0}^{+\infty} \subseteq \mathbb{R}^n$ such that $E(\mathbf{x}_i, \mathbf{x}_{i+1}) \triangleright 0$ for any $i \geq 0$. And the first element $\mathbf{x}_0$ in such infinity sequence is called a non-terminating point (or non-terminating input). Especially, we say $\mathbf{x}_0$ is a terminating point (or terminating input), if $\mathbf{x}_0$ is not a non-terminating point.*

Note that the definition of nontermination of Program P can also be equivalently restated as: Program P is non-terminating over the reals, if there exists an infinite sequence $\{(\mathbf{x}_i, \mathbf{x}_{i+1})\}_{i=0}^{+\infty} \subseteq \mathbb{R}^{2n}$ such that $E(\mathbf{x}_i, \mathbf{x}_{i+1}) \triangleright 0$ for any $i \geq 0$. If such the infinity sequence $\{\mathbf{x}_i\}_{i=0}^{+\infty}$ (or $\{(\mathbf{x}_i, \mathbf{x}_{i+1})\}_{i=0}^{+\infty}$) does not exist, then we say Program P is terminating over the reals. Moreover, the definition of nontermination of P can be easily extended to the nonhomogeneous semi-algebraic program $\widetilde{P}$.

Let $\mathbb{R}$ be the field of real numbers. A semi-algebraic system is a set of equations, inequations and inequalities given by polynomials. And the coefficients of those polynomials are all real numbers. Let $\mathbf{v} = (v_1, ..., v_d)^T \in \mathbb{R}^d$, $\mathbf{x} = (x_1, ..., x_n)^T \in \mathbb{R}^n$. Next, we give the definition of semi-algebraic systems (SASs for short).

Definition 2 *(Semi-algebraic systems). A semi-algebraic system is a conjunctive polynomial formula of the form*

$$\begin{cases} p_1(\mathbf{v}, \mathbf{x}) = 0, ..., p_r(\mathbf{v}, \mathbf{x}) = 0, \\ g_1(\mathbf{v}, \mathbf{x}) \geq 0, ..., g_k(\mathbf{v}, \mathbf{x}) \geq 0, \\ g_{k+1}(\mathbf{v}, \mathbf{x}) > 0, ..., g_t(\mathbf{v}, \mathbf{x}) > 0, \\ h_1(\mathbf{v}, \mathbf{x}) \neq 0, ..., h_m(\mathbf{v}, \mathbf{x}) \neq 0, \end{cases} \tag{2}$$

where $r \geq 1$, $t \geq k \geq 0$, $m \geq 0$ and all p_i's, g_i's and h_i's are polynomials in $\mathbb{R}[\mathbf{v}, \mathbf{x}] \setminus \mathbb{R}$. An semi-algebraic system is called parametric if $d \neq 0$, otherwise constant, where d is the dimension of $\mathbf{v}$.

We next recall the definition of ranking functions.

Definition 3 *(Ranking functions). Given Program P defined as before, we say $\rho(\mathbf{x})$ is a ranking function for P, if the following formula is true over the reals,*

$$\forall \mathbf{x}, \mathbf{x}'.\left(E(\mathbf{x}, \mathbf{x}') \triangleright 0 \Rightarrow \rho(\mathbf{x}) \geq 0 \land \rho(\mathbf{x}) - \rho(\mathbf{x}') \geq 1\right). \tag{3}$$

It is well known that the existence of ranking functions for P implies that Program P is terminating. For convenience, the function ρ satisfying Formula (3) is also called a ranking function over the set $S = \{(\mathbf{x}, \mathbf{x}') : E(\mathbf{x}, \mathbf{x}') \triangleright 0\}$.

3 Termination of Homogeneous Semi-algebraic Programs

In the section, we consider the termination of the homogeneous semi-algebraic program P as defined in (1). We will give some conditions under which Program P is nonterminating over the reals if and only if a certain semi-algebraic system has real solutions. Some useful lemmas will be introduced first.

Lemma 1 *[14]. If $\mathcal{H} : \mathbb{R}^k \to \mathbb{R}^d$ is continuous and $\mathfrak{S} \subseteq \mathbb{R}^k$ is bounded and closed, then the image $\mathcal{H}(\mathfrak{S})$ of $\mathfrak{S}$ under the continuous mapping $\mathcal{H}$ is bounded and closed too.*

Lemma 2 *[14]. If $\mathcal{H} : \mathbb{R}^k \to \mathbb{R}^d$ is continuous and $\mathfrak{S} \subseteq \mathbb{R}^k$ is connected, then the image $\mathcal{H}(\mathfrak{S})$ of $\mathfrak{S}$ under the continuous mapping $\mathcal{H}$ is connected too.*

Given the homogeneous semi-algebraic Program $P \triangleq P(E(\mathbf{x}, \mathbf{x}'))$ defined as in (1), let $S = \{(\mathbf{x}, \mathbf{x}') \in \mathbb{R}^{2n} : E(\mathbf{x}, \mathbf{x}') \triangleright 0\}$ and $\bar{S} = \{(\mathbf{x}, \mathbf{x}') \in \mathbb{R}^{2n} : E(\mathbf{x}, \mathbf{x}') \geq 0\}$. Clearly, $S \subseteq \bar{S}$. Since the algebraic expressions in $E(\mathbf{x}, \mathbf{x}')$ are all homogeneous polynomials, the set $\bar{S}$ is a subset of $\mathbb{R}^{2n}$ formed by rays starting from the origin. Therefore, $\bar{S}$ is a closed cone.

Let $\mathbf{U}(\mathbf{x}, \mathbf{x}') = \frac{\mathbf{x}'}{|\mathbf{x}'|} - \frac{\mathbf{x}}{|\mathbf{x}|}$. Let $\mathbf{U}(\bar{S} \setminus \{(0, 0)\}) = \{\mathbf{U}(\mathbf{x}, \mathbf{x}') : (\mathbf{x}, \mathbf{x}') \in \bar{S} \setminus \{(0, 0)\}\} = \{\mathbf{u} \in \mathbb{R}^n : \mathbf{u} = \frac{\mathbf{x}'}{|\mathbf{x}'|} - \frac{\mathbf{x}}{|\mathbf{x}|} \land (\mathbf{x}, \mathbf{x}') \in \bar{S} \setminus \{(0, 0)\}\}$. Define

$$\mathcal{H}_\lambda(\mathbf{x}, \mathbf{x}') = \mathbf{x}' - \lambda \mathbf{x}, \ (\lambda > 0).$$

Clearly, for any fixed positive number λ^*, $\mathcal{H}_{\lambda^*}(\mathbf{x}, \mathbf{x}')$ is a linear mapping from $\mathbb{R}^{2n}$ to $\mathbb{R}^n$. Let $\mathcal{H}_\lambda(\bar{S}) = \{\mathcal{H}_\lambda(\mathbf{x}, \mathbf{x}') : (\mathbf{x}, \mathbf{x}') \in \bar{S}\} = \{\mathbf{u} \in \mathbb{R}^n : \mathbf{u} = \mathbf{x}' - \lambda\mathbf{x} \wedge (\mathbf{x}, \mathbf{x}') \in \bar{S}\}$. and let $\mathcal{T} = \bigcup_{\lambda>0} \mathcal{H}_\lambda(\bar{S})$. It is not difficult to see that

$$\mathcal{T} = \{\mathbf{u} \in \mathbb{R}^n : \mathbf{u} = \mathbf{x}' - \lambda\mathbf{x} \wedge \lambda > 0 \wedge (\mathbf{x}, \mathbf{x}') \in \bar{S}\}.$$

Proposition 1. *With the above notion. Suppose that the function $\mathbf{U}(\mathbf{x}, \mathbf{x}')$ is continuous on $\bar{S} \setminus \{(0, 0)\}$. We have $\mathbf{U}(\bar{S} \setminus \{(0, 0)\}) \subseteq \mathcal{T}$.*

Proof. Because $\mathbf{U}(\mathbf{x}, \mathbf{x}')$ is continuous on $\bar{S}\setminus\{(0, 0)\}$, for any $(\mathbf{x}, \mathbf{x}') \in \bar{S}\setminus\{(0, 0)\}$, we have $|\mathbf{x}'| \neq 0 \wedge |\mathbf{x}| \neq 0$. Take arbitrarily an element $\mathbf{u}$ from $\mathbf{U}(\bar{S} \setminus \{(0, 0)\})$. By the definition of $\mathbf{U}(\bar{S} \setminus \{(0, 0)\})$, there must exist $(\mathbf{x}, \mathbf{x}') \in \bar{S} \setminus \{(0, 0)\}$ such that $\mathbf{u} = \mathbf{U}(\mathbf{x}, \mathbf{x}') = \frac{\mathbf{x}'}{|\mathbf{x}'|} - \frac{\mathbf{x}}{|\mathbf{x}|}$. And

$$\mathbf{u} = \frac{\mathbf{x}'}{|\mathbf{x}'|} - \frac{\mathbf{x}}{|\mathbf{x}|} = \left(\frac{\mathbf{x}'}{|\mathbf{x}'|}\right) - \frac{|\mathbf{x}'|}{|\mathbf{x}|} \cdot \left(\frac{\mathbf{x}}{|\mathbf{x}'|}\right).$$

Set $\lambda^* = \frac{|\mathbf{x}'|}{|\mathbf{x}|} > 0$. It is easy to see that the point $(\frac{\mathbf{x}}{|\mathbf{x}'|}, \frac{\mathbf{x}'}{|\mathbf{x}'|}) \in \bar{S}$, since $\bar{S}$ is a cone whose apex is the origin and the point $(\mathbf{x}, \mathbf{x}') \in \bar{S} \wedge \mathbf{x}' \neq 0 \wedge \mathbf{x} \neq 0$. Therefore, by the definition of $\mathcal{H}_\lambda(\bar{S})$, we get $\mathbf{u} \in \mathcal{H}_{\lambda^*}(\bar{S}) \subseteq \mathcal{T}$, since $(\frac{\mathbf{x}}{|\mathbf{x}'|}, \frac{\mathbf{x}'}{|\mathbf{x}'|}) \in \bar{S}$ and $\lambda^* > 0$. $\qquad\square$

Remark 1. Notice that $\mathcal{T}$ can be obtained by removing the quantifier prefix from the given formula,

$$\exists\mathbf{x}\exists\mathbf{x}'\exists\lambda.(\lambda > 0 \wedge E(\mathbf{x}, \mathbf{x}') \geq 0 \wedge \mathbf{u} = \mathbf{x}' - \lambda\mathbf{x}).$$

Now, let us construct the semi-algebraic system as follows.

$$sysm \triangleq \mathbf{x}' = \lambda\mathbf{x} \wedge \lambda > 0 \wedge \mathbf{x}' \neq 0 \wedge \mathbf{x} \neq 0 \wedge E(\mathbf{x}, \mathbf{x}') \triangleright 0. \qquad (4)$$

Let $T_{\mathbf{x}=0} = \{(\mathbf{x}, \mathbf{x}') \in \mathbb{R}^{2n} : \mathbf{x} = 0, \mathbf{x}' \neq 0\}$ and let $T_{\mathbf{x}'=0} = \{(\mathbf{x}, \mathbf{x}') \in \mathbb{R}^{2n} : \mathbf{x} \neq 0, \mathbf{x}' = 0\}$. Let $T_{\neq} = \{(\mathbf{x}, \mathbf{x}') \in \mathbb{R}^{2n} : \mathbf{x} \neq 0, \mathbf{x}' \neq 0\}$. Obviously, $\mathbb{R}^{2n} = T_{\mathbf{x}=0} \cup T_{\mathbf{x}'=0} \cup T_{\neq} \cup \{(0, 0)\}$. Denote by $\partial(\bar{S})$ the set of all boundary points of $\bar{S}$.

Lemma 3. *Given Program P defined as in (1), let $\odot = \{(\mathbf{x}, \mathbf{x}') \in \mathbb{R}^{2n} : |(\mathbf{x}, \mathbf{x}')| = 1\}$ be the unit sphere. If $\bar{S} \setminus \{(0, 0)\}$ is connected, then $\odot \cap \bar{S} \setminus \{(0, 0)\}$ is a connected subset of $\mathbb{R}^{2n}$.*

Proof. Given any two points $\hat{\mathbf{z}}_0 = (\hat{\mathbf{x}}, \hat{\mathbf{x}}'), \hat{\mathbf{z}}_1 = (\hat{\mathbf{y}}, \hat{\mathbf{y}}') \in \bar{S} \setminus (0, 0) \cap \odot$. Since $\bar{S}$ is a cone of $\mathbb{R}^{2n}$ and $\hat{\mathbf{z}}_0, \hat{\mathbf{z}}_1 \in \bar{S}$, we have $\lambda\hat{\mathbf{z}}_0, \lambda\hat{\mathbf{z}}_1 \subseteq \bar{S}$ for any $\lambda > 0$. That is, there must exist two rays $\ell_{\widehat{o\hat{\mathbf{z}}_0}}, \ell_{\widehat{o\hat{\mathbf{z}}_1}}$ in $\bar{S}$, which start from the origin O and pass through $\hat{\mathbf{z}}_0, \hat{\mathbf{z}}_1$, respectively. Obviously, the rays $\ell_{\widehat{o\hat{\mathbf{z}}_0}}, \ell_{\widehat{o\hat{\mathbf{z}}_1}}$ intersect with the unit sphere $\odot$ at $\hat{\mathbf{z}}_0$ and $\hat{\mathbf{z}}_1$, respectively. Take arbitrarily two nonzero points $\mathbf{z}_0, \mathbf{z}_1$ from the rays $\ell_{\widehat{o\hat{\mathbf{z}}_0}}, \ell_{\widehat{o\hat{\mathbf{z}}_1}}$, respectively. Since $\mathbf{z}_0, \mathbf{z}_1 \in \bar{S} \setminus \{(0, 0)\}$ and $\bar{S} \setminus \{(0, 0)\}$ is connected, there exists a path $L(\mathbf{z}_0, \mathbf{z}_1) \subseteq \bar{S} \setminus (0, 0)$ from $\mathbf{z}_0$ to $\mathbf{z}_1$. Let $\mathbf{z} \in L(\mathbf{z}_0, \mathbf{z}_1)$.

Clearly, $\mathbf{z} \neq (0,0)$ and for any $\mathbf{z} \in L(\mathbf{z}_0, \mathbf{z}_1)$, the ray $\ell_{\overrightarrow{oz}} \subseteq \bar{S}$. Denote by $\hat{\mathbf{z}}$ the intersection point of the ray $\ell_{\overrightarrow{oz}}$ and $\odot$. When $\mathbf{z}$ continuously moves from $\mathbf{z}_0$ to $\mathbf{z}_1$ along the path $L(\mathbf{z}_0, \mathbf{z}_1)$, the ray $\ell_{\overrightarrow{oz}}$ also continuously moves from $\ell_{\overrightarrow{oz_0}}$ to $\ell_{\overrightarrow{oz_1}}$. Obviously, the set of all the intersection points $\hat{\mathbf{z}}$ of the ray $\ell_{\overrightarrow{oz}}$ and $\odot$ forms a path $\hat{L}(\hat{\mathbf{z}}_0, \hat{\mathbf{z}}_1) \subseteq \odot \cap (\bar{S} \setminus \{(0,0)\})$ connecting $\hat{\mathbf{z}}_0$ and $\hat{\mathbf{z}}_1$. Since $\hat{\mathbf{z}}_0, \hat{\mathbf{z}}_1$ are taken arbitrarily from $(\bar{S} \setminus \{(0,0)\}) \cap \odot$, we have for any two points $\hat{\mathbf{z}}_0, \hat{\mathbf{z}}_1 \in \odot \cap (\bar{S} \setminus \{(0,0)\})$, there must exist a path $\hat{L}(\hat{\mathbf{z}}_0, \hat{\mathbf{z}}_1)$ in $\odot \cap (\bar{S} \setminus \{(0,0)\})$, which connects $\hat{\mathbf{z}}_0$ and $\hat{\mathbf{z}}_1$. Hence, $\odot \cap (\bar{S} \setminus \{(0,0)\})$ is connected. $\qquad \square$

Remark 2. This lemma indicates that if the set $\bar{S} \setminus \{(0,0)\}$ is connected, then the intersection of $\bar{S}$ and the unit sphere $\odot$ is also connected. This result will be used in the proof of Theorem 1.

Next, we will give our main result about the termination of Program P.

Theorem 1. *Let Program P be as in (1), and let the following conditions hold:*

(A) $\bar{S} \cap (T_{\mathbf{x}=0} \bigcup T_{\mathbf{x}'=0}) = \emptyset$, *i.e.,* $\bar{S} \subseteq T_{\neq} \cup \{(0,0)\}$,
(B) the formula

$$\mathbf{x}' = \lambda \mathbf{x} \wedge \lambda > 0 \wedge \mathbf{x}' \neq 0 \wedge \mathbf{x} \neq 0 \wedge (\mathbf{x}, \mathbf{x}') \in \partial(\bar{S}) \tag{5}$$

has no real solutions,
(C) $\bar{S} \setminus \{(0,0)\}$ *is a connected set,*
(D) there exist a closed convex polyhedral cone $\mathcal{C} = \{\mathbf{u} \in \mathbb{R}^n : A\mathbf{u} \geq 0\}$ and a hyperplane $\mathcal{L} = \{\mathbf{u} : \mathcal{L}(\mathbf{u}) = \mathbf{a}^T \mathbf{u} = 0\}$, such that $\mathbf{U}(\bar{S} \setminus \{(0,0)\}) \subseteq \mathcal{C}$ and $\mathcal{C} \cap \mathcal{L} = \{0\}$.

Then P is non-terminating over the reals iff (4) has a real solution.

Proof. **(I)** Suppose that the semi-algebraic system (4) has real solutions. That is, there exist $(\hat{\mathbf{x}}, \hat{\mathbf{x}}') \in \mathbb{R}^{2n}$ and $\lambda \in \mathbb{R}$, such that

$$\hat{\mathbf{x}}' = \lambda \hat{\mathbf{x}} \wedge \lambda > 0 \wedge \hat{\mathbf{x}}' \neq 0 \wedge \hat{\mathbf{x}} \neq 0 \wedge E(\hat{\mathbf{x}}, \hat{\mathbf{x}}') \rhd 0$$

Then, we can construct the following infinite sequence,

$$\mathbf{x}_0 = \hat{\mathbf{x}}, \mathbf{x}_1 = \lambda \mathbf{x}_0 = \hat{\mathbf{x}}', \mathbf{x}_2 = \lambda^2 \mathbf{x}_0, ..., \mathbf{x}_n = \lambda^n \mathbf{x}_0, ...$$

It is easy to see that

$$(\mathbf{x}_n, \mathbf{x}_{n+1}) = (\lambda^n \mathbf{x}_0, \lambda^{n+1} \mathbf{x}_0) = \lambda^n (\mathbf{x}_0, \lambda \mathbf{x}_0) = \lambda^n (\hat{\mathbf{x}}, \hat{\mathbf{x}}').$$

It immediately follows that $E(\mathbf{x}_n, \mathbf{x}_{n+1}) \rhd 0$ for any nonnegative integer $n \geq 0$, since $E(\hat{\mathbf{x}}, \hat{\mathbf{x}}') \rhd 0$, $\lambda > 0$ and the polynomials in $E(\mathbf{x}, \mathbf{x}')$ are all homogeneous polynomials in $\mathbf{x}, \mathbf{x}'$. Therefore, by the definition of nontermination of Program P stated in Sect. 1, Program P is non-terminating over the reals and $\mathbf{x} = \hat{\mathbf{x}}$ is a nonterminating point for P.

(II) Let $\pi_{\mathbf{x}} : \mathbb{R}^{2n} \to \mathbb{R}^n$ be a projection mapping. Define $\pi_{\mathbf{x}}(\bar{S}) = \{\mathbf{x} \in \mathbb{R}^n :$ $(\mathbf{x}, \mathbf{x}') \in \bar{S}\}$. The remaining task is to claim that if the four hypotheses (A), (B), (C), (D) of the theorem are all satisfied, then the semi-algebraic system (4) has no real solutions implies that Program P is terminating over the reals. Our approach to prove this is through the construction of a ranking functions over $\bar{S} \setminus \{(0,0)\}$.

First, by the hypothesis (A), we get that $\bar{S} \setminus \{(0,0)\} \subseteq T_{\neq}$. Let $\bar{S}^* = \bar{S} \setminus \{(0,0)\}$. Clearly, $\bar{S}^*$ consists of all the rays with the exclusion of the origin $(0,0)$, lying in the cone $\bar{S}$. Therefore, for any $(\mathbf{x}, \mathbf{x}') \in \bar{S}^*$ and any $\lambda > 0$, we have $\lambda(\mathbf{x}, \mathbf{x}') \in \bar{S}^*$. Furthermore, by the hypothesis (B), we know that (5) has no real solutions is equivalent to the quantified formula below holds,

$$\forall \mathbf{x}, \mathbf{x}', \lambda.(\lambda > 0 \wedge \mathbf{x}' \neq 0 \wedge \mathbf{x} \neq 0 \wedge (\mathbf{x}, \mathbf{x}') \in \partial(\bar{S}) \Rightarrow \mathbf{x}' \neq \lambda\mathbf{x}).$$

This further indicates that the formula

$$\forall \mathbf{x}, \mathbf{x}', \lambda.(\lambda > 0 \wedge (\mathbf{x}, \mathbf{x}') \in \partial(\bar{S}) \setminus \{(0,0)\} \Rightarrow \mathbf{x}' \neq \lambda\mathbf{x}) \tag{6}$$

is true, since $\bar{S} = \{(0,0)\} \cup \bar{S}^*$ and $\bar{S}^* \subseteq T_{\neq}$ by the hypothesis (A). At the same time, as the semi-algebraic system (4) has no real solutions, we have the formula below

$$\forall \mathbf{x}, \mathbf{x}', \lambda.(\lambda > 0 \wedge (\mathbf{x}, \mathbf{x}') \in S \Rightarrow \mathbf{x}' \neq \lambda\mathbf{x}), \tag{7}$$

is true, where $S = \{(\mathbf{x}, \mathbf{x}') \in \mathbb{R}^{2n} : E(\mathbf{x}, \mathbf{x}') \rhd 0\}$. Since $\bar{S}^* \cup \{(0,0)\} = \bar{S} = S \cup \partial\bar{S} = S \cup (\partial(\bar{S}) \setminus \{(0,0)\}) \cup \{(0,0)\}$, it follows that $S \cup (\partial(\bar{S}) \setminus \{(0,0)\}) = \bar{S}^*$. Therefore, by (6) and (7), we get that the following quantified formula

$$\forall \mathbf{x}, \mathbf{x}', \lambda.(\lambda > 0 \wedge (\mathbf{x}, \mathbf{x}') \in \bar{S}^* \Rightarrow \mathbf{x}' \neq \lambda\mathbf{x}) \tag{8}$$

is true. Next, we will construct a ranking function $\rho(\mathbf{x})$ over $\bar{S}^*$ such that $\rho(\mathbf{x})$ satisfies the following formulas

$$\forall \mathbf{x}, \mathbf{x}'.((\mathbf{x}, \mathbf{x}') \in \bar{S}^* \Rightarrow \rho(\mathbf{x}) \geq 0),$$
$$\forall \mathbf{x}, \mathbf{x}'.((\mathbf{x}, \mathbf{x}') \in \bar{S}^* \Rightarrow \rho(\mathbf{x}) - \rho(\mathbf{x}') \geq 1). \tag{9}$$

Let $\mathbf{U}(\mathbf{x}, \mathbf{x}') = \frac{\mathbf{x}'}{|\mathbf{x}'|} - \frac{\mathbf{x}}{|\mathbf{x}|}$. Since $\bar{S}^* \subseteq T_{\neq}$, $\mathbf{U}(\mathbf{x}, \mathbf{x}') : \mathbb{R}^{2n} \to \mathbb{R}^n$ is a continuous mapping over $\bar{S}^*$. Let $\mathbf{U}(\bar{S}^*) = \{\mathbf{u} \in \mathbb{R}^n : \mathbf{u} = \frac{\mathbf{x}'}{|\mathbf{x}'|} - \frac{\mathbf{x}}{|\mathbf{x}|} \wedge (\mathbf{x}, \mathbf{x}') \in \bar{S}^*\}$. By Formula (8), we get $0 \notin \mathbf{U}(\bar{S}^*)$. (This is because, if $0 \in \mathbf{U}(\bar{S}^*)$, then by the definition of $\mathbf{U}(\bar{S}^*)$, there exists $(\mathbf{x}, \mathbf{x}') \in \bar{S}^*$ such that $0 = \frac{\mathbf{x}'}{|\mathbf{x}'|} - \frac{\mathbf{x}}{|\mathbf{x}|}$. This clearly implies that there exists a positive number $\lambda = \frac{|\mathbf{x}'|}{|\mathbf{x}|}$, such that $\mathbf{x}' = \lambda\mathbf{x}$, which contradicts with Formula (8)). Hence, for any $(\mathbf{x}, \mathbf{x}') \in \bar{S}^*$,

$$0 < |\mathbf{U}(\mathbf{x}, \mathbf{x}')| \leq \left|\frac{\mathbf{x}'}{|\mathbf{x}'|}\right| + \left|\frac{\mathbf{x}}{|\mathbf{x}|}\right| \leq 2. \tag{10}$$

Therefore, the image $\mathbf{U}(\bar{S}^*)$ of $\bar{S}^*$ under the mapping $\mathbf{U}(\mathbf{x}, \mathbf{x}')$ is bounded. Additionally, for any $(\mathbf{x}, \mathbf{x}') \in T_{\neq}$ and any $\lambda > 0$, we have $\mathbf{u} = \frac{\mathbf{x}'}{|\mathbf{x}'|} - \frac{\mathbf{x}}{|\mathbf{x}|} = \frac{\lambda\mathbf{x}'}{|\lambda\mathbf{x}'|} - \frac{\lambda\mathbf{x}}{|\lambda\mathbf{x}|}$.

This indicates that the images of any two points $(\mathbf{x}, \mathbf{x}'), (\lambda\mathbf{x}, \lambda\mathbf{x}')$ on each ray of $\bar{S}^*$ under the mapping $\mathbf{U}(\mathbf{x}, \mathbf{x}')$ are the same. In other words, for any two points $(\mathbf{x}, \mathbf{x}'), (\mathbf{y}, \mathbf{y}') \in \bar{S}^*$, if $(\mathbf{y}, \mathbf{y}') = \lambda(\mathbf{x}, \mathbf{x}')$ for a certain $\lambda > 0$, then $\mathbf{U}(\mathbf{x}, \mathbf{x}') = \mathbf{U}(\mathbf{y}, \mathbf{y}')$. This enables us to take $(\mathbf{y}, \mathbf{y}') = \frac{1}{|(\mathbf{x}, \mathbf{x}')|}(\mathbf{x}, \mathbf{x}')$ for any point $(\mathbf{x}, \mathbf{x}') \in \bar{S}^*$, since $\mathbf{U}(\mathbf{y}, \mathbf{y}') \equiv \mathbf{U}(\mathbf{x}, \mathbf{x}')$. It is easy to see that $|(\mathbf{y}, \mathbf{y}')| = 1$, i.e., $(\mathbf{y}, \mathbf{y}') \in \odot$. With the above arguments, for any point $(\mathbf{x}, \mathbf{x}') \in \bar{S}^*$, taking $\lambda = \frac{1}{|(\mathbf{x}, \mathbf{x}')|}$, we have

$$
\begin{aligned}
\mathbf{U}(\bar{S}^*) &= \{\mathbf{u} \in \mathbb{R}^n : \mathbf{u} = \frac{\mathbf{x}'}{|\mathbf{x}'|} - \frac{\mathbf{x}}{|\mathbf{x}|} \wedge (\mathbf{x}, \mathbf{x}') \in \bar{S}^*\} \\
&= \{\mathbf{u} \in \mathbb{R}^n : \mathbf{u} = \frac{\lambda\mathbf{x}'}{|\lambda\mathbf{x}'|} - \frac{\lambda\mathbf{x}}{|\lambda\mathbf{x}|} \wedge (\lambda\mathbf{x}, \lambda\mathbf{x}') \in \bar{S}^* \wedge |\lambda(\mathbf{x}, \mathbf{x}')| = 1\} \\
&= \{\mathbf{u} \in \mathbb{R}^n : \mathbf{u} = \frac{\mathbf{y}'}{|\mathbf{y}'|} - \frac{\mathbf{y}}{|\mathbf{y}|} \wedge (\mathbf{y}, \mathbf{y}') \in \bar{S}^* \wedge |(\mathbf{y}, \mathbf{y}')| = 1\} \\
&= \{\mathbf{u} \in \mathbb{R}^n : \mathbf{u} = \frac{\mathbf{y}'}{|\mathbf{y}'|} - \frac{\mathbf{y}}{|\mathbf{y}|} \wedge (\mathbf{y}, \mathbf{y}') \in \bar{S} \wedge |(\mathbf{y}, \mathbf{y}')| = 1\} \\
&= \mathbf{U}(\bar{S}^* \cap \odot) = \mathbf{U}(\bar{S} \cap \odot).
\end{aligned}
\tag{11}
$$

The last equality is guaranteed by $\bar{S} \cap \odot = (\bar{S}^* \cup \{(0,0)\}) \cap \odot = \bar{S}^* \cap \odot$. Since both $\bar{S}$ and $\odot$ are closed sets, $\bar{S} \cap \odot$ is a closed set. This implies $\bar{S}^* \cap \odot$ is also a closed set. By the hypothesis (C), Lemma 3 and the above statements, we get that $\bar{S}^* \cap \odot$ is a bounded, closed and connected set. Since $\mathbf{U}(\mathbf{x}, \mathbf{x}')$ is continuous over $\bar{S}^*$, according to Lemmas 1 and 2, $\mathbf{U}(\bar{S}^* \cap \odot)$ is also a bounded, closed and connected set, since $\bar{S}^* \cap \odot$ a bounded, closed and connected set. This clearly indicates that $\mathbf{U}(\bar{S}^*)$ is also a bounded, closed and connected set, since $\mathbf{U}(\bar{S}^*) = \mathbf{U}(\bar{S}^* \cap \odot)$ by Formula (11). Because $0 \notin \mathbf{U}(\bar{S}^*)$, we get that $0 \notin \mathbf{U}(\bar{S}^* \cap \odot)$.

By Formula (10), since $0 < |\mathbf{U}(\mathbf{x}, \mathbf{x}')| \leq 2$ for any $(\mathbf{x}, \mathbf{x}') \in \bar{S}^*$, it follows that $0 < |\mathbf{U}(\mathbf{x}, \mathbf{x}')| \leq 2$ for any $(\mathbf{x}, \mathbf{x}') \in \bar{S}^* \cap \odot$. Since $\mathbf{U}(\bar{S}^* \cap \odot)$ is bounded, closed and connected and $|\circ| : \mathbb{R}^n \to \mathbb{R}$ is a continuous mapping, by properties of continuous functions, we get that there exists a positive number c, such that

$$
0 < c \leq |\mathbf{U}(\mathbf{x}, \mathbf{x}')| \leq 2,
\tag{12}
$$

for any $(\mathbf{x}, \mathbf{x}') \in \bar{S}^* \cap \odot$.

By the hypothesis (D), we know that there exists a closed polyhedral cone $\mathcal{C}$ and a hyperplane $\mathcal{L}$ such that $\mathbf{U}(\bar{S}^*) \subseteq \mathcal{C}$ and $\mathcal{C} \cap \mathcal{L} = \{0\}$. Hence, $\mathbf{U}(\bar{S}^* \cap \odot) \subseteq \mathcal{C}$. Since $0 \notin \mathbf{U}(\bar{S}^* \cap \odot)$, $\mathbf{U}(\bar{S}^* \cap \odot) \subseteq \mathcal{C}$ and $\mathcal{C} \cap \mathcal{L} = \{0\}$, we get that $\{\mathbf{u} \in \mathbb{R}^n : \mathcal{L}(\mathbf{u}) = 0\} \cap \mathbf{U}(\bar{S}^* \cap \odot) = \emptyset$. That is, for any $\mathbf{u} \in \mathbf{U}(\bar{S}^* \cap \odot)$, $\mathcal{L}(\mathbf{u}) = \mathbf{a}^T\mathbf{u} \neq 0$. Also, by Formula (11), since $\mathbf{U}(\bar{S}^*) = \mathbf{U}(\bar{S}^* \cap \odot)$, we have $\mathcal{L}(\mathbf{u}) = \mathbf{a}^T\mathbf{u} \neq 0$, for any $\mathbf{u} \in \mathbf{U}(\bar{S}^*)$. Furthermore, since $\mathbf{U}(\bar{S}^*)$ is a bounded, closed and connected set and $\mathcal{L}(\mathbf{u}) = \mathbf{a}^T\mathbf{u} \neq 0$, for any $\mathbf{u} \in \mathbf{U}(\bar{S}^*)$, there will be two cases to consider.

Case 1. For any $\mathbf{u} \in \mathbf{U}(\bar{S}^*)$, $\mathcal{L}(\mathbf{u}) = \mathbf{a}^T\mathbf{u} < 0$.

Case 2. For any $\mathbf{u} \in \mathbf{U}(\bar{S}^*)$, $\mathcal{L}(\mathbf{u}) = \mathbf{a}^T\mathbf{u} > 0$.

(i) Consider **Case 1.** Since $\mathbf{U}(\bar{S}^*)$ is a bounded, closed and connected set and the continuous function $\mathcal{L}(\mathbf{u}) = \mathbf{a}^T\mathbf{u} < 0$ for any $\mathbf{u} \in \mathbf{U}(\bar{S}^*)$, by properties of continuous functions, we have that there exists a positive number $c_1 > 0$, such that $\mathcal{L}(\mathbf{u}) = \mathbf{a}^T \mathbf{u} \le -c_1 < 0$ for any $\mathbf{u} \in \mathbf{U}(\bar{S}^*)$. Therefore, by the definition of $\mathbf{U}(\bar{S}^*)$, we obtain that for any $(\mathbf{x}, \mathbf{x}') \in \bar{S}^*$,

$$\mathbf{a}^T \left(\frac{\mathbf{x}'}{|\mathbf{x}'|} - \frac{\mathbf{x}}{|\mathbf{x}|} \right) \le -c_1 < 0, \tag{13}$$

that is,

$$\frac{\mathbf{a}^T}{c_1} \left(\frac{\mathbf{x}}{|\mathbf{x}|} - \frac{\mathbf{x}'}{|\mathbf{x}'|} \right) \ge 1 > 0, \tag{14}$$

for all $(\mathbf{x}, \mathbf{x}') \in \bar{S}^*$. Let $\mu(\mathbf{x}) = \frac{\mathbf{x}}{|\mathbf{x}|}$ and $\mu(\bar{S}^*) = \{\mathbf{u} \in \mathbb{R}^n : \mathbf{u} = \mu(\mathbf{x}) \wedge (\mathbf{x}, \mathbf{x}') \in \bar{S}^*\}$. Obviously, $\mu(\mathbf{x})$ is a continuous function over $\pi_{\mathbf{x}}(\bar{S}^*)$ (or $\bar{S}^*$) by the hypothesis (A). And for any $(\mathbf{x}, \mathbf{x}') \in \bar{S}^*$, we have $|\mu(\mathbf{x})| \equiv |\mu(\pi_{\mathbf{x}}(\mathbf{x}, \mathbf{x}'))| = 1$. This implies that for any $\mathbf{u} \in \mu(\bar{S}^*)$, $|\mathbf{u}| = 1$. Hence, the set $\mu(\bar{S}^*)$ is bounded. Similar to Formula (11), for any point $(\mathbf{x}, \mathbf{x}') \in \bar{S}^*$, taking $\lambda = \frac{1}{|(\mathbf{x}, \mathbf{x}')|}$ and setting $(\mathbf{y}, \mathbf{y}') = \lambda(\mathbf{x}, \mathbf{x}')$, we get

$$\begin{aligned}
\mu(\bar{S}^*) &= \{\mathbf{u} \in \mathbb{R}^n : \mathbf{u} = \frac{\mathbf{x}}{|\mathbf{x}|} \wedge (\mathbf{x}, \mathbf{x}') \in \bar{S}^*\} \\
&= \{\mathbf{u} \in \mathbb{R}^n : \mathbf{u} = \frac{\lambda \mathbf{x}}{|\lambda \mathbf{x}|} \wedge (\lambda \mathbf{x}, \lambda \mathbf{x}') \in \bar{S}^* \wedge |\lambda(\mathbf{x}, \mathbf{x}')| = 1\} \\
&= \{\mathbf{u} \in \mathbb{R}^n : \mathbf{u} = \frac{\mathbf{y}}{|\mathbf{y}|} \wedge (\mathbf{y}, \mathbf{y}') \in \bar{S}^* \wedge |(\mathbf{y}, \mathbf{y}')| = 1\} \qquad (15) \\
&= \{\mathbf{u} \in \mathbb{R}^n : \mathbf{u} = \frac{\mathbf{y}}{|\mathbf{y}|} \wedge (\mathbf{y}, \mathbf{y}') \in \bar{S} \wedge |(\mathbf{y}, \mathbf{y}')| = 1\} \\
&= \mu(\bar{S}^* \cap \odot) = \mu(\bar{S} \cap \odot).
\end{aligned}$$

Since $\bar{S}^* \cap \odot$ is a bounded and closed set and $\mu(\mathbf{x})$ is continuous on $\bar{S}^*$, $\mu(\bar{S}^*)$ is bounded and closed. Therefore, the linear function $\mathcal{Z}(\mathbf{u}) = \frac{\mathbf{a}^T}{c_1} \mathbf{u}$ has the minimum value over $\mu(\bar{S}^*)$, i.e., there exists $c_2 \in \mathbb{R}$, such that $\mathcal{Z}(\mathbf{u}) = \frac{\mathbf{a}^T}{c_1} \mathbf{u} \ge c_2$ for all $\mathbf{u} \in \mu(\bar{S}^*)$. By the definition of $\mu(\bar{S}^*)$, we obtain that for all $(\mathbf{x}, \mathbf{x}') \in \bar{S}^*$,

$$\frac{\mathbf{a}^T}{c_1} \cdot \frac{\mathbf{x}}{|\mathbf{x}|} - c_2 \ge 0. \tag{16}$$

Let $\rho(\mathbf{x}) = \frac{\mathbf{a}^T}{c_1} \cdot \frac{\mathbf{x}}{|\mathbf{x}|} - c_2$. By the definition of ranking function in Definition 3, and Formula (14)and (16), $\rho(\mathbf{x})$ is a ranking function over $\bar{S}^*$. And it immediately follows that $\rho(\mathbf{x})$ is also a ranking function over S, since $S \subsetneq \bar{S}^*$. This implies that Program P is terminating.

(ii) Consider **Case 2**: for any $\mathbf{u} \in \mathbf{U}(\bar{S}^*)$, $\mathcal{L}(\mathbf{u}) = \mathbf{a}^T \mathbf{u} > 0$. Since $\mathbf{U}(\bar{S}^*)$ is a bounded, closed and connected set and the continuous function $\mathcal{L}(\mathbf{u}) = \mathbf{a}^T \mathbf{u} > 0$ for any $\mathbf{u} \in \mathbf{U}(\bar{S}^*)$, by properties of continuous functions, we have that there exists a positive number $d_1 > 0$, such that $\mathcal{L}(\mathbf{u}) = \mathbf{a}^T \mathbf{u} \ge d_1 > 0$ for any $\mathbf{u} \in \mathbf{U}(\bar{S}^*)$. By the definition of $\mathbf{U}(\bar{S}^*)$, we obtain that for any $(\mathbf{x}, \mathbf{x}') \in \bar{S}^*$,

$$\mathbf{a}^T \left(\frac{\mathbf{x}'}{|\mathbf{x}'|} - \frac{\mathbf{x}}{|\mathbf{x}|} \right) \ge d_1 > 0, \tag{17}$$

that is,

$$\frac{-\mathbf{a}^T}{d_1}\left(\frac{\mathbf{x}}{|\mathbf{x}|} - \frac{\mathbf{x}'}{|\mathbf{x}'|}\right) \geq 1 > 0. \tag{18}$$

Similar to the analysis in **Case 1**, we define $\mu(\mathbf{x}) = \frac{\mathbf{x}}{|\mathbf{x}|}$ and $\mu(\bar{S}^*) = \{\mathbf{u} \in \mathbb{R}^n : \mathbf{u} = \mu(\mathbf{x}) \wedge (\mathbf{x}, \mathbf{x}') \in \bar{S}^*\}$. Let $\mathcal{Z}(\mathbf{u}) = \frac{-\mathbf{a}^T}{d_1}\mathbf{u}$. Since $\bar{S}^*$ is a bounded, closed and connected set and $\mathcal{Z}(\mathbf{u})$ is a continuous function over $\bar{S}^*$, there must exist $d_2 \in \mathbb{R}$, such that $\mathcal{Z}(\mathbf{x}) = \frac{-\mathbf{a}^T}{d_1}\mathbf{u} \geq d_2$ for all $\mathbf{u} \in \mu(\bar{S}^*)$. By the definition of $\mu(\bar{S}^*)$, we have that for all $(\mathbf{x}, \mathbf{x}') \in \bar{S}^*$,

$$\frac{-\mathbf{a}^T}{d_1} \cdot \frac{\mathbf{x}}{|\mathbf{x}|} - d_2 \geq 0. \tag{19}$$

Let $\rho(\mathbf{x}) = \frac{-\mathbf{a}^T}{d_1} \cdot \frac{\mathbf{x}}{|\mathbf{x}|} - d_2$. By the definition of ranking functions, (18) and (19) $\rho(\mathbf{x})$ is a ranking function over $\bar{S}^*$. Also, $\rho(\mathbf{x})$ is a ranking function over S, since $S \subsetneq \bar{S}^*$. This clearly implies that Program P is terminating.

By the arguments presented in (I) and (II), we get that if the four hypotheses (A), (B), (C) and (D) are all satisfied, then the semi-algebraic system (4) has no real solutions implies that Program P is terminating over the reals, since a ranking function $\rho(\mathbf{x})$ over S can always be constructed under such hypotheses.   □

Especially, if $E(\mathbf{x}, \mathbf{x}') \triangleright 0$ is a linear semi-algebraic system, i.e.,

$$E(\mathbf{x}, \mathbf{x}') \triangleright 0 = E_{Lin}(\mathbf{x}, \mathbf{x}') \triangleright 0 = (C(\mathbf{x})^T > 0, F(\mathbf{x}, \mathbf{x}')^T \geq 0)^T,$$
$$C(\mathbf{x}) \triangleq A_C\mathbf{x}, \quad F(\mathbf{x}, \mathbf{x}') \triangleq A_F\mathbf{x} + A'_F\mathbf{x}', \tag{20}$$

where $A_C \in \mathbb{R}^{s \times n}, A_F, A'_F \in \mathbb{R}^{m \times n}$, then the conditions in Theorem 1 can be further relaxed, as follows.

Theorem 2. *Given Program P defined by $E_{Lin}(\mathbf{x}, \mathbf{x}')$, if the following conditions*

(A) $\bar{S} \cap (T_{\mathbf{x}=0} \cup T_{\mathbf{x}'=0}) = \emptyset$,
(B) the semi-algebraic system $\{\mathbf{x}' = \lambda\mathbf{x} \wedge \lambda > 0 \wedge \mathbf{x}' \neq 0 \wedge \mathbf{x} \neq 0 \wedge (\mathbf{x}, \mathbf{x}') \in \partial(\bar{S})\}$ has no real solutions,
(C) there exist a closed convex polyhedral cone $\mathcal{C} = \{\mathbf{u} \in \mathbb{R}^n : A\mathbf{u}^T \geq 0\}$ and a hyperplane $\mathcal{L} = \{\mathbf{u} : \mathcal{L}(\mathbf{u}) = 0\}$, such that $\mathbf{U}(\bar{S} \setminus \{(0, 0)\}) \subseteq \mathcal{C}$ and $\mathcal{C} \cap \mathcal{L} = \{0\}$,

are satisfied, then, Program P is non-terminating over the reals if and only if the semi-algebraic system (4) has real solutions.

Proof. Let $B = \begin{pmatrix} A_C \\ A_F \end{pmatrix}$ and $B' = \begin{pmatrix} 0 \\ A'_F \end{pmatrix}$. Then, $E_{Lin}(\mathbf{x}, \mathbf{x}') \triangleright 0 \triangleq B\mathbf{x} + B'\mathbf{x}' \triangleright 0$. The proof of the theorem is completely similar to that of Theorem 1. We just need to note that when $E(\mathbf{x}, \mathbf{x}') = E_{Lin}(\mathbf{x}, \mathbf{x}'), \bar{S} = \{(\mathbf{x}, \mathbf{x}') \in \mathbb{R}^{2n} : E_{Lin}(\mathbf{x}, \mathbf{x}') \triangleright 0\} =$

$\{(\mathbf{x}, \mathbf{x}') \in \mathbb{R}^{2n} : B\mathbf{x} + B'\mathbf{x}' \rhd 0\}$ is a convex polyhedral cone with vertex at the origin. It is very easy to see that $\bar{S} \backslash \{(0,0)\}$ is still a connected set, since $\bar{S} \backslash \{(0,0)\}$ is a convex set. Therefore, the hypothesis (C) in Theorem 1 is naturally satisfied, when $E(\mathbf{x}, \mathbf{x}') = E_{Lin}(\mathbf{x}, \mathbf{x}')$.   □

We now take the following extremely simple example to illustrate our method.

Example 1. Consider the homogeneous semi-algebraic loop below

$$\textbf{while } x_1 > 0 \wedge x_1 - x_2 > 0 \textbf{ do}$$
$$\{-x_1 + 4x'_2 \geq 0, x'_1 + 4x_2 + 7x_1 - x'_2 \geq 0, x'_1 + x'_2 \geq 0, -x'_1 \geq 0\} \quad (21)$$

Let $E(x_1, x_2, x'_1, x'_2) = (x_1, x_1 - x_2, -x_1 + 4x'_2, x'_1 + 4x_2 + 7x_1 - x'_2, x'_1 + x'_2, -x'_1)$ and $S = \{(x_1, x_2, x'_1, x'_2) \in \mathbb{R}^4 : E(x_1, x_2, x'_1, x'_2) \rhd 0\}$, where $\rhd = (>, >, \geq, \geq, \geq, \geq)$. Since $E(x_1, x_2, x'_1, x'_2) \rhd 0$ is linear semi-algebraic system, we next apply Theorem 2 to determine if the program terminates. Let $\bar{S} = \{(x_1, x_2, x'_1, x'_2) \in \mathbb{R}^4 : E(x_1, x_2, x'_1, x'_2) \geq 0\}$.

Step 1. To check if the hypothesis (A) holds. To check if $\bar{S} \bigcap (T_{\mathbf{x}=0} \bigcup T_{\mathbf{x}'=0}) = \emptyset$ is equivalent to check if $\bar{S} \bigcap T_{\mathbf{x}=0} = \emptyset$ and $\bar{S} \bigcap T_{\mathbf{x}'=0} = \emptyset$. This is equivalent to check if the two semi-algebraic systems $\bar{S}_{\mathbf{x}=0}$ and $\bar{S}_{\mathbf{x}'=0}$ have only zero solution on $\mathbf{x}'$ and $\mathbf{x}$, respectively. Where $\bar{S}_{\mathbf{x}=0}$(resp. $\bar{S}_{\mathbf{x}'=0}$) is obtained by substituting $\mathbf{x} = 0$ (resp. $\mathbf{x}' = 0$) into $\bar{S}$. In the example, $\bar{S}_{\mathbf{x}=0} = \{(x'_1, x'_2) \in \mathbb{R}^2 : 4x'_2 \geq 0, x'_1 - x'_2 \geq 0, x'_1 + x'_2 \geq 0, -x'_1 \geq 0\}$, $\bar{S}_{\mathbf{x}'=0} = \{(x_1, x_2) \in \mathbb{R}^2 : x_1 \geq 0, x_1 - x_2 \geq 0, -x_1 \geq 0, 4x_2 + 7x_1 \geq 0\}$. By verification, we find that $\bar{S}_{\mathbf{x}=0} = \{0\}$ and $\bar{S}_{\mathbf{x}'=0} = \{0\}$. Thus, the hypothesis (A) is satisfied.

Step 2. To check if the hypothesis (B) holds, we may check if the following formula

$$\{\mathbf{x}' = \lambda\mathbf{x} \wedge \lambda > 0 \wedge \mathbf{x}' \neq 0 \wedge \mathbf{x} \neq 0 \wedge (\mathbf{x}, \mathbf{x}') \in \bar{S}\} \quad (22)$$

has no real solutions, since $\partial(\bar{S}) \subseteq \bar{S}$. By the tool RegularChains, we get that Formula (22) indeed has no real solutions. Hence, the hypothesis (B) is met.

Step 3. Check if the hypothesis (C) holds is to checking if there exist a close convex polyhedral cone $\mathcal{C}$ and a hyperplane $\mathcal{L}$, such that $\mathbf{U}(\bar{S}^*) \subseteq \mathcal{C}$ and $\mathcal{C} \cap \mathcal{L} = \{0\}$. To do this, we first predefine the polyhedral $\mathcal{C} \triangleq A\mathbf{u} \geq 0$ and the hyperplane $\mathcal{L} \triangleq \mathbf{a}^T\mathbf{u} = 0$. Where $A = (a_{ij})_{n \times n}$ is a parametric matrix and $\mathbf{a}^T = (a_1, ..., a_n)$ is a parametric vector. Therefore, in the example, to check if the hypothesis (C) holds is equivalent to check if there exist $a_{11}, a_{12}, a_{21}, a_{22}, a_1, a_2$, such that the following quantified formula,

$$\forall\mathbf{x}\forall\mathbf{x}'\forall\mathbf{u}.(\mathbf{u} = \frac{\mathbf{x}'}{|\mathbf{x}'|} - \frac{\mathbf{x}}{|\mathbf{x}|} \wedge (\mathbf{x}, \mathbf{x}') \in \bar{S} \backslash \{(0,0)\} \Longrightarrow A\mathbf{u} \geq 0), \quad (23)$$

holds and the following system

$$\{A\mathbf{u} \geq 0 \wedge \mathbf{a}^T\mathbf{u} = 0\}, \quad (24)$$

has only zero solution. By the hypothesis (A), since $\bar{S}^* = \bar{S} \backslash \{(0,0)\} \subseteq T_{\neq}$, the function $\mathbf{U}(\mathbf{x}, \mathbf{x}')$ is a continuous function on $\bar{S}^*$. Hence, by Proposition 1, we

have $\mathbf{U}(\bar{S}^*) \subseteq \mathcal{T}$. Since $\mathbf{U}(\bar{S}^*) = \{\mathbf{u} \in \mathbb{R}^n : \mathbf{u} = \frac{\mathbf{x}'}{|\mathbf{x}'|} - \frac{\mathbf{x}}{|\mathbf{x}|} \wedge (\mathbf{x}, \mathbf{x}') \in \bar{S} \backslash \{(0, 0)\}\}$
and $\mathcal{T} = \{\mathbf{u} \in \mathbb{R}^n : \mathbf{u} = \mathbf{x}' - \lambda\mathbf{x} \wedge \lambda > 0 \wedge (\mathbf{x}, \mathbf{x}') \in \bar{S}\}$, to check if Formula (23) is true, we may check if the following quantified formula holds,

$$\forall\mathbf{x}\forall\mathbf{x}'\forall\mathbf{u}.(\mathbf{u} = \mathbf{x}' - \lambda\mathbf{x} \wedge \lambda > 0 \wedge (\mathbf{x}, \mathbf{x}') \in \bar{S} \Longrightarrow A\mathbf{u} \geq 0). \tag{25}$$

This is because, if Formula (25) is true, then Formula (23) is also true.

In addition, to decide if Formula (24) has only zero solution is equivalent to decide if the following quantified formula is true,

$$\forall\mathbf{x}\forall\mathbf{x}'\forall\mathbf{u}.(a_{11}u_1 + a_{12}u2 \geq 0 \wedge a_{21}u1 + a_{22}u2 \geq 0 \wedge (u_1 \neq 0 \vee u_2 \neq 0)$$
$$\Longrightarrow a_1u_1 + a_2u_2 \neq 0). \tag{26}$$

Eliminating $\mathbf{x}, \mathbf{x}', \mathbf{u}$ from Formula (25) and (26), we get a semi-algebraic system only on a_{ij}'s, denoted by $\Phi(a_{11}, ..., a_{22}, \mathbf{a}^T)$. Solving $\Phi(a_{11}, ..., a_{22}, \mathbf{a}^T)$, we obtain that $a_{11} = -2, a_{12} = 1, a_{21} = -1, a_{22} = 1, a_1 = -\frac{3}{2}, a_2 = 1$. Therefore, there indeed exist $a_{11} = -2, a_{12} = 1, a_{21} = -1, a_{22} = 1, a_1 = -\frac{3}{2}, a_2 = 1$, such that both Formula (25) and Formula (26) hold. This immediately implies that when $a_{11} = -2, a_{12} = 1, a_{21} = -1, a_{22} = 1, a_1 = -\frac{3}{2}, a_2 = 1$, Formula (23) holds and the system(24) has only zero solution. Therefore, the hypothesis (C) is satisfied. By Theorem 2, the program is non-terminating over the reals if and only if the semi-algebraic system (4) has real solutions. By computing, we find the semi-algebraic system (4) has no real solutions. This clearly indicates that the program is terminating.

4   The Non-homogeneous Case

Theorem 1 indicates that when some conditions are satisfied, the termination of the homogeneous semi-algebraic program P can be equivalently reduced to semi-algebraic systems solving. In the section, we will further show that the termination of a non-homogeneous semi-algebraic program $\tilde{P}$ is equivalent to that of a homogeneous program such as P. Consider the non-homogeneous semi-algebraic program $\tilde{P}$,

$$\tilde{P}: \textbf{ while } \tilde{C}(\mathbf{x}) > 0 \textbf{ do}$$
$$\{\tilde{F}(\mathbf{x}, \mathbf{x}') \geq 0\} \tag{27}$$
$$\textbf{endwhile}$$

where $\mathbf{x}, \mathbf{x}' \in \mathbb{R}^n$, $\tilde{C}(\mathbf{x}) = (\tilde{c}_1(\mathbf{x}), ..., \tilde{c}_s(\mathbf{x})) \in (\mathbb{R}[\mathbf{x}])^s$ and $\tilde{F}(\mathbf{x}, \mathbf{x}') = (\tilde{f}_1(\mathbf{x}, \mathbf{x}'), ..., \tilde{f}_m(\mathbf{x}, \mathbf{x}')) \in (\mathbb{R}[\mathbf{x}])^m$. Let $\tilde{E}(\mathbf{x}, \mathbf{x}') = (\tilde{C}(\mathbf{x})^T, \tilde{F}(\mathbf{x}, \mathbf{x}')^T)$. For $\tilde{P}$, we say that $\tilde{P}$ is non-terminating over the reals, if there exists an infinite sequence $\{\mathbf{x}_i\}_{i=0}^{+\infty} \subseteq \mathbb{R}^n$ such that $\tilde{E}(\mathbf{x}_i, \mathbf{x}_{i+1}) \triangleright 0$ for any $i \geq 0$, where $\triangleright$ is defined as before. More importantly, one always can homogenize Program $\tilde{P}$ by introducing two new program variables z, z' and adding three additional constraints on the z and z' to obtain the following homogeneous semi-algebraic program $P^{\mathbf{H}}$ such as P

$$P^{\mathbf{H}} : \textbf{while}\ \ C(\mathbf{x}, z) > 0 \wedge z > 0\ \ \textbf{do}$$
$$\{F(\mathbf{x}, \mathbf{x}', z) \geq 0, z' \geq z, -z' \geq -z\} \tag{28}$$
$$\textbf{endwhile,}$$

where $C(\mathbf{x}, z) = (c_1(\mathbf{x}, z), ..., c_s(\mathbf{x}, z)))$, $F(\mathbf{x}, \mathbf{x}', z) = (f_1(\mathbf{x}, \mathbf{x}', z), ..., f_m(\mathbf{x}, \mathbf{x}', z))$, $c_i(\mathbf{x}, z)$ and $f_j(\mathbf{x}, \mathbf{x}', z)$ are homogeneous polynomials, for $i = 1, ..., s$, $j = 1, ..., m$.

Example 2. consider the nonhomogeneous semi-algebraic program $\widetilde{Q}$: **while** $(x_1 - x_2^2 - 1 > 0 \wedge x_1^2 + 2x_2 + 1 > 0)\textbf{do}\{x_1' + x_2' - x_1 + 1 \geq 0, x_1^3 + x_2^4 - x_2' - 3 \geq 0\}$. In $\widetilde{Q}$, all the algebraic expressions $x_1 - x_2^2 - 1, x_1^2 + 2x_2 + 1, x_1' + x_2' - x_1 + 1, x_1^3 + x_2^4 - x_2' - 3$ are nonhomogeneous polynomials. By introducing additional variables z, z', $\widetilde{Q}$ can be converted to a homogeneous semi-algebraic program $Q^{\mathbf{H}}$: **while**$(x_1 \cdot z - x_2^2 - 1 \cdot z^2 > 0 \wedge x_1^2 + 2x_2 \cdot z + 1 \cdot z^2 > 0 \wedge z > 0)\textbf{do}\{x_1' + x_2' - x_1 + 1 \cdot z \geq 0, x_1^3 \cdot z + x_2^4 - x_2' \cdot z^3 - 3 \cdot z^4 \geq 0, z' \geq z, -z' \geq -z\}$. According to the following Theorem 3, we have the termination of $\widetilde{Q}$ is equivalent to that of $Q^{\mathbf{H}}$.

Note that the update statements on z, $\{z' \geq z, -z' \geq -z\}$ in (28), is equivalent to $z' = z$. For Program $P^{\mathbf{H}}$, denote by d_{c_i} and d_{f_j} the degrees of $c_i(\mathbf{x}, z)$ and $f_j(\mathbf{x}, \mathbf{x}', z)$. Let $\mathbf{y} = (\mathbf{x}, z)$ and $\mathbf{y}' = (\mathbf{x}', z')$. Let $\rhd_H = (\underbrace{>, ..., >}_{s+1\ \text{times}}, \underbrace{\geq, ..., \geq}_{m+2\ \text{times}})$

and $\rhd_{cf} = \rhd = (\underbrace{>, ..., >}_{s\ \text{times}}, \underbrace{\geq, ..., \geq}_{m\ \text{times}})$.

Define

$$E^{\mathbf{H}}(\mathbf{y}, \mathbf{y}') \rhd_H 0 \triangleq (\overbrace{C(\mathbf{x}, z)^T > 0, z > 0}^{s+1}, \overbrace{F(\mathbf{x}, \mathbf{x}', z)^T \geq 0, z' - z \geq 0, -z' + z \geq 0}^{m+2})^T,$$
$$E_{cf}^{\mathbf{H}}(\mathbf{y}, \mathbf{y}') \rhd_{cf} 0 \triangleq (C(\mathbf{x}, z)^T > 0, F(\mathbf{x}, \mathbf{x}', z)^T \geq 0)^T,$$
$$S^{\mathbf{H}} = \{(\mathbf{y}, \mathbf{y}') \in \mathbb{R}^{2n+2} : E^{\mathbf{H}}(\mathbf{y}, \mathbf{y}') \rhd_H 0\}. \tag{29}$$

Obviously, for any $(\mathbf{y}, \mathbf{y}') \in \mathbb{R}^{2n+2}$,

$$E^{\mathbf{H}}(\mathbf{y}, \mathbf{y}') \rhd_H 0 \Longrightarrow E_{cf}^{\mathbf{H}}(\mathbf{y}, \mathbf{y}') \rhd_{cf} 0. \tag{30}$$

The following theorem indicates that the termination of $\widetilde{P}$ is equivalent to that of $P^{\mathbf{H}}$.

Theorem 3. *The non-homogeneous Program $\widetilde{P}$ is non-terminating over the reals if and only if the homogeneous Program $P^{\mathbf{H}}$ is non-terminating over the reals.*

Proof. If the non-homogenous Program $\widetilde{P}$ does not terminate, say on input $\mathbf{x} = \mathbf{x}^* \in \mathbb{R}^n$, then the homogenous Program $P^{\mathbf{H}}$ does not terminate on input $\mathbf{y} = (\mathbf{x}, z) = (\mathbf{x}^*, 1)$.

For the converse, suppose that the homogenous Program $P^{\mathbf{H}}$ does not terminate on input $\mathbf{y} = \mathbf{y}^* = (\mathbf{x}^*, z^*)$. That is, there exists a infinite sequence

$$\{\mathbf{y}_i^*\}_{i=0}^{+\infty} = \{\mathbf{y}_0^* = (\mathbf{x}_0^*, z_0^*) = \mathbf{y}^* = (\mathbf{x}^*, z^*), \mathbf{y}_1^*, \mathbf{y}_2^*, ..., \mathbf{y}_i^* = (\mathbf{x}_i^*, z_i^*), ...\} \subseteq S^{\mathbf{H}},$$

which is produced by the input $\mathbf{y}^*$, such that

$$E^{\mathbf{H}}(\mathbf{y}_i^*, \mathbf{y}_{i+1}^*) \triangleright_H 0 \qquad (31)$$

for $i = 0, 1, ...$. Clearly, for all $i = 0, 1, ...$, we have $\pi_z(\mathbf{y}_i^*) = z^* = z_0^*$, since $\{z' \geq z, -z' \geq -z\}$ in (28) is equivalent to $z' = z$, where π_z is a projection mapping from $\mathbb{R}^{n+1}$ to $\mathbb{R}$. Therefore, Formula (31) is equivalent to

$$C(\mathbf{x}_i^*, z_i^*) > 0 \wedge z_i^* > 0 \wedge F(\mathbf{x}_i^*, \mathbf{x}_{i+1}^*, z_i^*) \geq 0 \wedge z_{i+1}^* = z_i^* = z^*, \qquad (32)$$

for all $i = 0, 1, ...$. Set $\mathbf{y}^{**} = (\frac{\mathbf{x}^*}{z^*}, 1)$. We next show that $P^{\mathbf{H}}$ does not terminate on $\mathbf{y} = \mathbf{y}^{**}$. Since $c_i(\mathbf{x}, z)$'s and $f_j(\mathbf{x}, \mathbf{x}', z)$'s are all homogeneous polynomials in $\mathbf{x}, \mathbf{x}', z$, we have

$$\begin{pmatrix} (z^*)^{d_{c_1}} \cdot c_1(\frac{\mathbf{x}_i^*}{z^*}, 1) > 0 \\ \vdots \\ (z^*)^{d_{c_s}} \cdot c_s(\frac{\mathbf{x}_i^*}{z^*}, 1) > 0 \end{pmatrix} \wedge \frac{z_i^*}{z^*} = 1 > 0 \wedge \begin{pmatrix} (z^*)^{d_{f_1}} \cdot F_1(\frac{\mathbf{x}_i^*}{z^*}, \frac{\mathbf{x}_{i+1}^*}{z^*}, 1) \geq 0 \\ \vdots \\ (z^*)^{d_{f_m}} \cdot F_m(\frac{\mathbf{x}_i^*}{z^*}, \frac{\mathbf{x}_{i+1}^*}{z^*}, 1) \geq 0 \end{pmatrix} \qquad (33)$$

for all $i = 0, 1, ...$. Since $z^* > 0$, we further get that

$$C\left(\frac{\mathbf{x}_i^*}{z^*}, 1\right) > 0 \wedge \frac{z_i^*}{z^*} = 1 > 0 \wedge F\left(\frac{\mathbf{x}_i^*}{z^*}, \frac{\mathbf{x}_{i+1}^*}{z^*}, 1\right) \geq 0 \qquad (34)$$

for all $i = 0, 1, ...$. Setting $\hat{\mathbf{y}}_i = (\hat{\mathbf{x}}_i, \hat{z}_i) = (\frac{\mathbf{x}_i^*}{z^*}, 1)$, we obtain that

$$E^{\mathbf{H}}(\hat{\mathbf{y}}_i, \hat{\mathbf{y}}_{i+1}) \triangleright_H 0,$$

for all $i = 0, 1, 2, ...$. Hence, Program $P^{\mathbf{H}}$ does not terminate on $\mathbf{y}^{**} = \hat{\mathbf{y}}_0 = (\frac{\mathbf{x}_0^*}{z^*}, 1) = (\frac{\mathbf{x}^*}{z^*}, 1)$. In other words, we can construct a infinite sequence

$$\{\hat{\mathbf{y}}_0 = \mathbf{y}^{**}, \hat{\mathbf{y}}_1, \hat{\mathbf{y}}_2, ..., \hat{\mathbf{y}}_i = (\frac{\mathbf{x}_i^*}{z^*}, 1), ..., \}$$

such that

$$E^{\mathbf{H}}(\hat{\mathbf{y}}_i, \hat{\mathbf{y}}_{i+1}) = E^{\mathbf{H}}\left(\frac{\mathbf{x}_i^*}{z^*}, 1, \frac{\mathbf{x}_{i+1}^*}{z^*}, 1\right) \triangleright_H 0$$

for any $i \geq 0$. Thus, for any $i \geq 0$, by Formula (30) we get

$$E_{cf}^{\mathbf{H}}(\hat{\mathbf{y}}_i, \hat{\mathbf{y}}_{i+1}) = E_{cf}^{\mathbf{H}}(\frac{\mathbf{x}_i^*}{z^*}, 1, \frac{\mathbf{x}_{i+1}^*}{z^*}, 1) \triangleright_{cf} 0.$$

This immediately implies that

$$\widetilde{E}\left(\frac{\mathbf{x}_i^*}{z^*}, \frac{\mathbf{x}_{i+1}^*}{z^*}\right) \triangleright 0$$

for any $i \geq 0$, since $\widetilde{E}(\mathbf{x}, \mathbf{x}') \equiv E_{cf}^{\mathbf{H}}(\mathbf{x}, 1, \mathbf{x}', 1)$ and $\triangleright = \triangleright_{cf}$. Therefore, Program $\widetilde{P}$ does not terminate on $\mathbf{x} = \frac{\mathbf{x}_0^*}{z^*} = \frac{\mathbf{x}^*}{z^*}$, according to the definition of nontermination of $\widetilde{P}$. □

5   Conclusion

We have analyzed the termination of a class of semi-algebraic loop programs. Some conditions are given such that under such conditions the termination problem of this kind of loop programs over the reals can be equivalently reduced to the satisfiability of a certain semi-algebraic system, i.e., when such conditions are satisfied, a semi-algebraic loop program is not terminating over the reals if and only if the constructed semi-algebraic system has a real solution. Especially, for the nonhomogeneous semi-algebraic program $\widetilde{P}$, one always can convert $\widetilde{P}$ to its homogeneous version $P^{\mathbf{H}}$ by introducing two auxiliary variables z, z' and adding three additional constraints $\{z > 0, z' \geq z, -z' \geq -z\}$ to the body of loop. The above constraints can also be rewritten as $\{z > 0, z' = z\}$. This particular constraints cause that the only λ satisfying the semi-algebraic system $sysm_{P^{\mathbf{H}}}$, which is associated with $P^{\mathbf{H}}$ and is similar to Formula (4), is 1,

$$sysm_{P^{\mathbf{H}}} \triangleq \mathbf{y}' = \lambda\mathbf{y} \wedge \lambda > 0 \wedge \mathbf{y}' \neq 0 \wedge \mathbf{y} \neq 0 \wedge E^{\mathbf{H}}(\mathbf{y}, \mathbf{y}') \triangleright_{\mathbf{H}} 0, \qquad (35)$$

where $\mathbf{y} = (\mathbf{x}, z), \mathbf{y}' = (\mathbf{x}', z')$. Therefore, since $\lambda = 1$, for each solution $(\mathbf{y}, \mathbf{y}')$ of Formula (35), we have $\mathbf{y} = \mathbf{y}'$.

Acknowledgments. The author would like to thank the anonymous reviewers for their helpful suggestions. This research is partially supported by the National Natural Science Foundation of China NNSFC (61572024, 61103110).

References

1. Bagnara, R., Mesnard, F.: Eventual linear ranking functions. In: Proceedings of the 15th Symposium on Principles and Practice of Declarative Programming. Madrid, Spain, pp. 229–238. ACM (2013)
2. Bagnara, R., Mesnard, F., Pescetti, A., Zaffanella, E.: A new look at the automatic synthesis of linear ranking functions. Inf. Comput. **215**, 47–67 (2012)
3. Ben-Amram, A.M., Genaim, S.: On multiphase-linear ranking functions. In: Majumdar, R., Kunčak, V. (eds.) CAV 2017. LNCS, vol. 10427, pp. 601–620. Springer, Cham (2017). doi:10.1007/978-3-319-63390-9_32
4. Ben-Amram, A., Genaim, S.: On the linear ranking problem for integer linear-constraint loops. In: Proceedings of the 40th Annual ACM SIGPLAN-SIGACT Symposium on Principles of Programming Languages (POPL 2013), Rome, Italy, pp. 51–62. ACM (2013)
5. Ben-Amram, A., Genaim, S.: Ranking functions for linear-constraint loops. J. ACM **61**(4), 1–55 (2014)
6. Bradley, A.R., Manna, Z., Sipma, H.B.: Linear ranking with reachability. In: Etessami, K., Rajamani, S.K. (eds.) CAV 2005. LNCS, vol. 3576, pp. 491–504. Springer, Heidelberg (2005). doi:10.1007/11513988_48
7. Bradley, A.R., Manna, Z., Sipma, H.B.: Termination of polynomial programs. In: Cousot, R. (ed.) VMCAI 2005. LNCS, vol. 3385, pp. 113–129. Springer, Heidelberg (2005). doi:10.1007/978-3-540-30579-8_8
8. Braverman, M.: Termination of integer linear programs. In: Ball, T., Jones, R.B. (eds.) CAV 2006. LNCS, vol. 4144, pp. 372–385. Springer, Heidelberg (2006). doi:10.1007/11817963_34

9. Chen, H.Y., Flur, S., Mukhopadhyay, S.: Termination proofs for linear simple loops. In: Miné, A., Schmidt, D. (eds.) SAS 2012. LNCS, vol. 7460, pp. 422–438. Springer, Heidelberg (2012). doi:10.1007/978-3-642-33125-1_28

10. Chen, Y., Xia, B., Yang, L., Zhan, N., Zhou, C.: Discovering non-linear ranking functions by solving semi-algebraic systems. In: Jones, C.B., Liu, Z., Woodcock, J. (eds.) ICTAC 2007. LNCS, vol. 4711, pp. 34–49. Springer, Heidelberg (2007). doi:10.1007/978-3-540-75292-9_3

11. Colóon, M.A., Sipma, H.B.: Synthesis of linear ranking functions. In: Margaria, T., Yi, W. (eds.) TACAS 2001. LNCS, vol. 2031, pp. 67–81. Springer, Heidelberg (2001). doi:10.1007/3-540-45319-9_6

12. Cook, B., See, A., Zuleger, F.: Ramsey vs. lexicographic termination proving. In: Piterman, N., Smolka, S.A. (eds.) TACAS 2013. LNCS, vol. 7795, pp. 47–61. Springer, Heidelberg (2013). doi:10.1007/978-3-642-36742-7_4

13. Cousot, P.: Proving program invariance and termination by parametric abstraction, lagrangian relaxation and semidefinite programming. In: Cousot, R. (ed.) VMCAI 2005. LNCS, vol. 3385, pp. 1–24. Springer, Heidelberg (2005). doi:10. 1007/978-3-540-30579-8_1

14. Duistermaat, J., Kolk, J.: Multidimensional Real Analysis. Cambridge University Press, Cambridge (2004)

15. Heizmann, M., Hoenicke, J., Leike, J., Podelski, A.: Linear ranking for linear lasso programs. In: Hung, D., Ogawa, M. (eds.) ATVA 2013. LNCS, vol. 8172, pp. 365–380. Springer, Cham (2013). doi:10.1007/978-3-319-02444-8_26

16. Leike, J., Heizmann, M.: Ranking templates for linear loops. In: Ábrahám, E., Havelund, K. (eds.) TACAS 2014. LNCS, vol. 8413, pp. 172–186. Springer, Heidelberg (2014). doi:10.1007/978-3-642-54862-8_12

17. Li, Y., Zhu, G., Feng, Y.: The L-depth eventual linear ranking functions for single-path linear constraints loops. In: 10th International Symposium on Theoretical Aspects of Software Engineering (TASE 2016), pp. 30–37. IEEE (2016)

18. Li, Y.: Witness to non-termination of linear programs. Theor. Comput. Sci. **681**, 75–100 (2017)

19. Liu, J., Xu, M., Zhan, N.J., Zhao, H.J.: Discovering non-terminating inputs for multi-path polynomial programs. J. Syst. Sci. Complex. **27**, 1284–1304 (2014)

20. Podelski, A., Rybalchenko, A.: A complete method for the synthesis of linear ranking functions. In: Steffen, B., Levi, G. (eds.) VMCAI 2004. LNCS, vol. 2937, pp. 239–251. Springer, Heidelberg (2004). doi:10.1007/978-3-540-24622-0_20

21. Tiwari, A.: Termination of linear programs. In: Alur, R., Peled, D.A. (eds.) CAV 2004. LNCS, vol. 3114, pp. 70–82. Springer, Heidelberg (2004). doi:10.1007/ 978-3-540-27813-9_6

22. Xia, B., Zhang, Z.: Termination of linear programs with nonlinear constraints. J. Symb. Comput. **45**(11), 1234–1249 (2010)

23. Xia, B., Yang, L., Zhan, N., Zhang, Z.: Symbolic decision procedure for termination of linear programs. Formal Aspects Comput. **23**(2), 171–190 (2011)

Computing Exact Loop Bounds for Bounded Program Verification

Tianhai Liu[1]([✉]), Shmuel Tyszberowicz[2,3], Bernhard Beckert[1],
and Mana Taghdiri[4]

[1] Karlsruhe Institute of Technology, Karlsruhe, Germany
liu@ira.uka.de
[2] RISE, Southwest University, Chongqing, China
[3] The Academic College Tel Aviv Yaffo, Tel Aviv, Israel
[4] Horus Software GmbH, Ettlingen, Germany

Abstract. Bounded program verification techniques verify functional properties of programs by analyzing the program for user-provided bounds on the number of objects and loop iterations. Whereas those two kinds of bounds are related, existing bounded program verification tools treat them as independent parameters and require the user to provide them. We present a new approach for automatically calculating *exact* loop bounds, i.e., the greatest lower bound and the least upper bound, based on the number of objects. This ensures that the verification is complete with respect to all the configurations of objects on the heap and thus enhances the confidence in the correctness of the analyzed program. We compute the loop bounds by encoding the program and its specification as a logical formula, and solve it using an SMT solver. We performed experiments to evaluate the precision of our approach in loop bounds computation.

1 Introduction

Bounded program verification techniques (e.g. [7,15,24]) verify functional properties of object-oriented programs, where loops are unrolled and the number of objects for each class is bounded. These techniques typically encode the program and the property of interest into a logical formula and check the satisfiability of the formula by invoking an SMT solver. They provide an attractive trade-off between automation and completeness. They automatically exhaustively analyze a program based on the user-provided bounds[1] and thus guarantee to find any bug (with respect to the analyzed property) within the bounds, but defects outside bounds may be missed. As a result, bounded program verification has becomes an increasingly attractive choice for gaining confidence in the correctness of software.

Existing bounded program verification techniques typically require the user to provide two kinds of bounds as separate parameters: (1) the *loop bounds* that

---

[1] They analyze a program based on *both* bounds—objects and loop iterations. Thus, not all object space within bounds is necessarily explored (as explain in what follows).

© Springer International Publishing AG 2017　　　•
K.G. Larsen et al. (Eds.): SETTA 2017, LNCS 10606, pp. 147–163, 2017.
https://doi.org/10.1007/978-3-319-69483-2_9

limit the number of iterations for each loop, and (2) the *class bounds* that limit the number of objects for each class. (The class bound for a primitive type, e.g., the Java `int`, is the size of integer bit-width.) These two kinds of bounds, however, are not independent and have to be chosen carefully; the class bounds can affect the number of loop iterations, and the loop bounds can influence the size of object space to be explored. To clarify, consider the following loop that traverses an acyclic singly-linked list in Java, starting from the header entry `l.head`: `e=l.head; while(e!=null){ e=e.next;}`.

Supposing the loop bound is 2, we unroll the loop twice and then add an `assume` clause which fails if it iterates more than 2 times. The code will be: `e=l.head; if(e!=null){ e=e.next; if(e!=null){ e=e.next;assume(e==null);}}`.

When the list has at most one element (implied, e.g., by the provided class bound), the second `if`-condition always evaluates to `false` and thus the following code is unreachable. On the other hand, if the list contains at least 5 elements (implied, e.g., by the specification or by the rest of the code), the encoding formula of the code evaluates to `false`, since the `assume` statement will never evaluate to `true`. Furthermore, when the list contains up to 5 elements, only the lists with at most 2 elements are analyzed and the other elements in the list are not treated.

When a loop is unrolled too many times (i.e., more than the least upper bound on the number of loop iterations), the unrolled program has many unreachable paths which may impede the performance of the underlying solver. The verification process may fail due to the solver being overloaded. When a loop is unrolled too few times (i.e., fewer than the greatest lower bound on the number of loop iterations), none of the program executions that reach the loop will be valid in the unrolled program, thus any property concerning the loop will vacuously hold in all runs. This is also the case for *infinite* loops; we consider a loop infinite when it does not terminate for any input. Selecting a loop bound that lies strictly between the greatest lower- and the least upper-bound causes the analysis to be incomplete (i.e., it explores only a part of object space).

Several approaches have been developed to compute loop *upper* bounds (e.g. [4,10,18,22]), and many of them (e.g. [4,10,18]) do not require a specific bound for objects; they compute loop upper bounds as functions, based on the input sizes. However, none of those approaches can handle arbitrary configuration of objects on the heap. They either focus only on primitive types [4,18,22] or support only particular configuration of objects [10]. Furthermore, none of those approaches considers specifications in computing loop bounds, and many of them compute a valid upper bound, which is not necessarily the *least* upper bound.

Incremental bounded model checkers, e.g. NBIS [11], also can be used to compute loop upper bounds. Starting from an initial number, they unroll a loop and check whether the loop condition still holds after the last unrolled iteration. If so, a new upper bound candidate is found and checked again iteratively. This approach, however, is imprecise in the presence of class bounds and specifications. It may compute upper bounds that are higher than the least upper bound, thus many unreachable paths arise in the unrolled program, and the verification may fail. To overcome this potential failure, the user may restart verification with smaller class bounds. Thus the confidence in the correctness of the code is reduced, as the number of relevant objects is smaller. Moreover, this approach does not compute bounds when the loops are *non-terminating*. A loop is considered non-terminating when it does not terminate for at least one input while it may terminate for other inputs. This is inconsistent with bounded program verification tool, which do analyze the terminating executions of a method and ignore non-terminating runs.

We present an approach that is meant to be used as a pre-processing phase in bounded program verification. It focuses on data-structure-rich programs and can handle arbitrary configurations for the objects in the heap. Given both a program method m selected for analysis and class bounds b, we compute both the *greatest* lower bound and the *least* upper bound for each loop that is reachable from m. Our approach, therefore, can provide the user with an insight on what loop bounds to consider in bounded program verification, and to enhance the confidence for program correctness with respect to the class bounds. When a method specification exists, we consider it as well. In addition to numerical bounds, we also output a pre-state (and an execution trace) that witnesses each computed bound, which guarantees that the computed bounds are feasible. We produce loop bounds even for a non-terminating loop, provided that the loop has at least one terminating execution; recall our definition of non-terminating. This is consistent with bounded program verification approaches. Besides, we can detect unreachable loops by analyzing the results of bound computation.

We compute loop lower- and upper-bounds for Java programs annotated in a subset of JML (Java Modeling Language) [13]. We translate the code, its precondition, and its sub-routines' specifications (if they exist) into a first-order formula, encoding the loops' effects as recursive uninterpreted functions. The resulting formula is solved for the *exact* greatest lower and least upper bounds for each loop. This is achieved by calling an SMT (satisfiability modulo theories) solver that is able to solve optimization problems. Given a formula f and an optimization objective o, the SMT solver finds a model for f to achieve the goal o. Several off-the-shelf SMT solvers have been extended to solve optimization problems, e.g., Z3 [19] with νZ [3] and SYMBA [14], MathSAT5 [5] with OptiMathSAT [21]. Besides, some SMT-based algorithms for solving optimization problems have been developed, e.g., the authors of [17] integrated an SMT solver with a classical incremental solving algorithm to solve generic optimization problems, and an algorithm in [20] aims to solve linear arithmetic problems. Our approach takes advantage of these recent advances in SMT solvers. Our target logic is undecidable, i.e., it is possible for the underlying solver to output '*unknown*'. However, for the small class bounds generally used in bounded program verification, the solver returns a definite answer.

We have implemented our approach and NBIS' approach in prototype tools *BoundJ* and *IncUnroll*, respectively. We compared the computed bounds using these tools. Our experiments reveal that in all cases *BoundJ* has computed precise loop bounds, while *IncUnroll* does not. On the other hand, *IncUnroll* can produces loop bounds with the increased class bounds, while *BoundJ* returns '*unknown*' for large class bounds.

2 Logical Formalism

We focus on analyzing object-oriented programs, and currently support a basic subset of Java—excluding floating-point numbers, strings, generics, and concurrency. We support class hierarchy without interfaces and abstract classes. Any method which is called by the analyzed method and has no specification is inlined into its call sites; otherwise it is replaced by its specification. Specifications are written in JML [13], and should not include exceptional behaviors and model fields. The constructs `requires` and `ensures` define, respectively, a method's pre- and postcondition. We support arbitrarily nested universal and existential quantifiers, and allow using the JML reachability construct.

We translate the given code and its specifications into a first-order SMT logic that consists of quantified bit-vectors, unbounded integers, and uninterpreted functions. We now describe our target logic. For this we use the SMT-LIB 2.0 syntax [2], in which expressions are given in a prefix notation. The command (`declare-fun f` (A_1 .. A_{n-1}) A_n) declares a function $f : A_1 \times .. \times A_{n-1} \to A_n$. Constants are functions that take no arguments. The command (`assert F`) asserts a formula F in the current logical context; multiple assert statements are assumed to be implicitly conjoined. The (`push`) command pushes an empty assertion set onto the assertion-set stack. and (`pop`) pops the top assertion set from the stack. The command (`check-sat`) triggers solving a conjunction of formulas. The operator `ite` denotes a ternary if-then-else expression. Basic formulas are combined using the boolean operators `and`, `or`, `not`, and `=>` (implies). Universal and existential quantifiers are denoted by the keywords `forall` and `exists`. The Z3 solver contains two extensions of SMT-LIB to express optimization objectives. The command (`maximize t`) instructs the solver to produce a model that maximizes the value of the integer term `t` and to return the assignment for `t` in the solution, if one exists. The (`minimize t`) command finds the smallest value of `t`.

Our translation uses the fixed-size bit-vectors theory, in which sorts are of the form (`_BitVec m`), where m is a non-negative integer denoting the size of the bit-vector. Bit-vectors of different sizes represent different sorts in SMT. This theory models the precise semantics of unsigned and of signed two-complements arithmetic, supporting a large number of logical and arithmetic operations on bit vectors. The translation also uses the unbounded integers theory, which contains only one sort `Int`, corresponding to integer numbers. It supports arithmetic operations to be applied to numerical constants or variables.

3 Motivating Example

We use two examples to illustrate that we compute precise loop bounds. Figure 1 shows a Java program for a check-in process in a youth hostel. The check-in

```
1  int guest = 0;
2  //@requires (\forall int i; 0<=i&&i<ages.length; 0<ages[i]&&ages[i]<=18);
3  void checkin(int[] ages) {
4      for (int i = 0; i < ages.length; i++) {
5          if (ages[i] <= 27) guest++;}
6      openRoomFor();}
7  //@ ensures 3<=\old(guest) && \old(guest)<=10;
8  native void openRoomFor();
```

Fig. 1. A simple program for youth hostel checkin. The method `openRoomFor`'s post-condition at line 7 constrains the range of values of `guest`.

process requires that the guests will be young (the precondition of the method `checkin`), and the group size is neither greater than 10 nor less than 3 (the postcondition[2] of the method `openRoomFor`). Carefully inspecting the code, we notice that the branch condition at line 5 is implied `true` by the precondition of the method `checkin`. Thus the number of loop (line 4) iterations equals to the number of young guests (the field `guest` when the loop terminates). Suppose the Java `int` bit-width is 6 (i.e., integer numbers ranging from -32 to 31). We evaluate the loop lower- and upper bounds to 3 and 10, respectively. However, using an approach that computes loop bounds by incrementally unrolling loops (Sect. 1), the loop upper bound is 31, since it does not consider the postcondition of the method `openRoomFor` when evaluating the loop condition. The incremental loop unrolling approach provides no lower bounds computation.

Figure 2 illustrates a Java implementation of a `copy` method for a singly-linked list of `Data` entries. Given an instance d of `Data`, the `copy` method deeply copies the receiver list, starting from the first occurrence (exclusive) of d. If d does not exist, nothing is copied. Bounded program verification techniques analyze programs with respect to specific bounds on the number of objects. Assume that the maximum number of objects of type `List`, `Entry`, and `Data` is 2, 26, and 1, respectively. The bounds for each loop are computed separately. For instance, when the upper bound of the second loop (*Loop2* at line 9) is computed, no specific bound for the preceding loop (*Loop1* at line 4) is assumed. Our technique computes the loop upper bounds considering all the cases in which the loops terminate. Loop1 may not terminate for some inputs, e.g., when the receiver list is a cyclic one. For all inputs for which Loop1 terminates our technique outputs 26 as the upper bound and generates a witness in which an acyclic list contains 26 entries, where the last one is followed by `null`. For all inputs for which Loop2 terminates, our technique outputs 12 as Loop2's upper bound and generates a witness where an acyclic list has 13 entries, where the first one has `Data` d. That makes sense, because the `copy` method deep copies an acyclic linked list, and fresh entries one-to-one correspond (excluding the entry containing d) to the entries in the receiver list. Thus Loop2 can allocate at most

---

[2] We intentionally refer to the postcondition rather than to a precondition that limits the number of guests in order to demonstrate our approach.

```
1   class List { Entry head;
2     List copy(Data d) {
3       Entry curr = head;
4       while (curr != null && curr.data != d) curr = curr.next; // Loop 1
5       List result = new List();
6       if (curr != null) {
7         curr = curr.next;
8         Entry last = null;
9         while (curr != null) {   // Loop 2
10          Entry e = new Entry(curr.data);
11          if (last == null) { result.head = e; }
12          else { last.next = e; }
13          last = e;
14          curr = curr.next;
15        }
16      }
17      return result;}}
18  class Entry {/*@ nullable */ Entry next; /*@ nullable */ Data data;
19      Entry(Data d) { data = d; next = null;}}
20  class Data {}
```

Fig. 2. A Java program to perform a deep copy on a linked list.

12 fresh objects, and a total of 25 ($=12*2+1$) **Entry** objects are used. If Loop2 iterates 13 times, e.g., a total 27 objects are needed, which is larger than the bound (26) on **Entry**.

As shown, the number of loop iterations heavily depends on both the specifications and the number of objects in the analyzed domain. Furthermore, the number of iterations of different loops is inter-dependent. Detecting those dependencies manually and computing the precise loop bounds can be prohibitively difficult, thus an automatic approach for computing exact loop bounds (in the presence of specifications and class bounds) can significantly enhance the bounded program verification engineers' confidence in the correctness of the analyzed programs.

4   Our Approach

Given a method p selected for analysis from a piece of code and a set of class bounds b, our technique computes for each loop l two numbers, GLB_l and LUB_l. They respectively denote the *greatest lower bound* and the *least upper bound* on the number of iterations of l, since we ensure that no valid execution of p can iterate the loop l less times than GLB_l or more times than LUB_l. In order to analyze only valid executions, we consider the whole code when computing the bounds for a loop l. For each computed bound the output also contains a witnessing pre-state and an execution trace.

We translates the given method p and its constraints (specifications) c (consisting of p's precondition and the annotations of the methods reachable from p)

into an SMT formula, based on a set of user-provided bounds b on the number of instances of the classes. Let T denote this translation; then $T[p, c, b]$ produces a tuple (s, f, N_l), where s is the pre-state of p, f is an SMT formula that encodes the control flow and dataflow of p and the additional constraints b and c, and N_l represents the number of times the loop condition has been checked for loop l. We distinguish various loops inside p using loop ids.

The LUB_l is computed by delegating a formula of the form $f \land exit(l, N_l) \land maximize(N_l)$ to an SMT solver that provides the functions to solve optimization problems. The function $exit$ means that the loop l exits after checking the loop condition N_l times. The $maximize$ command instructs the solver to find a model where N_l is the biggest compared to the values in other models. A satisfying solution to this formula represents a terminating execution of p in which the loop l is reachable when running p and the number of iterations of l is $N_l - 1$ ($N_l > 0$), and unreachable in case that $N_l = 0$. When the formula is unsatisfiable, it means that either the user-provided class bounds are too small or the methods are over-specified (e.g., the precondition of the analyzed method is `false`). The GLB_l is computed similarly to least upper bound computation, using the command $minimize$ instead of $maximize$.

Unbounded integers are used in our translation to encode loop iterations. Since (1) a loop may not terminate, and (2) even for terminating loops, the number of iterations is not known and thus cannot be bounded apriori. Hence, our target logic is undecidable, and it is possible for the solver to output 'unknown'. In such a case, our analysis terminates with no conclusive outcome.

4.1   Encoding Control Flow

We encode the control flow of the analyzed method using a *computation graph* [24]. Each node in this graph represents a control point in the program, and each edge represents either a program statement or a branch condition. There are exactly one node to entry the graph and one node to exit from the graph. If a loop in the analyzed method is triggered multiple times, due to either method invocations or that it is an inner loop, then multiple occurrences of the loop exist in the computation graph. We compute the loop bounds for each loop occurrence. This is consistent with bounded program verification.

Bounded program verification tools such as Jalloy [24] and InspectJ [15] also use computation graphs to encode control flow. However, they unroll loops and thus assume that the graph is acyclic. In that case, control flow can be encoded by simple boolean variables. Our approach, on the other hand, preserves loops as cycles in the graph and encodes their (cyclic) control flow via uninterpreted functions in the SMT logic. More precisely, similarly to previous approaches, we encode an edge that does not belong to any loop from node m to node n, using a boolean variable E_{mn}, whose truth value denotes whether the edge is traversed or not. When an edge belongs to a loop, the encoding must clarify in which loop iterations the edge is traversed. Therefore, when an edge from m to n belongs to loop l, we encode it using a boolean-valued, uninterpreted function $E_{mn} : Int^{>0} \rightarrow Bool$ ($Int^{>0}$ denotes positive integers). The expression $E_{mn}(i)$

evaluates to true if the edge is traversed in the ith iteration of l. The exit edge of l is traversed only once the loop condition is not fulfilled for the $(N_l)^{th}$ iteration of the loop. We encode the exit edge of a loop l as the expression $E_{mn}(N_l)$, where $N_l > 0$ and $N_l = 1 + K$, where K is the number of iterations of the loop l.

We use the term *entry edge* (*exit edge*) to denote an edge that leads to the entry node (exits from the exit node) of a loop but does not belong to that loop. We use the term *head edge* (*tail edge*) to denote the first (the last) edge of a loop. The control flow for the computation of the loop bounds is encoded using the following four general rules. (1) The first edge of the computation graph must be traversed. (2) If an edge E_{mn} is traversed, at least one of the outgoing edges of node n must be traversed (dataflow constraints prevent more than one outgoing edge from being traversed). If node n belongs to a loop l, the iteration index must be considered. In particular, if n is the loop's head node, then (3) if a head edge is traversed, either the first iteration starts or the loop exits before the first iteration, and (4) if a tail edge at the ith iteration of the loop is traversed, then either the $(i + 1)^{th}$ iteration starts or the loop exits before this iteration.

Figure 3 provides an example. Figure 3(a) shows a Java method that sets the field data of all list elements to the input value, provided that this value is not null. Figure 3(b) gives the corresponding computation graph. The edge labels denote the statements and branch conditions in a special SSA-like format as described in Sect. 4.2. Figure 3(c) presents our encoding of the control flow. In this example, E_0_1, E_1_2, E_2_3, E_2_8, E_3_4, E_4_5, and E_8_9 are boolean variables, while E_5_6, E_6_7, E_7_5, and E_5_8 are boolean-valued functions. The edge E_4_5 is an entry edge, E_5_6 is a head edge, E_7_5 is a tail edge, and E_5_8 is an exit edge. The numbers preceding the constraints correspond to the four encoding rules presented above. For each loop l in the computation graph, we introduce an integer variable N_l to represent the number of times that the loop condition has been checked; e.g., N_{L0} and N_{L1} in Fig. 3(b) and N_L_1 (that encodes N_{L1}) in Fig. 3(c).

4.2   Encoding Dataflow

We now provide an overview of our encoding of Java statements, which is based on the InspectJ approach [15]. We focus on how loops affect the encoding. The types that are accessed in the analyzed code are encoded using bit-vectors in the SMT logic. That is, if a Java type T is bounded by the user-provided number n, we encode T as a bit-vector of size $\lceil log(n + 1) \rceil$ (including the *null* value). In the following description, we use $BV[T]$ to represent the bit-vector of a Java type T.

In an acyclic computation graph, all variables and fields of the program are renamed so that they are assigned at most once along each path of the graph. Since our computation graphs can be cyclic, renaming cannot be achieved by enumerating all paths. We rename variables and fields of the program assuming that each loop constructs a separate naming context (similar to a called method, e.g.). This separates the naming of variables (fields) in one loop from the others,

```
class Entry { Entry next;
   Data data;
   boolean assign(Data d){
      boolean r = false;
      if(d != null){
         r = true;
         Entry e = this;
         while(e != null){
            e.data = d;
            e = e.next;
         }
      }
      return r;}}
class Data {}
```

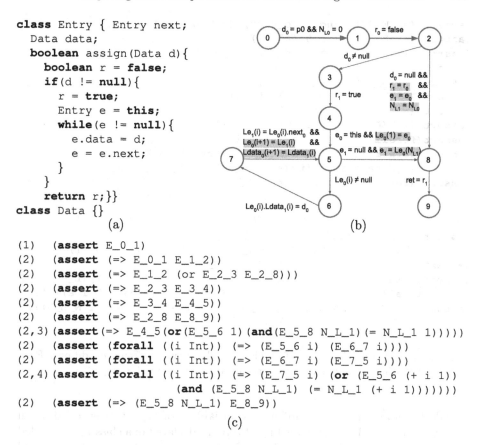

(a)   (b)

(1)   (**assert** E_0_1)
(2)   (**assert** (=> E_0_1 E_1_2))
(2)   (**assert** (=> E_1_2 (or E_2_3 E_2_8)))
(2)   (**assert** (=> E_2_3 E_3_4))
(2)   (**assert** (=> E_3_4 E_4_5))
(2)   (**assert** (=> E_2_8 E_8_9))
(2,3) (**assert** (=> E_4_5 (or (E_5_6 1) (**and** (E_5_8 N_L_1) (= N_L_1 1)))))
(2)   (**assert** (**forall** ((i Int)) (=> (E_5_6 i) (E_6_7 i))))
(2)   (**assert** (**forall** ((i Int)) (=> (E_6_7 i) (E_7_5 i))))
(2,4) (**assert** (**forall** ((i Int)) (=> (E_7_5 i) (**or** (E_5_6 (+ i 1))
                          (**and** (E_5_8 N_L_1) (= N_L_1 (+ i 1)))))))
(2)   (**assert** (=> (E_5_8 N_L_1) E_8_9))

(c)

Fig. 3. (a) sample Java code, (b) its computation graph, (c) our control flow encoding. The character 'L' in (b) denotes loop id. The variable N_{L0} represents the number of times that the loop condition has been checked. After exiting the loop it is renamed to N_{L1}, and is encoded in (c) as the SMT variable N_L_1.

which makes it easier to support complex loop structures. More precisely, renaming variables (fields) involves the following steps. (1) Starting from the innermost loop l, we give any variable (field) that may be updated by l an initial name, and then perform renaming within the body of l as for an acyclic computation graph. (2) We collapse the cycle (loop) l of the computation graph into a single node, denoting the initial and the final names of the variables updated in l. (3) We repeat step 1, considering the collapsed loops. Hence, any time a collapsing node is visited, adequate conditions are produced to ensure that the variables (fields) of the current context hold the same values as the initial/final variables (fields) of the collapsed loop. In the example in Fig. 3(b), d_0, N_{L0}, N_{L1}, e_0, e_1, r_0, r_1, and $data_0$ belong to the outer context, whereas Le_0, Le_1, $Ldata_0$, and $Ldata_1$ belong to the loop context. Since the loop does not update the $next_0$ field and the constants, e.g., $this$, $p0$ and ret, both contexts share them. Data accesses

```
(assert (=> E_0_1 (and (= d_0 p0) (= N_L_0 0))))
(assert (=> E_1_2 (= r_0 false)))
(assert (=> E_2_3 (not (= d_0 null_Data))))
(assert (=> E_3_4 (= r_1 true)))
(assert (=> E_4_5 (= e_0 this)))
(assert (=> E_2_8 (= d_0 null_Data)))
(assert (=> E_8_9 (= ret r_1)))
(assert (=> (E_5_8 N_L_1) (= e_1 null_Entry)))
(assert(forall((i Int)) (=>(E_5_6 i) (not(=(L_e_0 i)null_Entry)))))
(assert(forall((i Int)) (=>(E_6_7 i) (forall((e Entry))
      (=(L_data_1 i e) (ite(= e (L_e_0 i))d_0(L_data_0 i e)))))))
(assert (forall ((i Int)) (=> (E_7_5 i)
                           (= (L_e_1 i) (next_0 (L_e_0 i))))))
```

(a)

```
(assert(=> E_4_5 (and (= (L_e_0 1) e_0)
                      (= (L_data_0 1) e_0) (data_0 e_0))))
(assert (=> (E_5_8 N_L_1) (= (L_e_0 N_L_1) e_1)))
(assert(forall((i Int)) (=>(E_7_5 i) (and(=(L_e_0(+ i 1))(L_e_1 i))
      (forall((e Entry)) (=(L_data_0(+ i 1) e)(L_data_1 i e)))))))
(assert (=> E_2_8 (and (= r_1 r_0) (= e_1 e_0) (= N_L_1 N_L_0))))
```

(b)

Fig. 4. Dataflow Encoding SMT formulas: (a) dataflow (b) frame conditions. Each class has a distinct null value, e.g., the null_Data and the null_Entry.

outside loops are encoded as follows: A variable v of type T is encoded as an SMT variable $v : BV[T]$, and a field f of type T_2 declared in a class T_1 is encoded as a function $f : BV[T_1] \to BV[T_2]$. However, if a variable or field is updated within a loop, one needs to know the updates performed in each loop iteration. A variable v_l of type T that may be modified within a loop l is encoded as a function $v_l : Int^{>0} \to BV[T]$, where $v_l(i)$ denotes the value of v in the ith iteration of l. Similarly, a field f of type T_2 declared in a class of type T_1, that may be modified within a loop l, is encoded as a function $f_l : Int^{>0} \times BV[T_1] \to BV[T_2]$, $f_l(i, o)$ denotes the value of $o.f$ in the ith iteration of the loop. Figure 4(a) shows the dataflow formulas for Fig. 3(b). The first 8 formulas correspond to the edges outside the loop. The last 3 encode the dataflow in each loop iteration.

Frame Conditions. Frame conditions are used to avoid underspecification of nodes with multiple incoming edges, and to ensure the correctness of the dataflow. The highlighted expressions in Fig. 3(b) are the frame conditions for the corresponding dataflow expressions. Lets take the merge node 8 for example. Since the variables r, e, and N_L are updated (and thus renamed) only in the path $2 \to 3 \to 4 \to 5 \to 8$, frame conditions for these variables are required when the path $2 \to 8$ is taken; the last formula in Fig. 4(b) is the relevant frame condition. Special frame conditions are required when the merge node is a loop's head node, e.g., node 5 in Fig. 3(b). Before the first iteration, $Le_0(1)$ equals e_0 and $Ldata_0(1)$ equals $data_0$ (encoded as the first formula in Fig. 4(b)). After the last iteration, e_1 equals $Le_0(N_l)$ (encoded as the second formula in Fig. 4(b)). Furthermore, in each new iteration $i+1$, $Le_0(i+1)$ equals $Le_1(i)$ and $Ldata_0(i+1)$ equals $Ldata_1(i)$ (encoded as the third formula in Fig. 4(b)). It should be noted that the frame condition for N_l ensures that the variable N_L_1 equals to the number of times that the loop condition has been checked. Without the frame condition we may get wrong loop bounds, due to traversing spurious paths.

Nested Loops. If a loop l_2 is nested in a loop l_1, the iterations of l_2 depend on the iterations of l_1. Therefore, we encode those variables (fields) that are updated in the inner loop l_2 by adding an additional iteration column to the SMT functions that represent those variables (fields). That is, if a variable v^{l_2} of type T is modified within l_2, we declare an SMT function $v^{l_2} : Int^{>0} \times Int^{>0} \rightarrow BV[T]$, where $v^{l_2}(i_1, i_2)$ denotes the value of v^{l_2} in the $(i_2)^{th}$ iteration of l_2 while in the $(i_1)^{th}$ iteration of l_1. Updated fields are encoded in a similar way by adding an additional column. Moreover, the edge variables encoding the control flow of l_2 will also get an additional column corresponding to the iteration number of the outer loop. It works the same way for any depth of nesting.

Encoding Specifications. Computing loop bounds does not require any user-provided specifications or annotations; the user only provides bounds on the number of elements of each type. However, if the precondition of the analyzed method or the method contracts of the called methods are provided, our analysis will take them into account. That is, if the user provides method contracts for an invoked method, they will be used to substitute any call to the method. Otherwise, the method body will be inlined in its call sites. More details can be found in our previous work [15].

4.3   Computing Loop Bounds

In order to compute the bounds of a loop l, we constrain l to terminate, that is, its exit edge to be traversed when its loop condition has been checked N_l times, if l is reachable from the analyzed method. We also trigger the solver to find the model where N_l has the maximal assignment. We solve the conjunction of all the control flow, dataflow, frame conditions, and specification formulas. If this formula is satisfiable, N_l is assigned a value in the satisfying solution, where $N_l > 0$ denotes the loop l is reachable and its bound is N_l, $N_l = 0$ denotes l is not reachable. If the formula is unsatisfiable, then either the user-provided class bounds are not large enough or the user-provided specifications are not consistent by themselves. The following formulas give the SMT commands that computes the least upper bound for the example of Fig. 3.

```
(push) (assert (=> (> N_L_1 0) (E_5_8 N_L_1))))
(maximize N_L_1) (check-sat) (get-model) (pop)
```

To compute a loop lower bound, we just replace the *maximize* command by *minimize*. It is possible for the solver to output 'unknown' because our logic is undecidable, and then our analysis terminates with no conclusive outcome.

5   Experiments

Our prototype tool (*BoundJ*) uses: the Jimple 3-address intermediate representation provided by the Soot optimization framework [23] to preprocess Java program code; the Common JML Tools package (ISU) [13] to preprocess JML specifications; and Z3 version 4.4.2 [19] as the underlying SMT solver. We also have considered the approach used in NBIS [11]to evaluate the precision of our approach. Since NBIS targets C and C++ code, and does not consider specifications or class bounds, we implemented that approach in a prototype tool (*IncUnroll*) that targets Java and accepts the same inputs that *BoundJ* does. We report on a collection of benchmarks, selected from InspectJ [15] (a bounded program verification system), KeY [1] (a program verification system), JDK (Java Development Kit), and TPDB (Termination Programs Data Base) [25]. All the experiments[3] have been performed on an Intel Core 2.50 GHz with 4 GB of RAM using Linux 64bit.

Table 1. Results of computing loop bounds using *BoundJ* and *IncUnroll*.

Method	Scope Size	GLB (*BoundJ*)	LUB (*BoundJ*)	LUB (*IncUnroll*)
BinaryHeap deleteMin	3	X	X	3
	4	1	1	4
	5	1	1	5
	6	1	2	6
KeYList. removeDup	4	5, 0	5, 0	7, 1
	5	5, 0	6, 0	15, 2
	6	5, 0	7, 1	31, 3
	7	5, 0	8, 2	63, 4
	8	5, 0	9, 3	127, 5
OurList. copy	3	0, 0	3, 1	?, ?
	4	0, 0	4, 1	?, ?
	5	0, 0	5, 2	?, ?
	6	0, 0	6, 2	?, ?
JDKList. add	3	1, 1	1, 2	1, 2
	4	1, 1	3, 4	3, 4
	5	1, 1	7, 8	7, 8
	7	1, 1	21, ?	21, 32
	10	1, 1	?, ?	175, 256
NonTerm. fibonacci	3	0	9	?
	4	0	10	?
	5	0	10	?
	6	0	10	?
NonTerm. gause	6	0	63	63
	7	0	?	127
	9	0	?	508

The evaluation results are shown in Table 1. The *Method* column shows the names of the entry methods of the analyzed programs. There are in total 6 pro-

---

[3] The complete benchmarks can be found at http://asa.iti.kit.edu/478.php.

grams have been analyzed. To increase the complexity of the specifications, we also added special method contracts for the sub-routines (if exist). The method *deleteMin* of the class *BinaryHeap* (that calls another 3 methods, 109 LOC) effectively extracts the minimum element in a *min heap*[4] and restores the properties of min heap. The *removeDups* method of the class *KeYList* (3 methods, 33 LOC) removes the duplicate elements from a queue. The *add* method (2 methods, 39 LOC) is classical implementation in JDK 1.7. All those methods have complex preconditions, i.e., quantifiers have been involved. The methods *deleteMin* and *add* also use JML reachability expressions to constrain the heap configurations. The *copy* method is the method presented in Sect. 3. In addition to these data-structure-based benchmarks, we have also used benchmarks that involve only primitive types. Such benchmarks are typical for the loop bound computation and the non-termination detection communities. They (the methods `fibnacci` and `gause`) do not have any specification and there are around 10 LOC in average in each method, however, we have selected these benchmarks for the following two reasons: (i) The number of iterations for many of these loops is non-linearly distributed. Therefore, computing their loop bounds is particularly challenging in many existing approaches. (ii) To validate that our approach indeed computes loop bounds for the methods that contain at least one terminating path. For each analyzed program, we have exploited each tool to compute ~ 4 loop upper bounds for different *class bounds*, and in total 25 loop upper bounds computations have been done using each tool. Besides, we also used *BoundJ* to compute the loop lower bounds. Thus in total we have done 75 computations.

The *Scope Size* column shows the bounds on the number of objects of each class and on the size of the integer bit-width. For a scope size n, the analyses of *deleteMin*, *removeDups*, and *copy* methods have to explore data spaces of size $(n + 1)^{11} * 2^{9n}$, $(n + 1)^9 * 2^n$, and $(n + 1)^5 * 2^n$, respectively. (The numbers are calculated based on the number of accessed classes, fields, and parameters. Computations are skipped for space reasons.) The columns *GLB* and *LUB* represent the computed lower and upper bounds, respectively. When a method contains more than one loop, the bounds are shown as a sequence of numbers separated by commas. The symbol X denotes that the loop is not reachable from the method. Question marks (?) mean that no definite answer is achieved after the timeout limit of 20 min.

In Table 1 we observe that: (i) *BoundJ* computed *exact* loop lower-/upper-bounds for all data structure-rich methods. A careful inspection of the code reveals that all computed loop lower- and upper bounds are greatest and least, respectively. (ii) *IncUnroll* does not always compute precise loop bounds as *BoundJ* does. Since *IncUnroll* does not consider the whole code in loop bounds computation, on average its computed loop upper bounds are 4.2 times (median 4, maximum 13) greater than the ones *BoundJ* computed. In addition, *IncUnroll* failed to compute the loop upper bounds for the non-terminating methods `copy` and `fibonacci` because of timeout, while *BoundJ* still produced the loop

---

[4] A min heap is a binary heap where the values that are stored in the children nodes are greater than the value stored in the parent node.

upper bounds for the methods since it considers all program executions that terminate. (iii) Nevertheless, *IncUnroll* can compute loop upper bounds with increased class bounds, while *BoundJ* timeouts for 2 (of 25) cases. According to the *small scope hypothesis* [12], bounded program verification systems aim to analyze the program with respect to a small scope. Furthermore, *BoundJ* always produces the *exact* loop bounds based on the user-provided class bounds, thus it guarantees any bounded program verification is complete for the given class bounds and enhances the confidence in the correctness of the analyzed program.

6 Related Work

Various techniques have been developed to compute loop bounds in real-time systems. To estimate loop bounds, they either require annotations [8] or perform a numerical analysis to achieve a numerical interval of loop upper bounds [6]. To achieve better analysis performance and more precise loop bounds, the technique described in [16] employs a combination of abstract interpretation, inter-procedural program slicing, and inter-procedural dataflow analysis. Unlike our technique that requires explicit user-provided class bounds, the techniques described in [4,9,22] generate symbolic bounds (functions) in terms of loop inputs. All these approaches, however, focus on numerical loops; some of them (e.g. [4,6]) even require well-structured loops with no branches inside them. Our approach can work on arbitrary loops and target data structure-rich programs.

To compute bounds for complex loops in C++ code, SPEED [10] generates computational complexity bound functions that contain well-implemented abstract data structures. For loops that access data structures, it generates symbolic bound expressions in terms of numerical properties of the data structures and user-defined quantitative functions. Generating symbolic bounds depends on the generation of loop invariants.

SPEED, however, requires user-provided specifications, does not support complex heap configurations (e.g., transitive reachability) in the precondition of the analyzed method, and in some cases outputs only an approximate loop bound functions.

The incremental bounded model checker NBIS [11] can be used to compute loop upper bounds. It instruments every loop in a given C/C++ code with an unrolling assertion that checks whether the loop can iterate beyond the current loop bound. The encoding of the code along with the unrolling assertions is checked for satisfiability. A satisfiable instance denotes an execution trace in which a loop iterates more times than its current bound. That loop is then unrolled according to the newly-found bound and the process starts over. However, such an approach has the following three drawbacks. (1) It does not terminate in case of non-terminating loops. (2) It does not always return the least upper bound, because the trace corresponding to a satisfying instance stops the execution when exiting the loop with a new bound, hence the code following the loop is ignored.[5] Preventing the trace from stopping is not possible, since it requires an encoding that does not depend on the loops being unrolled (since the needed number of unrolling is unknown prior to invoking the satisfiability procedure),[6] (3) This approach reduces the necessary confidence in the correctness of the analyzed code. The over-approximated loop upper bounds result in many unreachable paths after loop unrolling. Hence, the formulas that are translated from the unreachable paths may overload the underlying solver and cause the verification process to fail. It might be due to the computed upper bound of the loop is too high or the user-provided class bounds are too big. When the verification engineer uses smaller class bounds than those in the previous run to re-compute the loop upper bounds, the new loop upper bounds still may be too high, since the computation ignores the code and specifications following the loop under consideration. If the engineer arbitrarily selects smaller loop upper bounds, not all inputs concerning the class bounds are completely analyzed, thus the correctness of the code is not guaranteed for the class bounds. Consequently, although this iterative approach is successful for terminating loops in the absence of class bounds and specifications, it is not applicable in the context of bounded program verification since the class bounds are necessary and the confidence in the correctness of the code is not guaranteed for the class bounds.

7   Conclusion

We have presented an approach for computing *exact* loop bounds of a given loop for bounded inputs. Such an analysis is particularly useful for bounded program verification in which the user has to provide bounds on both the size of objects and the number of loop unrollings. Our approach provides the user with insight on what loop bounds to consider and enhances the confidence in the correctness of the analyzed programs. We focus on data-structure-rich programs and support arbitrary configurations of the objects on a heap. We translate the Java code and its JML specifications (excluding the postconditions of the entry method) into an SMT formula and solve it using an SMT solver that can solve optimization problems. We compared our approach with another one that incrementally unrolls a loop and checks whether the loop condition still holds after the last unrolled iteration. Experiments show that our method indeed produces the exact bounds, whereas the other method computes lower and upper

---

[5] For instance, when each iteration of the loop allocates one instance of class A and the code after the loop allocates 2 objects of type A. If $bound(A) = 5$, no valid execution of the code (with respect to class bounds) can iterate the loop more than 3 times, whereas the computed upper bound will be 5 when ignoring the code after the loop.

[6] An alternative checking whether the trace is valid with respect to the class bounds by executing (symbolically or dynamically) the whole code. Invalidity of the current instance, however, does not necessarily mean that the newly-found loop bound is impossible; it may still be that another satisfying instance can be valid and gives a higher loop bound. Thus, in the worst case, such a validity check requires enumerating all possible satisfying instances, which makes the approach impractical.

bounds, which not necessarily are the exact ones. Our approach can assist the bounded program verification engineers to obtain complete confidence in the verification process. Although our analysis is not guaranteed to produce a definite outcome (due to the undecidability of the target logic), our experiments show that in practice the unknown outcome occurs for higher input bounds and not for the small bounds that are typically used in bounded program verification.

Bounded verification techniques unroll not only loops but also recursive methods. Currently we handle loops only; computing bounds for recursion is left for future work. To improve scalability, we will optimize the encoding formulas using quantify elimination techniques, e.g., symmetry breaking and pattern matching. We will also study the application of our approach to other areas, e.g., worst-case execution time and dynamic heap consumption program analysis.

References

1. Ahrendt, W., Beckert, B., Bubel, R., Hähnle, R., Schmitt, P.H., Ulbrich, M. (eds.): Deductive Software Verification – The KeY Book: From Theory to Practice. LNCS, vol. 10001. Springer, Heidelberg (2016). doi:10.1007/978-3-319-49812-6
2. Barrett, C., Fontaine, P., Tinelli, C.: The SMT-LIB standard: Version 2.5. Technical report, The University of Iowa (2015)
3. Bjørner, N., Phan, A.-D., Fleckenstein, L.: vZ - an optimizing SMT solver. In: Baier, C., Tinelli, C. (eds.) TACAS 2015. LNCS, vol. 9035, pp. 194–199. Springer, Heidelberg (2015). doi:10.1007/978-3-662-46681-0_14
4. Blanc, R., Henzinger, T.A., Hottelier, T., Kovács, L.: ABC: algebraic bound computation for loops. In: Clarke, E.M., Voronkov, A. (eds.) LPAR 2010. LNCS, vol. 6355, pp. 103–118. Springer, Heidelberg (2010). doi:10.1007/978-3-642-17511-4_7
5. Cimatti, A., Griggio, A., Schaafsma, B.J., Sebastiani, R.: The MathSAT5 SMT solver. In: Piterman, N., Smolka, S.A. (eds.) TACAS 2013. LNCS, vol. 7795, pp. 93–107. Springer, Heidelberg (2013). doi:10.1007/978-3-642-36742-7_7
6. Cullmann, C., Martin, F.: Data-flow based detection of loop bounds. In: WCET. OASICS, vol. 6. Schloss Dagstuhl (2007)
7. Dennis, G.D.: A Relational Framework for Bounded Program Verification. Ph.D. thesis, MIT (2009)
8. Gulavani, B.S., Gulwani, S.: A numerical abstract domain based on *expression abstraction* and *max operator* with application in timing analysis. In: Gupta, A., Malik, S. (eds.) CAV 2008. LNCS, vol. 5123, pp. 370–384. Springer, Heidelberg (2008). doi:10.1007/978-3-540-70545-1_35
9. Gulwani, S., Jain, S., Koskinen, E.: Control-flow refinement and progress invariants for bound analysis. In: PLDI, pp. 375–385. ACM (2009)
10. Gulwani, S., Mehra, K.K., Chilimbi, T.M.: SPEED: precise and efficient static estimation of program computational complexity. In: POPL, pp. 127–139. ACM (2009)
11. Günther, H., Weissenbacher, G.: Incremental bounded software model checking. In: SPIN, pp. 40–47. ACM (2014)
12. Jackson, D.: Software Abstractions: Logic, Language and Analysis. MIT, Cambridge (2016)
13. Leavens, G.T., Baker, A.L., Ruby, C.: Preliminary design of JML: a behavioral interface specification language for java. ACM SIGSOFT Softw. Eng. Notes **31**(3), 1–38 (2006)

14. Li, Y., Albarghouthi, A., Kincaid, Z., Gurfinkel, A., Chechik, M.: Symbolic optimization with SMT solvers. In: POPL, pp. 607–618. ACM (2014)
15. Liu, T., Nagel, M., Taghdiri, M.: Bounded program verification using an SMT solver: a case study. In: ICST, pp. 101–110. IEEE (2012)
16. Lokuciejewski, P., Cordes, D., Falk, H., Marwedel, P.: A fast and precise static loop analysis based on abstract interpretation, program slicing and polytope models. In: CGO, pp. 136–146. IEEE (2009)
17. Ma, F., Yan, J., Zhang, J.: Solving generalized optimization problems subject to SMT constraints. In: Snoeyink, J., Lu, P., Su, K., Wang, L. (eds.) AAIM/FAW -2012. LNCS, vol. 7285, pp. 247–258. Springer, Heidelberg (2012). doi:10.1007/978-3-642-29700-7_23
18. Michiel, M.D., Bonenfant, A., Cassé, H., Sainrat, P.: Static loop bound analysis of C programs based on flow analysis and abstract interpretation. In: RTCSA, pp. 161–166. IEEE (2008)
19. Moura, L., Bjørner, N.: Z3: an efficient SMT solver. In: Ramakrishnan, C.R., Rehof, J. (eds.) TACAS 2008. LNCS, vol. 4963, pp. 337–340. Springer, Heidelberg (2008). doi:10.1007/978-3-540-78800-3_24
20. Sebastiani, R., Tomasi, S.: Optimization in SMT with LA (Q) cost functions. In: Gramlich, B., Miller, D., Sattler, U. (eds.) IJCAR 2012. LNCS, vol. 7364, pp. 484–498. Springer, Heidelberg (2012). doi:10.1007/978-3-642-31365-3_38
21. Sebastiani, R., Trentin, P.: OptiMathSAT: a tool for optimization modulo theories. In: Kroening, D., Păsăreanu, C.S. (eds.) CAV 2015. LNCS, vol. 9206, pp. 447–454. Springer, Cham (2015). doi:10.1007/978-3-319-21690-4_27
22. Shkaravska, O., Kersten, R., van Eekelen, M.: Test-based inference of polynomial loop-bound functions. In: PPPJ, pp. 99–108. ACM (2010)
23. Vallée-Rai, R., Co, P., Gagnon, E., Hendren, L.J., Lam, P., Sundaresan, V.: Soot - a Java bytecode optimization framework. In: CASCON, p. 13. IBM (1999)
24. Vaziri, M.: Finding Bugs in Software with a Constraint Solver. Ph.D. thesis, MIT (2004)
25. Termination problems data base (TPDB). http://termination-portal.org/wiki/TPDB. Accessed June 2017

AndroidLeaker: A Hybrid Checker for Collusive Leak in Android Applications

Zipeng Zhang[1,2] and Xinyu Feng[1,2(✉)]

[1] School of Computer Science and Technology,
University of Science and Technology of China, Hefei, China
zhangzpp@mail.ustc.edu.cn, xyfeng@ustc.edu.cn
[2] Suzhou Institute for Advanced Study,
University of Science and Technology of China, Suzhou, China

Abstract. Android phones often carry sensitive personal information such as contact books or physical locations. Such private data can be easily leaked by buggy applications by accident or by malicious applications intentionally. Much work has been proposed for privacy protection in Android systems, but there still lacks effective approaches to prevent information leak caused by Inter-Component Communication (ICC).

We present AndroidLeaker, a new hybrid analysis tool of privacy protection based on taint analysis for Android applications to prevent the privacy leak caused by multiple application cooperation. Our approach combines static analysis and dynamic checking. Static analysis is used to check the information leak in the individual applications and dynamic checking at runtime is responsible for preventing the information leak caused by cooperation of multiple applications. Such a combination may effectively reduce the runtime overhead of pure dynamic checking, and reduce false alarms in pure static analysis.

Keywords: Taint analysis · Security · Information flow control · Android

1 Introduction

With the growth of smart phone applications, more and more private information, such as contact books, physical locations etc., are stored in the phone. Users are now faced with the high risk of these applications stealing their privacy. There is already a lot of work to detect the privacy leak in Android applications, but there are two problems that have not been solved very well:

1. The definition of privacy leak is over-restricted. Current tools [1,4,5,13,14,16, 20,21] generally regard all behaviors that send privacy outside the device as the information leak, such as sending contact books to the Internet. However,

---

This work is supported in part by grants from National Natural Science Foundation of China (NSFC) under Grant Nos. 61632005 and 61379039.

© Springer International Publishing AG 2017
K.G. Larsen et al. (Eds.): SETTA 2017, LNCS 10606, pp. 164–180, 2017.
https://doi.org/10.1007/978-3-319-69483-2_10

only the privacy sending that a user disallows should be recognized as privacy leak. For example, a taxi booking application like Uber may send your location to its server to help send the nearest vehicle for you. Such behavior should not be viewed as the leakage.

In addition, those tools [1,9,21] that analyze each application separately may regard all program points that send the privacy to another application as the leak points, even though the privacy may not leave the system. It prevents benign cooperation among applications.

2. The lack of effective approaches to prevent collusive leak. Generally, there are two kinds of mechanisms to prevent collusive leak:

   (a) Analyzing the application interactions statically. For example, Epicc [11] infers the possible interactions among applications by analyzing the configuration files *AndroidManifest.xml* of Android applications. The components that can receive messages from another application or be invoked by another application must be declared in the configuration files explicitly. Analysis of the application code may find the program points that perform ICC. By combining these two kinds of information, an interaction graph among applications is established. But this kind of graph may include lots of interactions that may never happen, thus introduce lots of false alarms.

   (b) Tracking the contents of messages sent among applications dynamically. TaintDroid [4] and TaintART [14] add the support for tracking information flow in Android operating system. Therefore the privacy accessed by an application or sent from each other can be detected at runtime. But they do not track the implicit channel, such as the information flow dependent on the control flow. Moreover, since the operating system needs to perform the information flow tracking for each instruction, especially when even the data contains no privacy, considerable time and memory overheads are introduced.

To address the *first* problem, we extend the existing Android permission model by introducing a new kind of permission, *send permission*. Send permission is a collection of private information that an application is allowed to send outside of the device. For example, if an application declares its send permission as {Location}, then sending the location outside is not considered as information leak. Conversely, sending the contact book to the Internet is recognized as leak because of the lack of such send permission.

In contrast, the existing Android permission model is based on access control and is more coarse-grained. If an application needs to access some private information, then it needs to declare the request of the corresponding permission in the configuration file. The permissions are granted at the time of installation, if the user agrees. However, if the user does not grant it the permissions, for the concern that it may disclose the privacy, then the corresponding functionalities cannot be used. For example, an input method may need to access the contact book to provide hints when the user types his friend's name. If the input method is not authorized to read the contact book, then the user is not able to use this

```
1   // action is "com.test.action.MY_ACTION"
2   public class MyReceiver extends BroadcastReceiver{
3     public void onReceive(Context context, Intent intent){
4       String ip = intent.getStringExtras("ip");
5       String number = getContact("Alice");
6       try{
7         Socket socket = new Socket(ip, "8080");
8         OutputStream out = socket.getOutputStream();
9         DataOutputStream wrt = DataOutputStream(out);
10        wrt.writeUTF(number);
11      } catch(IOException e){...}
12    }
13
14    private String getContact(String name){
15    // get the phone number of a given name
16    }
17  }
```

(a)

```
1   public class MyActivity extends Activity{
2     public void onCreate(Bundle savedInstanceState){
3       ...
4       Intent intent = new Intent();
5       intent.setAction("com.test.action.MY_ACTION");
6       intent.putExtra("ip", "192.168.1.1");
7       sendBroadcast(intent);
8     }
9   }
```

(b)

Fig. 1. An example showing collusive leak

convenient feature. We can use the send permission to address this dilemma. An input method may declare its send permission as {}, *i.e.*, it does not send any private information outside. Then the user can safely grant the access permission of contact book to it, without worrying that their privacy information is leaked.

Before the discussion of the solution of the second problem, we first show an example in Fig. 1, where an untrusted application sends the privacy out with the help of another trusted application. Application (a) is a trusted application, and it is authorized to access the contact book. It receives an IP address from another application and then sends the contact book to that address. Here we assume that it only sends out the phone number of Alice. Application (b) may be a malicious application, and it is not allowed to access any confidential data. It sends a non-confidential message to application (a), which contains an IP address and a port number. With the help of application (a), application (b) sends the contact book to the specific address successfully, although it does not access any privacy directly. The static analysis tools before [1,3,11] do not detect this kind of leakage, because application (b) itself never access any confidential data, let alone leak the privacy. The dynamic analysis tools [4,14] cannot find this leak either, because they just inform the user that application (a) sends the contact book to the network. Since application (a) is trusted, so the user will not prevent this privacy sending.

To address this problem, we propose a combination of static and dynamic checking, which can reduce the runtime overhead of pure dynamic checking and reduce the false alarms in pure static analysis. In the static analysis, we check the information leak generated by individual applications, and collect some information for dynamic analysis. In the dynamic analysis, communications among applications are checked and those which may cause privacy leak are prevented.

Figure 2 shows the overall architecture of our method. During the static phase, our tool takes the Android application package (APK) as the input, and read the permissions from its configuration file. Then it analyzes the possible privacy leaks in the application and outputs the leak points if they are found. If no leak is found, it generates the security labels of the outgoing messages and a *dependent send permission*, which are taken as the input by the instrumentation phase. The dynamic checking is performed by instrumentation. The communications among applications ICC is inspected at runtime, and the interaction that may result in the privacy leak is prevented. The leak behavior is also recorded.

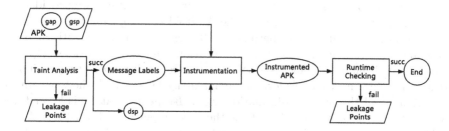

Fig. 2. The overall architecture

Our work makes the following contributions:

1. We introduce a novel permission, *send permission*, to describe the collection of private information that an application is allowed to send out. On the one hand, the user can allow the application to access the confidential data without worrying about privacy leak. On the other hand, a secure application that must access confidential data can declare the send permission as low as possible to show it is indeed trustworthy.
2. We develop a new hybrid analysis tool *AndroidLeaker* to prevent the information leak caused by the cooperation of multiple applications. We can reduce false-alarms in pure static analysis and reduce the runtime overhead of pure dynamic checking.

The rest of this paper is organized as follows. Section 2 presents the new permissions and gives the key ideas of the static analysis and dynamic analysis.

Section 3 discusses the implementation of AndroidLeaker and shows some important details about the taint analysis. Section 4 shows the evaluate results for AndroidLeaker. Section 5 compares our method with related work and concludes.

2   Our Approach

We first give the extension of the Android permission model, and then introduce the static checking rules and dynamic checking based on the extension.

2.1   The Policy

Each application is associated with a security policy θ, which represents the permissions to access and spread the privacy. As shown in Fig. 3, a security policy is a triple (gap, gsp, dsp). gap is the permission to access the privacy, similar with the access permission in Android. Unlike in Android, gap not only limits the access to privacy directly through system calls, but also restricts the application from getting privacy by inter-component communication. gsp is the permission to send the privacy out. Only the private information described in gsp are allowed to be sent outside. gsp is usually the subset of gap, because a privacy can be sent out only when it can be accessed first. gap and gsp are provided by the programmer, and are declared in the Android configuration file. The third part dsp is the dependent send permission. It is generated in the static analysis phase and used to help check the communications among applications. gap, gsp and dsp are all labels. A label is a set of variables, each representing a kind of private information. All labels form a partial lattice $(L, \sqsubseteq)$, with the top element $\top$ containing all the private information. $L_1 \sqsubseteq L_2$ means that L_1 is lower than L_2, $i.e.$, L_1 is a subset of L_2. $L_1 \sqcup L_2$ is the least upper bound of L_1 and L_2.

Dependent send permission dsp specifies the private information that an application may send out when requested by other applications. It is a refinement of the send permission gsp. There is a dependent send behavior in Application (a) in Fig. 1, because whether it sends the contact book to the network depends on the messages from other applications. So the dependent send permission of Application (a) is {ContactBook}. On the contrary, there is no risk to allow an untrusted application to communicate with a privileged application with empty dependent send permission. Generally, the dependent send permission dsp is lower than the send permission gsp, ($i.e.$, $dsp \sqsubseteq gsp$) — An application disallowed to send out the private information cannot send out that privacy for others.

$$
\begin{array}{llll}
(Labels) & L, \pi, gap, gsp & \in & \mathcal{P}(Vars) \\
(Policy) & \theta & ::= & (gap, gsp, dsp)
\end{array}
\qquad
\begin{array}{l}
L_1 \sqsubseteq L_2 \text{ iff } L_1 \subseteq L_2 \vee L_2 = \top \\
L_1 \sqcup L_2 = \begin{cases} \top & \text{if } L_1 = \top \text{ or } L_2 = \top \\ L_1 \cup L_2 & \text{otherwise} \end{cases}
\end{array}
$$

Fig. 3. The label and policy

As shown in Fig. 1, the security policy of Application (a) is declared as ({ContactBook}, {ContactBook}, {ContactBook}), which means it may read the contact book, send it out, and also send it out for other applications. The security policy of Application (b) is ({}, {}, {}). Because it cannot access any privacy, its send permission and dependent send permission must be {} too.

In the following parts, we present the key checking rules for static analysis and dynamic analysis.

2.2   Static Analysis

Representative taint analysis rules are shown in Fig. 4. The judgement is in the form of $\pi, \theta \vdash \{\Gamma\}\ c\ \{\Gamma'\}$. The pc label π represents the security label of the program counter, which is used to check the implicit information channel dependent on control flows. θ is the security policy. Γ is the type environment which is a partial mapping from variables to labels, i.e., $\Gamma \in Vars \rightharpoonup Labels$. It stores the private information that may be contained in different variables. Γ and Γ' represent the type environments of variables before and after the execution of the instruction c respectively. Here variables are divided into two classes: the local variable p and shared variable s. The former is variables declared in an application, and the latter represents the private information maintained globally by the Android system and shared across applications. The judgement $\Gamma \vdash e : L$ means that the label of expression e is L under the given type environment Γ. The judgement $\Gamma \vdash b : L$ for the boolean expression b is similar. Note that the dependent send permission dsp in the policy θ is not used in the static checking. It is *inferred* in the static checking phase and used in the dynamic checking.

To simplify the discussion, we use some primitives to model the actual operations in Android (shown in Fig. 5). Primitives **recv** and **send** are used to model the inter-component communication operations. An application can send a broadcast to another application, or start an activity or a service in another application. All these can be viewed as a simple message-passing style inter-process communication. The message is called *Intent* in Android. $p := recv()$

$$\frac{}{\pi, (gap, gsp, \_) \vdash \{\Gamma\}\ p := \mathbf{recv}()\ \{\Gamma[p \mapsto gap]\}}\ (\text{Recv})$$

$$\frac{}{\pi, \theta \vdash \{\Gamma\}\ \mathbf{send}(e_1, e_2)\ \{\Gamma\}}\ (\text{Send}) \qquad \frac{\Gamma \vdash e : L_e \quad \pi \sqcup L_e \sqsubseteq gsp}{\pi, (gap, gsp, \_) \vdash \{\Gamma\}\ \mathbf{out}(e)\ \{\Gamma\}}\ (\text{Out})$$

$$\frac{s \in gap}{\pi, (gap, gsp, \_) \vdash \{\Gamma\}\ p := \mathbf{get}(s)\ \{\Gamma[p \mapsto \pi \sqcup \{s\}]\}}\ (\text{Get})$$

$$\frac{\Gamma \vdash b : L_b \quad \pi \sqcup L_b, \theta \vdash \{\Gamma\}\ \{c_i\}\ \{\Gamma'_i\} \quad i = 1, 2}{\pi, \theta \vdash \{\Gamma\}\ \mathbf{if}\ b\ \mathbf{then}\ c_1\ \mathbf{else}\ c_2\ \{\Gamma_1 \sqcup \Gamma_2\}}\ (\text{If})$$

Fig. 4. Selected rules for taint analysis

Primitives	Operations
$p := \textbf{recv}()$	onReceive(), getIntent(), onActivityResult(), ...
$\textbf{send}(e_1, e_2)$	sendBroadcast(), startActivity(), startActivityForResult(), ...
$\textbf{out}(e)$	sendTextMessage(), openConnection(), connect(), ...
$p := \textbf{get}(s)$	managedQuery(), getDeviceId(), getLatitude(), ...

Fig. 5. The primitives

gets the intent from another application and stores it in p. $\textbf{send}(e_1, e_2)$ sends an intent, represented by expression e_2 here, to an application identified by expression e_1. Primitive $\textbf{out}(e)$ is used to model the system calls to send the information e outside the device through Internet or text messages, etc. It is the only way that the information can leave the system. Primitive $p := \textbf{get}(s)$ is used to model the operations to read the private information (modeled as global variables, as explained before), such as the contact book, location, short messages etc. It reads the private information s into the local variable p.

As shown in Fig. 4, the RECV rule assigns the label gap to the variable p at the end of the receive command. This is a conservative estimation since the actual label of the incoming message is unknown statically. It is safe because the dynamic checking for **send** ensures that the label of the outgoing message should be lower than the receiver's gap. The checking for **send** primitive is done at runtime (explained below), so the static checking does nothing (see the SEND rule). The OUT rule says that, to output the private information, an application needs to have the corresponding send permission. The GET rule says that the access of the private information represented as the shared variable s requires the corresponding access permission, i.e., $s \in gap$. It assigns the label $\pi \sqcup \{s\}$ to the local variable p at the end to record the implicit information flow through the control flow. The IF rule shows how the implicit information flow is tracked. The label of the condition b is joined with the pc label π. The resulting type environment is the union of the resulting type environments of two branches. The union of two type environments $\Gamma_1 \sqcup \Gamma_2$ is defined as $\{p \rightsquigarrow (\Gamma_1(p) \sqcup \Gamma_2(p)) \mid p \in dom(\Gamma_1)\}$ (note $dom(\Gamma_1) = dom(\Gamma_2)$). The checking rules for other statements are standard and are omitted here.

Dependent Send Permission. Other than the taint analysis, we also infer the dependent send permission in the static phase. Each application declares the methods which can be invoked by other applications in its configuration file. The output of private information in these exposed methods should be declared in the dependent send permission. As an example, the *onReceive* function of Application (a) in Fig. 1 is one such method. To infer the dependent send permission, we traverse these exposed methods only and apply the following rule for each **out** command:

$$\frac{\Gamma \vdash e : L_e}{\pi, \Gamma \vdash \{dsp\}\ \textbf{out}(e)\ \{dsp \sqcup L_e \sqcup \pi\}}$$

We start from an empty label $\{\}$ as the initial *dsp*. The type environment Γ and the pc label π is computed in the taint analysis previously.

2.3 The Instrumentation and the Dynamic Analysis

The **send** primitive is the only program point which needs to be dynamically checked. The dynamic checking is implemented by code instrumentation, which inserts permission checking code before each corresponding system call. The system call is executed only if the checking is passed. The following rule shows the checking for each **send** primitive.

$$\frac{\Gamma \vdash e_2 : L_2 \quad L = L_2 \sqcup \pi \quad \Theta(\texttt{cid}) = (gap_a, gsp_a, dsp_a) \quad \Theta(e_1) = (gap_b, gsp_b, dsp_b)}{\pi, \Gamma, \Theta \vdash \textbf{send}(e_1, e_2)}$$

$$L \sqsubseteq gap_b \quad L \not\sqsubseteq gsp_a \Rightarrow L \not\sqsubseteq gsp_b \quad dsp_b \sqsubseteq dsp_a$$

π and Γ are the pc label and the type environment derived from the static taint analysis. Θ is a mapping from application identifiers to the policy set, *i.e.*, $\Theta ::= \{t_1 \leadsto \theta_1, \ldots, t_n \leadsto \theta_n\}$. It stores policies of all applications and can be globally accessed. (gap_a, gsp_a, dsp_a) and (gap_b, gsp_b, dsp_b) are the policies of the sender and the receiver respectively, which are acquired from Θ. The first line in the premise computes the label L of the outgoing message. It is the union of the label L_2 of message e_2 and the pc label π. The second line shows the constraints for **send** primitive.

To show the validity of the checking, we first analyze the possible ways that one application may leak privacy with the help of another application.

Figure 6 shows four cases in which the communication between two applications may result in the privacy leak. The three sets above the applications are their policies. The policies of application A and application B are referred as (gap_a, gsp_a, dsp_a) and (gap_b, gsp_b, dsp_b) respectively in the following explanation, and label L represents the label of the message. In case (a), A has the permission to access and send out the contact book, while B has neither permissions. A

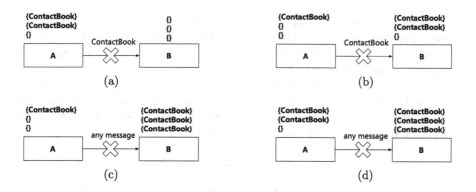

(a)   (b)

(c)   (d)

Fig. 6. Cases of forbidden send

should not send contact book to B, or B will access the privacy it is not allowed to access. So all messages an application receives should not contain any privacy which are not declared in its access permission. The first restriction is

$$L \sqsubseteq gap_b.$$

In case (b), A has the permission to access the contact book, but is not allowed to send the contact book out, while B can send the contact book out. If A sends contact book to B, then it is possible that B may send out contact book for A. This case should be forbidden. The second restriction is

$$L \not\sqsubseteq gsp_a \Rightarrow L \not\sqsubseteq gsp_b.$$

It means that if the message cannot be sent out by A, then it is disallowed to be sent to an application which may send it out.

In case (c), A is not authorized to send out the contact book, and B may send out the contact book depending on the message it receives. Application (a) in Fig. 1 shows such a case. Any messages should be forbidden to be sent from A to B, when even the message does not contain any privacy. Therefore, the third restriction is

$$dsp_b \sqsubseteq gsp_a.$$

This restriction forbids A from leaking confidential data by invoking B. Because A can only call B to send the confidential data which is described in dsp_b, so A must have the permission to send the data which dsp_b describes.

The above three constraints, however, are not sufficient to prevent leakage. Consider the following example:

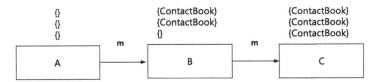

A sends a non-confidential data **m** to B, and then B just passes **m** to C directly. The communications between A and B, and that between B and C, do not violate these three restrictions above, but A sends a message to C indirectly, which violates the restriction shown in case (c).

This example explains why sending message from A to B in case (d) in Fig. 6 should be forbidden. We require that if A sends messages to B which may send confidential data dependently, then A must declare the corresponding dependent send permission. This is our fourth constraint:

$$dsp_b \sqsubseteq dsp_a$$

From the fourth condition and the property that $dsp_a \sqsubseteq gsp_a$, we can derive the third condition.

In summary, the condition to perform the **send** operation successfully is

$$L \sqsubseteq gap_b \quad \text{and} \quad L \not\sqsubseteq gsp_a \Rightarrow L \not\sqsubseteq gsp_b \quad \text{and} \quad dsp_b \sqsubseteq dsp_a$$

This is what we need to check at runtime.

3   The Implementation

AndroidLeaker is based on the Soot framework [15], which converts Android Dalvik bytecode into the call graph and provides usefull basic data flow analysis framework. We use Heros [2], a generic implementation of inter-procedural, finite, distributive subset(IFDS) problems [12] solver, to perform the inter-procedural data flow analysis. In the following parts, we introduce the concrete implementation of static and dynamic analysis, and then give some important details about the actual taint analysis and the instrumentation.

3.1   The Overall Architecture

The overall architecture of our static analysis is shown in Fig. 7. Before the taint analysis, we need to collect the following information:

1. Sources and sinks: Android APIs, including those serving as sources and sinks, change frequently with the upgrade of Android. To adapt to the change, we allow users to customize the source and sink functions. The component **SourceSinkParser** parses the user-defined sources and sinks.
2. Callback functions: Callback functions are not invoked directly in the program. To analyze the call back functions, we treat them as entries of the program. The procedure **CallbackAnalysis** is used to find all the callbacks the program registered. There are two kinds of callbacks: those registered in the program and those declared in the configuration files.

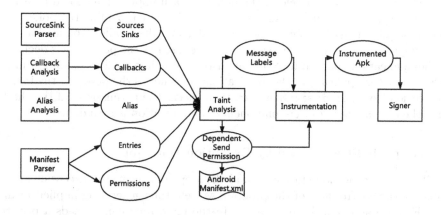

Fig. 7. The concrete architecture of static analysis

3. Alias: Alias information is needed in order to handle information flow through object fields. `Alias Analysis` traverses the program and generates the alias information.
4. Entries and permissions: The entry points and required permissions of the application are declared in the configuration file *AndroidManifest.xml*, which are extracted by the component `ManifestParser`. The callbacks return by `Call Analysis` are also considered as entries. The permissions include the access permissions and the send permissions.

The information collected above are fed into `TaintAnalysis`. To improve the precision, our analysis is flow-sensitive, context-sensitive, field-sensitive and object-sensitive. The `TaintAnalysis` generates the security labels of the outgoing messages and the dependent send permission, which are used by the phase `Instrumentation` to instrument the dynamic checking code into the application. The `TaintAnalysis` also records the dependent send permission in the configuration file *AndroidManifest.xml*. Finally, the instrumented application is sent to the android application signer to obtain a signed version, which can be installed in the phone.

Fig. 8. The architecture of dynamic analysis

The dynamic checking model is shown in Fig. 8. To check the permissions at the inter-component communication program points, both permissions of the sender and the receiver should be known (recall the dynamic checking for **send** primitive in Sect. 2.3). Therefore, the policy of each application should be available at runtime. `PermissionManager` is responsible for providing the policies of specific applications. It collects the permissions of all installed applications by inquiring the operating system, since the system stores the permissions declared in the application configuration file at installation time. Before sending a message to another application, the application first queries the policy of the target application from `PermissionManager`, then checks the permissions. If the check fails, the corresponding inter-component communication is abandoned.

3.2   Implicit Flow Caused by Exception Handling

Similar to the condition statement and the loop statement, the exceptions may influence the control flow of the program as well, thus generating implicit channels. Figure 9 shows such an example. In the main function, it reads a private data into variable x first, and then calls function f with the parameter x. The

```
1   class A {
2     public static void main(){
3       int x=getPrivacy(), y=0;
4       try {
5         f(x);
6       } catch(Exception e){
7         y=1;
8       }
9       int w=0;
10    }
11
12    public void f(int x)
13            throws Exception{
14      if (x > 0) {
15        throw new Exception();
16      } else {
17        return;
18      }
19    }
20  }
```

(a)

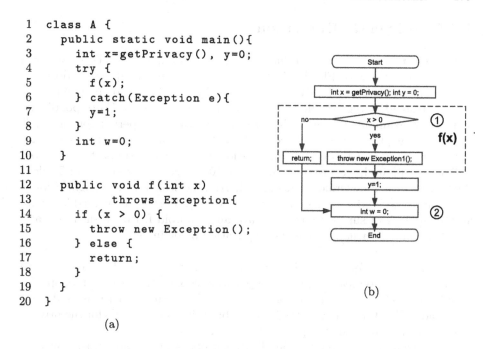

(b)

Fig. 9. An example for implicit flow due to exception

function f decides whether to throw an exception or return normally depending on the value of x. After the execution of function f, if f throws an exception, then y is set to 1, otherwise y remains 0. Although x is not assigned to y directly, we can know whether x is higher than 0 by testing the value of y.

Part (b) in Fig. 9 shows the control flow graph of program (a). The control flow graph of function f is expanded and embedded into that of the function main. In the expanded control flow graph, it is clear to see that the value of y depends on the value of x.

To prevent this kind of implicit channels, when a branch node is analyzed, we compute its immediate post-dominator [7], and push a new pc label on the stack, along with the node's immediate post-dominator. When the immediate post-dominator is reached, the pc label is popped. Since the inter-procedural analysis is too time-consuming, we use on-the-fly intra-procedural static analysis of immediate post-dominator. If the immediate post-dominator of a branch node inside its function is null, then there are two cases: (a) if this function does not throw an exception, then its immediate post-dominator should be the exit point of the function; (b) otherwise, its immediate post-dominator should be the immediate post-dominator of the calling point. For example, the immediate post-dominator of node ① in Fig. 9 is node ②, because function f, to which node ① belongs, throws an exception and the immediate post-dominator of the calling point of f, *i.e.*, line 5 in program (a), is node ②.

4   Experimental Evaluation

Precision for individual applications. We test 82 examples from DroidBench, a test suite released with FlowDroid [1] to assess the taint analyses. It does not contain any test cases with privacy leak caused by communication between applications. Instead, it regards all communication points as sinks. Correspondingly, to test the examples in AndroidLeaker, we set the send permission of applications to be {} by default. Since Android lifecycle and Java threading have not been considered in our implementation yet, and the support of Android callbacks is limited, we just test the other 10 categories, as shown in Table 1. Compared with FlowDroid (version 1.5), AndroidLeaker gets a slightly lower precision (82%), and a slightly higher recall (69%). Most of the missed leaks are due to the lack of support for some Java and Android features, such as static initialization, unreachable code and application modeling etc.

Inter-Component Communication. We do not know any standard test suite to test the information leak caused by application cooperation, so we develop four sample applications to test the communication between applications, and compare AndroidLeaker with FlowDroid [1]. Table 2 shows the results for the privacy leak test caused by inter-component communication.

Each example contains two applications: one sends a broadcast (by calling **sendBroadcast**, a representative inter-component communication method in Android), and the other receives it. In the first two examples, the sender A sends the device ID d (private data) to the receiver B, and B sends anything it

Table 1. DroidBench test results

App category	AndroidLeaker			FlowDroid		
	$\otimes$	$*$	$\circ$	$\otimes$	$*$	$\circ$
EmulatorDetection	3	0	0	2	0	1
InterComponentCommunication	10	0	7	14	1	3
GeneralJava	13	1	9	12	1	8
Reflection	3	0	1	1	0	3
Aliasing	0	1	0	0	1	0
ArraysAndLists	3	4	0	2	4	1
ImplicitFlows	3	0	1	0	0	4
AndroidSpecific	7	0	3	7	0	3
FieldAndroidObjectSensitivity	2	4	0	2	0	0
InterAppCommunication	2	0	1	3	0	0
Total	46	10	21	43	7	23
Precision=$\otimes/(\otimes + *)$	82%			86%		
Recall= $\otimes/(\otimes + \circ)$	69%			65%		

$\otimes$ = correct warning $*$ = false warnings $\circ$ = missed leaks.

Table 2. The privacy leak test for inter-component communication

Example	Sender's policy	Receiver's policy	AndroidLeaker	FlowDroid	Expect
broadcastTest1	({d},{d}, {})	({d},{d},{})	No leak	Leak	No leak
broadcastTest2	({d},{d}, {})	({},{}, {})	Leak	Leak	Leak
broadcastTest3	({},{},{})	({d},{d}, {d})	Leak	No leak	Leak
broadcastTest4	({},{},{})	({},{}, {})	No leak	No leak	No leak

receives to the Internet. Whether there is information leak or not depends on the policies of these two applications. If B has the permission to access and send out the device ID, then there is no information leak, as in the broadcastTest1 example. Otherwise the send operation of B should be regarded as information leak, as in the broadcastTest2 example. Since FlowDroid considers all inter-component communication as sinks, it reports information leak in both examples. AndroidLeaker can distinguish the two situations and reports the right results.

The third example is similar with the program in Fig. 1. The sender A sends an IP address and a port number to the receiver B. B reads the deviceID, and then sends it to the IP address and the port number it receives. Since B may send the device ID out for A but A itself has no permission to send, this communication may result in privacy leak. The forth example is a variation of the third example and the only difference is that the receiver B does nothing. So it is safe to allow A to send messages to B. Since the FlowDroid detects nothing private sent from A to B, it reports no leak for the last two examples. AndroidLeaker can distinguish these two cases and report the right results. Compared with FlowDroid, AndroidLeaker gets lower false positive alarms for leak caused by inter-component communication.

Runtime Overhead. We test the runtime overhead for inter-component communication on an Android 4.4 phone with the cpu of Qualcomm MSM 8226 (4 cores, 1.2 GHz, 1 G Memory). Since the instrumented checking for all communication APIs are the same, we take the system call `sendBroadcast` as an example to test the overhead, because it is asynchronous and directly returns without user interaction. It takes 0.4 s to run the `sendBroadcast` for 1000 times, and 6 s to run the instrumented version. That is AndroidLeaker adds 5.6 ms for each inter-component communication system call. The overall overhead depends on the number of the communication system calls when an application runs. That is, the total overhead is $\frac{T+N \times 5.6\ \mathrm{ms}}{T} - 1$, where T is the execution time of the original application, and N is the number of the communication system calls. By contrast, TaintDroid introduces 14% CPU overhead on average for each application [4]. If communication happens rarely, AndroidLeaker gets better performance.

5   Related Work and Conclusion

Various static analysis techniques have been proposed to detect privacy leak in Android [1,5,6,8,9,11,17,18,20,21]. CHEX [8] is a tool to locate one kind of vulnerable applications (*hijacking vulnerable applications*). An application is *hijacking vulnerable* if it exposes an interface to perform some privileged action such that another application without the permission may access the confidential data by calling it. In our method, this kind of vulnerable applications can be also detected, because one application is forbidden to access or send confidential data on behalf of another unprivileged application.

Leakminer [19] and FlowDroid [1] are both based on the soot framework and implement the Android lifecycle, but they do not consider the cooperation among applications and focus on the analysis of one application. Epicc [11] performs a study of ICC (inter-component communication) vulnerabilities, including sending an Intent that may be intercepted by a malicious component, or exposing components to be launched by a malicious Intent. This is similar with our concern with ICC. It analyzes the sensitive data transmission in Android by analyzing the configuration file, *i.e.*, AndroidManifest.xml, to infer the communication. Since Epicc performs a static analysis and adopts a conservative strategy while we perform dynamic checking when the communication occurs, we may achieve less false positive.

ComDroid [3] detects the implicit Intents possibly intercepted by malicious applications and the components exposed to external applications which may be utilized by the malicious applications to acquire confidential data or perform some unauthorized actions. It does not concern whether the private information is contained in Intents or returned by the components.

There are also some works [4,10,14] based on dynamic checking or dynamic taint analysis. TaintDroid [4] tracks the flow of privacy sensitive data through third party applications and automatically taints data from privacy-sensitive sources and tracks the labels when the sensitive data propagates. Aquifer [10] prevents accidental information inside one application from being stolen by another application. Both of them do not track the implicit channel depending on the control flow.

Conclusion. In this paper, we present a new hybrid analysis for Android applications to prevent the information leak caused by application cooperations. The combination of the static checking and the dynamic checking can reduce the overhead at runtime effectively.

References

1. Arzt, S., Rasthofer, S., Fritz, C., Bodden, E., Bartel, A., Klein, J., Le Traon, Y., Octeau, D., McDaniel, P.: FlowDroid: precise context, flow, field, object-sensitive and lifecycle-aware taint analysis for android apps. In: PLDI 2014, pp. 259–269 (2014)
2. Bodden, E.: Inter-procedural data-flow analysis with IFDS/IDE and soot. In: SOAP 2012, pp. 3–8 (2012)
3. Chin, E., Felt, A.P., Greenwood, K., Wagner, D.: Analyzing inter-application communication in android. In: Proceedings of the 9th International Conference on Mobile Systems, Applications, and Services, pp. 239–252 (2011)
4. Enck, W., Gilbert, P., Chun, B.G., Cox, L.P., Jung, J., McDaniel, P., Sheth, A.N.: TaintDroid: an information-flow tracking system for realtime privacy monitoring on smartphones. In: OSDI 2010, pp. 1–6 (2010)
5. Gibler, C., Crussell, J., Erickson, J., Chen, H.: AndroidLeaks: automatically detecting potential privacy leaks in android applications on a large scale. In: Katzenbeisser, S., Weippl, E., Camp, L.J., Volkamer, M., Reiter, M., Zhang, X. (eds.) Trust 2012. LNCS, vol. 7344, pp. 291–307. Springer, Heidelberg (2012). doi:10.1007/978-3-642-30921-2_17
6. Kim, J., Yoon, Y., Yi, K., Shin, J.: SCANDAL: static analyzer for detecting privacy leaks in android applications. In: MoST 2012 (2012)
7. Lengauer, T., Tarjan, R.E.: A fast algorithm for finding dominators in a flowgraph. ACM Trans. Program. Lang. Syst. 1(1), 121–141 (1979)
8. Lu, L., Li, Z., Wu, Z., Lee, W., Jiang, G.: CHEX: statically vetting android apps for component hijacking vulnerabilities. In: CCS 2012, pp. 229–240 (2012)
9. Mann, C., Starostin, A.: A framework for static detection of privacy leaks in android applications. In: Proceedings of the 27th Annual ACM Symposium on Applied Computing, SAC 2012, pp. 1457–1462 (2012)
10. Nadkarni, A., Enck, W.: Preventing accidental data disclosure in modern operating systems. In: CCS 2013, pp. 1029–1042 (2013)
11. Octeau, D., McDaniel, P., Jha, S., Bartel, A., Bodden, E., Klein, J., Le Traon, Y.: Effective inter-component communication mapping in android with Epicc: an essential step towards holistic security analysis. In: SEC 2013, pp. 543–558 (2013)
12. Reps, T., Horwitz, S., Sagiv, M.: Precise interprocedural dataflow analysis via graph reachability. In: POPL 1995, pp. 49–61 (1995)
13. Sakamoto, S., Okuda, K., Nakatsuka, R., Yamauchi, T.: DroidTrack: tracking information diffusion and preventing information leakage on android. In: Park, J.J.J.H., Ng, J.K.-Y., Jeong, H.Y., Waluyo, B. (eds.) Multimedia and Ubiquitous Engineering. LNEE, vol. 240, pp. 243–251. Springer, Dordrecht (2013). doi:10.1007/978-94-007-6738-6_31
14. Sun, M., Wei, T., Lui, J.C.: TaintART: a practical multi-level information-flow tracking system for android runtime. In: CCS 2016, pp. 331–342 (2016)
15. Vallée-Rai, R., Co, P., Gagnon, E., Hendren, L., Lam, P., Sundaresan, V.: Soot: a java bytecode optimization framework. In: CASCON 2010, pp. 214–224 (2010)
16. Xia, M., Gong, L., Lyu, Y., Qi, Z., Liu, X.: Effective real-time android application auditing. In: S&P 2015, pp. 899–914 (2015)
17. Xiao, X., Tillmann, N., Fahndrich, M., de Halleux, J., Moskal, M.: User-aware privacy control via extended static-information-flow analysis. In: ASE 2012, pp. 80–89 (2012)

18. Xu, R., Saïdi, H., Anderson, R.: Aurasium: Practical policy enforcement for android applications. In: Security 2012, pp. 27–27 (2012)
19. Yang, Z., Yang, M.: LeakMiner: detect information leakage on android with static taint analysis. In: WCSE 2012, pp. 101–104 (2012)
20. Yang, Z., Yang, M., Zhang, Y., Gu, G., Ning, P., Wang, X.S.: AppIntent: analyzing sensitive data transmission in android for privacy leakage detection. In: CCS 2013, pp. 1043–1054 (2013)
21. Zhao, Z., Osorio, F.C.C.: TrustDroid: preventing the use of smartphones for information leaking in corporate networks through the used of static analysis taint tracking. In: MALWARE 2012, pp. 135–143 (2012)

Modeling and Verification

Remark on Some π Variants

Jianxin Xue[1,2(✉)], Huan Long[2], and Yuxi Fu[2]

[1] College of Computer and Information Engineering,
Shanghai Polytechnic University, Shanghai, China
jxxue@sspu.edu.cn

[2] BASICS, MOE-MS Key Laboratory for Intelligent Computing and Intelligent
Systems, Department of Computer Science, Shanghai Jiao Tong University,
Shanghai, China
longhuan@sjtu.edu.cn, fu-yx@cs.sjtu.edu.cn

Abstract. Two π variants that restrict the use of received names are studied. For either variant the external characterization of the absolute equality is given using a family of bisimulations; the expressive completeness of the calculus is established; and a complete equational proof system is constructed. The relative expressiveness between the two variants and their relationship to π are revealed in terms of subbisimilarity.

1 Introduction

The π-calculus of Milner, Parrow and Walker [MPW92] has proved to be a versatile and robust programming language. Being a prime model of interaction, it accommodates the λ-calculus [Mil92, CF11] and is capable of explaining a wide range of phenomena where dynamic reconfiguration is a fundamental property [Wal95]. The name-passing communication mechanism of π is so strong that most of its variations are able to simulate each other both operationally and observationally to a considerable degree [San93, Tho95, San96a]. It has been an interesting topic to investigate different π-variants from different viewpoints [Pal03, Gor08, FL10]. The results obtained so far are important in that they help improve our understanding of the interaction models at both technical and conceptual levels [Fu16].

There are a number of ways to restrain the power of the π-calculus. Different variants are obtained by considering different forms of the choice operator and the recursion operator. Based on the results on CCS [BGZ03, BGZ04, GSV04], the relative expressive power of these variants have been examined [Pal03, Gor08], the most recent results being reported in [FL10]. Less close relatives are obtained by restricting the usage of the output prefix operator. In the private π-calculus [San96a], denoted by π^P, the exported names are always local names. Apart from its ability to code up higher order processes, the expressive power of π^P is almost unknown. The difficulty is partly attributed to the fact that it has a different set of action labels than the other π-variants. In the asynchronous π-calculus studied in [HT91a, HT91b, Bou92, ACS96] an output prefix is detached from any continuation. The reason why this simple syntactic manipulation provides a semantic modeling of asynchrony is explained

© Springer International Publishing AG 2017
K.G. Larsen et al. (Eds.): SETTA 2017, LNCS 10606, pp. 183–199, 2017.
https://doi.org/10.1007/978-3-319-69483-2_11

in [Fu10] using the theory by process. It is our personal view that the theory of the asynchronous π-calculus is best seen as an example of the theories defined by the processes [Fu10]. More distant cousins of the π-calculus are the process-passing calculi [Tho95, San93]. It is shown in [Fu16] however that these models are too weak from the viewpoint of computation, despite of the result proved in [LPSS08]. To achieve completeness we need to turn the higher order model into abstraction passing calculi [San93, XYL15, Fu17].

Instead of restraining the power of the operators, the calculi that impose conditions on the usage of the received names are the truly proper variants of the name-passing calculus [FZ15]. In the local π-calculus introduced in [Mer00, MS04] a received name can never be used as an input channel. A piece of program that defines a local subroutine with name say f can be rest assured that no other programs or subroutines share the name f. This is certainly a useful safety property and the π^L-calculus is defined in this way. Symmetrically the π^R-calculus imposes the condition that a received name cannot appear as an output channel. A host that intends to send a piece of information to another site may send a local channel name to the site and upload the information using the private channel. In this way the host makes sure that it is not at the receiving end of anything from any aliens. In another variant, called π^S-calculus, a process is not allowed to pass a received name to a third party. This appears as a more fundamental restriction on the name-passing mechanism. In this model the dynamic reconfiguration of the local communication topology is regional. One can never know a secret about you from a third party. Despite of their practical significance these three important variants have not been systematically studied.

Which aspects of π^L, π^R, π^S should we look into? We cannot claim to understand the calculi if we do not know the relative expressiveness between them and the π-calculus. This brings up the issue of model independence since the expressiveness relationship must be defined irrespectively of any particular model. As it turned out, the majority results in process theory are about particular models [Hoa78, Mil89, SW01]. The lack of the emphasis on model independence has been a blocking factor for the development of the process theory. Theory of Interaction proposed in [Fu16] is an attempt to provide a theory for all interaction models. The fundamental part of Theory of Interaction, the theory of equality, the theory of expressiveness, and the theory of completeness have been outlined, and a number of foundational results have been revealed. The applications of this general approach to the value-passing calculus and the name-passing calculus are reported in [Fu13, FZ15]. We will apply the general methodology of Theory of Interaction to π^L, π^R and π^S. The observational theory developed in this manner will help to construct an equational proof system for each of the variants. As a fallout, we will be able to say more about the theory of π^P.

Section 2 defines the semantics of the π-calculus and the three variants. Section 3 characterizes for π^L and π^R the absolute equality in terms of external bisimulation. Section 4 discusses the relative expressiveness of π^L, π^R and π-calculus. Section 5 confirms that all the three variants are legitimate models of interaction. Section 6 describes the equational proof systems for π^L and π^R. Section 7 takes a look at the private π-calculus. Section 8 concludes.

2   Pi and the Variants

Our definition of the π-calculus follows the presentation given in [FZ15]. Throughout the paper, we adopt the following notational conventions.

- The small letters a, b, c, d, e, f, g, h from the beginning of the alphabet range over the infinite set $\mathcal{N}$ of the *names*.
- The lowercase letters u, v, w, x, y, z towards the end of the alphabet range over the infinite set $\mathcal{N}_v$ of the *name variables*.
- The letters l, m, n, o, p, q in the middle of the alphabet range over $\mathcal{N} \cup \mathcal{N}_v$.

We often write $\widetilde{c}$ for a name sequence $c_1, \ldots, c_n$ and similarly $\widetilde{x}$ for $x_1, \ldots, x_n$.

In the π-calculus a name received in a communication can be used as either an output channel, or an input channel, or the content of a further communication. The π-terms are defined by the following grammar:

$$T := \mathbf{0} \mid \sum_{i \in I} n(x).T_i \mid \sum_{i \in I} \overline{n}m_i.T_i \mid T \mid T' \mid (c)T \mid [p{=}q]T \mid [p{\neq}q]T \mid !n(x).T \mid !\overline{n}m.T.$$

In the π^L-calculus a received name cannot be used as an input channel. The π^L-terms are inductively generated by following grammar:

$$T := \mathbf{0} \mid \sum_{i \in I} a(x).T_i \mid \sum_{i \in I} \overline{n}m_i.T_i \mid T \mid T' \mid (c)T \mid [p{=}q]T \mid [p{\neq}q]T \mid !a(x).T \mid !\overline{n}m.T.$$

In the π^R-calculus a received name cannot appear as an output channel. Its terms are produced by the following grammar:

$$T := \mathbf{0} \mid \sum_{i \in I} n(x).T_i \mid \sum_{i \in I} \overline{a}m_i.T_i \mid T \mid T' \mid (c)T \mid [p{=}q]T \mid [p{\neq}q]T \mid !n(x).T \mid !\overline{a}m.T.$$

Finally in the π^S-calculus a received name is not allowed to be transmitted to another process. The grammar is

$$T := \mathbf{0} \mid \sum_{i \in I} n(x).T_i \mid \sum_{i \in I} \overline{n}c_i.T_i \mid T \mid T' \mid (c)T \mid [p{=}q]T \mid [p{\neq}q]T \mid !n(x).T \mid !\overline{n}c.T.$$

$\sum_{i \in I} n(x).T_i$ is an input choice term and $\sum_{i \in I} \overline{n}m_i.T_i$ an output choice term. The binder $n(x)$ is an input prefix and $\overline{n}m_i$ an output prefix. The components $n(x).T_i$ and $\overline{n}m_i.T_i$ are called summands. In $n(x).T_i$ the name variable x is bound. A name variable is free if it is not bound. We use guarded replications $!n(x).T$ and $!\overline{n}m.T$. The conditional operator $[p{=}q]$ is a match and $[p{\neq}q]$ a mismatch. The restriction term $(c)T$ is in localization form, where the name c is local. A name is global if it is not local. We will write $gn(\_)$ for the function that returns the set of the global names. The derived prefix operator $\bar{a}(c).T$ abbreviates $(c)\bar{a}c.T$. We assume α-convention, meaning that no misuse of names/name variables ever occurs. The application of a substitution $\sigma = \{n_1/x_1, \ldots, n_i/x_i\}$ to a term is denoted by $T\sigma$. A term is open if it contains free name variables; it is closed otherwise. A process is a closed term. For each π-variant π', the set of the π'-processes is denoted by $\mathcal{P}_{\pi'}$ and is ranged over by L, M, N, O, P, Q.

The semantics of π, π^L, π^R, π^S is defined by the same labeled transition system. The observable action set is $\mathcal{L} = \{ab, \bar{a}b, \bar{a}(c) \mid a, b, c \in \mathcal{N}\}$. The action set $\mathcal{L} \cup \{\tau\}$ is ranged over by λ. The semantic rules are given below.

Action

$$\sum_{i \in I} a(x).T_i \xrightarrow{ac} T_i\{c/x\} \qquad \sum_{i \in I} \bar{a}c_i.T_i \xrightarrow{\bar{a}c_i} T_i$$

Composition

$$\frac{T \xrightarrow{\lambda} T'}{S \mid T \xrightarrow{\lambda} S \mid T'} \qquad \frac{S \xrightarrow{ac} S' \quad T \xrightarrow{\bar{a}c} T'}{S \mid T \xrightarrow{\tau} S' \mid T'} \qquad \frac{S \xrightarrow{ac} S' \quad T \xrightarrow{\bar{a}(c)} T'}{S \mid T \xrightarrow{\tau} (c)(S' \mid T')}$$

Localization

$$\frac{T \xrightarrow{\bar{a}c} T'}{(c)T \xrightarrow{\bar{a}(c)} T'} \qquad \frac{T \xrightarrow{\lambda} T'}{(c)T \xrightarrow{\lambda} (c)T'} c \notin gn(\lambda)$$

Condition

$$\frac{T \xrightarrow{\lambda} T'}{[a=a]T \xrightarrow{\lambda} T'} \qquad \frac{T \xrightarrow{\lambda} T'}{[a \neq b]T \xrightarrow{\lambda} T'}$$

Replication

$$!\bar{a}c.T \xrightarrow{\bar{a}c} T \mid !\bar{a}c.T \qquad !a(x).T \xrightarrow{ac} T\{c/x\} \mid !a(x).T$$

The notation $\Longrightarrow$ denotes the reflexive and transitive closure of $\xrightarrow{\tau}$.

3   Observational Theory

The first fundamental relationship in process theory is the equality relationship. It is argued in [Fu16] that from the point of view of both computation and interaction, there is only one equality that satisfies the following conditions:

- it is model independent;
- it is an equality for self-evolving and interactive objects.

Self-evolution is the feature of computation and interaction is what a process is supposed to do. We refer the reader to [Fu16] for a detailed argument and convincing technical support for the above remarks and the principles behind the argument. In this short paper we simply repeat the relevant definitions.

Definition 1. *A symmetric relation $\mathcal{R}$ on processes is a bisimulation if it validates the following bisimulation property:*

- *If $Q\mathcal{R}P \xrightarrow{\tau} P'$ then one of the following statements is valid:*
  (i) $Q \Longrightarrow Q'$ for some Q' such that $Q'\mathcal{R}P$ and $Q'\mathcal{R}P'$.
  (ii) $Q \Longrightarrow Q''\mathcal{R}P$ for some Q'' such that $\exists Q'.Q'' \xrightarrow{\tau} Q'\mathcal{R}P'$.

It is codivergent *if the following codivergence property is satisfied:*

- *If $Q\mathcal{R}P \xrightarrow{\tau} P_1 \xrightarrow{\tau} \ldots \xrightarrow{\tau} P_i \ldots$ is an infinite computation, then $\exists Q'.\exists i \geqslant 1.Q \xRightarrow{\tau} Q'\mathcal{R}P_i$.*

It is extensional *if the following extensionality property holds:*

1. *if $M\mathcal{R}N$ and $P\mathcal{R}Q$ then $(M \mid P)\mathcal{R}(N \mid Q)$;*
2. *if $P\mathcal{R}Q$ then $(a)P\mathcal{R}(a)Q$ for every $a \in \mathcal{N}$.*

It is equipollent *if $P \Downarrow \Leftrightarrow Q \Downarrow$ whenever $P\mathcal{R}Q$, where $P \Downarrow$, meaning that P is observable, if and only if $P \Longrightarrow \xrightarrow{\lambda} P'$ for some P' and some $\lambda \neq \tau$.*

The bisimulation of the above definition is what van Glabbeek and Weijland called branching bisimulation [vGW89, Bae96]. Codivergence is Priese's eventually progressing property [Pri78]. Equipollence is the most abstract form of Milner and Sangiorgi's barbness condition [MS92]. All the properties introduced in Definition 1 are model independent. Their combination imposes a minimal requirement from the point of computation as well as interaction and a maximal condition from the point of model independence.

Definition 2. *The absolute equality $=$ is the largest relation on processes validating the following statements:*

1. *The relation is reflexive;*
2. *The relation is equipollent, extensional, codivergent and bisimilar.*

The approach to extend the absolute equality from the processes to the terms is standard. The model independence necessarily means that the absolute equality is difficult to work with. If there is a single technical lemma that helps reason about $=$, it must be the Bisimulation Lemma stated next.

Lemma 1. *If $P \Longrightarrow P' = Q$ and $Q \Longrightarrow Q' = P$, then $P = Q$.*

The property stated in Lemma 1 is called X-property by De Nicola, Montanari and Vaandrager [DNMV90].

Once we have defined the absolute equality, we can distinguish two kinds of internal actions. We will write $T \xrightarrow{\iota} T'$ if $T \xrightarrow{\tau} T' \neq T$, and $T \to T'$ if $T \xrightarrow{\tau} T' = T$.

The observational theory discusses model specific characterizations of the absolute equality. A model dependent counterpart of $=$ is often far more tractable. An external bisimilarity is a Milner-Park style bisimulation [Mil89, Par81] in which every action is explicitly bisimulated. The external characterization of $=$ for the π-calculus is given in [Fu16]. For the π^L-calculus we can give an external counterpart in terms of a family of relations in the style of Sangiorgi's open bisimulations [San96b]. In the following definition $\subseteq_f$ stands for the finite subset relationship.

Definition 3. *A π^L-bisimulation is a family $\{\mathcal{R}_\mathcal{F}\}_{\mathcal{F} \subseteq_f \mathcal{N}}$ of codivergent symmetric bisimulations on $\mathcal{P}_{\pi^L}$ if the following statements are valid whenever $P\mathcal{R}_\mathcal{F}Q$:*

1. *If $P \xrightarrow{ab} P'$ then $Q \Longrightarrow Q'' \xrightarrow{ab} Q'\mathcal{R}_\mathcal{F}P'$ and $P\mathcal{R}_\mathcal{F}Q''$ for some Q',Q''.*
2. *If $P \xrightarrow{\bar{a}b} P'$ and $a \notin \mathcal{F}$ then $Q \Longrightarrow Q'' \xrightarrow{\bar{a}b} Q'\mathcal{R}_\mathcal{F}P'$ and $P\mathcal{R}_\mathcal{F}Q''$ for some Q',Q''.*
3. *If $P \xrightarrow{\bar{a}(c)} P'$ and $a \notin \mathcal{F}$ then $Q \Longrightarrow Q'' \xrightarrow{\bar{a}(c)} Q'\mathcal{R}_{\mathcal{F}\cup\{c\}}P'$ and $P\mathcal{R}_\mathcal{F}Q''$ for some Q',Q''.*

We write $\left\{\simeq_\mathcal{F}^{\pi^L}\right\}_{\mathcal{F} \subseteq_f \mathcal{N}}$ for the largest π^L-bisimulation, each $\simeq_\mathcal{F}^{\pi^L}$ is called the $\mathcal{F}$-π^L-bisimilarity. The $\simeq^{\pi^L}$-bisimilarity $\simeq^{\pi^L}$ is the $\emptyset$-π^L-bisimilarity.

The idea of Definition 3 is that an indexing set $\mathcal{F}$ records all the local names that have been opened up as it were by bound output actions. If $P \simeq_\mathcal{F}^{\pi^L} Q$ and $a \in \mathcal{F}$ then the action $\bar{a}b$ say need not be bisimulated for the reason that no environments that have received the local name a will ever do any input actions at a. A similar idea motivates the following definition.

Definition 4. *A π^R-bisimulation is a family $\{\mathcal{R}_\mathcal{F}\}_{\mathcal{F} \subseteq_f \mathcal{N}}$ of codivergent symmetric bisimulations on $\mathcal{P}_{\pi^R}$ if the followings are valid whenever $P\mathcal{R}_\mathcal{F}Q$:*

1. *If $P \xrightarrow{ab} P'$ and $a \notin \mathcal{F}$ then $Q \Longrightarrow Q'' \xrightarrow{ab} Q'\mathcal{R}_\mathcal{F}P'$ and $P\mathcal{R}_\mathcal{F}Q''$ for some Q',Q''.*
2. *If $P \xrightarrow{\bar{a}b} P'$ then $Q \Longrightarrow Q'' \xrightarrow{\bar{a}b} Q'\mathcal{R}_\mathcal{F}P'$ and $P\mathcal{R}_\mathcal{F}Q''$ for some Q',Q''.*
3. *If $P \xrightarrow{\bar{a}(c)} P'$ then $Q \Longrightarrow Q'' \xrightarrow{\bar{a}(c)} Q'\mathcal{R}_{\mathcal{F}\cup\{c\}}P'$ and $P\mathcal{R}_\mathcal{F}Q''$ for some Q',Q''.*

We write $\left\{\simeq_\mathcal{F}^{\pi^R}\right\}_{\mathcal{F} \subseteq_f \mathcal{N}}$ for the largest π^R-bisimulation, each $\simeq_\mathcal{F}^{\pi^R}$ is called the $\mathcal{F}$-π^R-bisimilarity. The $\simeq^{\pi^R}$-bisimilarity $\simeq^{\pi^R}$ is the $\emptyset$-π^R-bisimilarity.

The proof of the next lemma is routine.

Lemma 2. *Both $\simeq^{\pi^L}$ and $\simeq^{\pi^R}$ are equivalence and congruence relations.*

The next is another useful technical lemma.

Lemma 3. *The following statements are valid:*

1. *If $P \simeq_{\mathcal{F}\cup\{c\}}^{\pi^L} Q$ and c is not a global output channel in $P \mid Q$ then $P \simeq_\mathcal{F}^{\pi^L} Q$;*
2. *If $P \simeq_{\mathcal{F}\cup\{c\}}^{\pi^R} Q$ and c is not a global input channel in $P \mid Q$ then $P \simeq_\mathcal{F}^{\pi^R} Q$;*
3. *If $P \simeq_{\mathcal{F}\cup\{c\}}^{\pi'} Q$ for $\pi' \in \{\pi^L, \pi^R\}$ then $(c)(P \mid A) \simeq_\mathcal{F}^{\pi'} (c)(Q \mid A)$ for each A.*

Without further ado, we come to the main result of this section.

Theorem 1. *The following are valid:*

1. *The π^L-bisimilarity $\simeq^{\pi^L}$ coincides with $=^{\pi^L}$.*
2. *The π^R-bisimilarity $\simeq^{\pi^R}$ coincides with $=^{\pi^R}$.*

Proof. Lemma 2 implies both $\simeq^{\pi^L}\,\subseteq\,=^{\pi^L}$ and $\simeq^{\pi^R}\,\subseteq\,=^{\pi^R}$. The proof of the reverse inclusions is a modification of a proof in [Fu05]. For the present proof we only have to mention the part that differs from the previous proofs. (2) Let $\{\mathcal{R}_{\mathcal{F}}\}_{\mathcal{F}\subseteq_f\mathcal{N}}$ be defined in the following manner.

$$
\mathcal{R}_{\{c_1,\dots,c_n\}} \stackrel{\text{def}}{=} \left\{ (P,Q) \,\middle|\, \begin{array}{l} \{a_1,\dots,a_n\}\cap gn(P\mid Q)=\emptyset, \\ (c_1,\dots,c_n)(\overline{a_1}c_1\mid\dots\mid\overline{a_n}c_n\mid P) =^{\pi^R} \\ (c_1,\dots,c_n)(\overline{a_1}c_1\mid\dots\mid\overline{a_n}c_n\mid Q) \end{array} \right\}.
$$

We prove that $\{\mathcal{R}_{\mathcal{F}}\}_{\mathcal{F}\subseteq_f\mathcal{N}}$ is a π^R-bisimulation. Suppose $A = B$, where

$$
A \stackrel{\text{def}}{=} (c_1,\dots,c_n)(\overline{a_1}c_1\mid\dots\mid\overline{a_n}c_n\mid P),
$$
$$
B \stackrel{\text{def}}{=} (c_1,\dots,c_n)(\overline{a_1}c_1\mid\dots\mid\overline{a_n}c_n\mid Q),
$$

such that $\{a_1,\dots,a_n\}\cap gn(P\mid Q)=\emptyset$. Consider $P\stackrel{\overline{a}(c)}{\longrightarrow}P'$ for some $c\notin\{c_1,\dots,c_n\}$. Let d,f,a_{n+1} be fresh and let D be defined by

$$
D \stackrel{\text{def}}{=} a(x).(\overline{a_{n+1}}x\mid[x\notin gn(P\mid Q)]f)+a(x).d.
$$

Now

$$
A\mid D\stackrel{\tau}{\longrightarrow}(c)(A'\mid\overline{a_{n+1}}c\mid[c\notin gn(P\mid Q)]f)
$$

must be bisimulated by

$$
B\mid D\Longrightarrow B''\mid D\stackrel{\tau}{\longrightarrow}(c)(B'\mid\overline{a_{n+1}}c\mid[c\notin gn(P\mid Q)]f).
$$

Since $B''\mid D\stackrel{\tau}{\longrightarrow}(c)(B'\mid\overline{a_{n+1}}c\mid[c\notin gn(P\mid Q)]f))$ must be a change-of-state, it must be the case that $A\mid D=^{\pi^R}B''\mid D$. It follows easily from Bisimulation Lemma that $B\Longrightarrow B''\stackrel{\overline{a}(c)}{\longrightarrow}B'=^{\pi^R}A'$. Clearly,

$$
A' \equiv (c_1,\dots,c_n)(\overline{a_1}c_1\mid\dots\mid\overline{a_n}c_n\mid P'),
$$
$$
B' \equiv (c_1,\dots,c_n)(\overline{a_1}c_1\mid\dots\mid\overline{a_n}c_n\mid Q'),
$$
$$
B'' \equiv (c_1,\dots,c_n)(\overline{a_1}c_1\mid\dots\mid\overline{a_n}c_n\mid Q'')
$$

for some P',Q',Q''. It follows from $B\Longrightarrow B''\stackrel{\overline{a}(c)}{\longrightarrow}B'$ that $Q\Longrightarrow Q''\stackrel{\overline{a}(c)}{\longrightarrow}Q'$. Moreover $P\mathcal{R}_{\{c_1,\dots,c_n\}}Q''$ and $P'\mathcal{R}_{\{c_1,\dots,c_n,c\}}Q'$ by definition. We can symmetrically deal with (1). $\qquad\square$

4   Relative Expressiveness

The second fundamental relationship in process theory is the expressiveness relationship between process calculi. For this relationship to make sense at all, model independence has to be a born property. A theory of expressiveness is developed in [Fu16]. The philosophy of the theory is that the expressiveness relationship is the generalization of the absolute equality from one model to two models. Again we will simply repeat the definition here.

Definition 5. *Suppose* $\mathbb{M}, \mathbb{N}$ *are two* π-*variants. A binary relation* $\mathfrak{R} \subseteq \mathcal{P}_{\mathbb{M}} \times \mathcal{P}_{\mathbb{N}}$ *is a* subbisimilarity *if it validates the following statements.*

1. $\mathfrak{R}$ *is reflexive in the following sense:*
    (a) $\mathfrak{R}$ *is total, meaning that* $\forall M \in \mathcal{P}_{\mathbb{M}}.\exists N \in \mathcal{P}_{\mathbb{N}}.M\mathfrak{R}N.$
    (b) $\mathfrak{R}$ *is sound, meaning that* $N_1\mathfrak{R}^{-1}M_1 =_{\mathbb{M}} M_2\mathfrak{R}N_2$ *implies* $N_1 =_{\mathbb{N}} N_2.$
2. $\mathfrak{R}$ *is equipollent, extensional, codivergent and bisimilar.*

We say that $\mathbb{M}$ is subbisimilar to $\mathbb{N}$, notated by $\mathbb{M} \sqsubseteq \mathbb{N}$, if there is a subbisimilarity from $\mathbb{M}$ to $\mathbb{N}$. We write $\mathbb{M} \sqsubset \mathbb{N}$ if $\mathbb{M} \sqsubseteq \mathbb{N}$ and $\mathbb{N} \not\sqsubseteq \mathbb{M}$. Intuitively $\mathbb{M} \sqsubseteq \mathbb{N}$ means that $\mathbb{N}$ is at least as expressive as $\mathbb{M}$.

Theorem 2. *Suppose* $\mathbb{M}, \mathbb{N} \in \{\pi, \pi^L, \pi^R\}$. *If* $\mathbb{M}, \mathbb{N}$ *are distinct then* $\mathbb{M} \not\sqsubseteq \mathbb{N}$.

Proof. Suppose $\mathfrak{F}$ is a subbisimilarity from $\mathbb{M}$ to $\mathbb{N}$. The proof of Theorem 4.23 in [Fu16] essentially shows that, for all P, Q such that $P\mathfrak{F}Q$, the following *Global Bisimulation* property holds:

- If $P \xrightarrow{ac} P'$ then $Q \Longrightarrow Q'' \xrightarrow{ac} Q'\mathfrak{F}^{-1}P'$ and $P\mathfrak{F}Q''$ for some $Q'', Q'.$
- If $Q \xrightarrow{ac} Q'$ then $P \Longrightarrow P'' \xrightarrow{ac} P'\mathfrak{F}^{-1}Q'$ and $P''\mathfrak{F}Q$ for some $Q'', Q'.$
- If $P \xrightarrow{\bar{a}c} P'$ then $Q \Longrightarrow Q'' \xrightarrow{\bar{a}c} Q'\mathfrak{F}^{-1}P'$ and $P\mathfrak{F}Q''$ for some $Q'', Q'.$
- If $Q \xrightarrow{\bar{a}c} Q'$ then $P \Longrightarrow P'' \xrightarrow{\bar{a}c} P'\mathfrak{F}^{-1}Q'$ and $P''\mathfrak{F}Q$ for some $Q'', Q'.$

Using the above property the following crucial fact is established in [Fu16]:

Self Interpretation: $A\mathfrak{F}A$ whenever A contains no replication operator.

The Global Bisimulation property, the extensionality and Theorem 1 are necessary to guarantee the Self Interpretation property.

Now we can argue as follows:

- It should be clear that $a(x).x \ \mathfrak{F} \ a(x).x$ implies $\pi \not\sqsubseteq \pi^L$, $a(x).\bar{x} \ \mathfrak{F} \ a(x).\bar{x}$ implies $\pi \not\sqsubseteq \pi^R$.
- It follows from the Self Interpretation property that

$$(d)(\bar{a}d \,|\, \bar{d}.\bar{e}) \ \mathfrak{F} \ (d)(\bar{a}d \,|\, \bar{d}.\bar{e}), \tag{1}$$

$$(d)(\bar{a}d \,|\, \bar{d}) \ \mathfrak{F} \ (d)(\bar{a}d \,|\, \bar{d}). \tag{2}$$

The equality $(d)(\bar{a}d \,|\, \bar{d}.\bar{e}) = (d)(\bar{a}d \,|\, \bar{d})$ holds in π^L. It holds in none of π, π^R. Therefore (1) and (2) imply $\pi^L \not\sqsubseteq \pi$ and $\pi^L \not\sqsubseteq \pi^R$.

– Similarly the Self Interpretation property implies

$$(d)(\overline{a}d \mid d.\overline{e}) \; \mathfrak{F} \; (d)(\overline{a}d \mid d.\overline{e}),$$
$$(d)(\overline{a}d \mid d) \; \mathfrak{F} \; (d)(\overline{a}d \mid d),$$

from which $\pi^R \not\sqsubseteq \pi$ and $\pi^R \not\sqsubseteq \pi^L$ follow.

We are done. $\qquad\qquad\qquad\qquad\qquad\qquad\qquad\qquad\qquad\qquad\qquad\quad$ □

5   Expressive Completeness

The variants π^L, π^R, π^S would not be interesting if they are not complete. For interaction models completeness means that the computable functions can be encoded as interactive processes. There are many notions of Turing completeness. In this paper we adopt the definition introduced in [Fu16]. The idea is to formalize the following interactive version of the Church-Turing Thesis:

Axiom of Completeness. $\mathbb{C} \sqsubseteq \mathbb{M}$ for all models of interaction $\mathbb{M}$.

Here $\mathbb{C}$ is the Computability Model, which is basically the interaction model of computable function. The reader is referred to [Fu16] for the definition of $\mathbb{C}$ and what it means for $\mathbb{M}$ to satisfy $\mathbb{C} \sqsubseteq \mathbb{M}$. It is sufficient to say that a proof of completeness boils down to showing how the natural numbers are defined and how the recursive functions are translated into processes that can input natural numbers and output the computing results. To avoid confusion we will write $\underline{0}, \underline{1}, \underline{2}, \ldots, \underline{i}, \ldots$ for the natural numbers. We will use the following notations for the recursive functions defined in [Rog87]:

– $\mathsf{s}(x)$ is the successor function.
– $\underline{i}(x_1, \ldots, x_n)$ is the n-ary constant function with value $\underline{i}$.
– $\mathsf{p}_n^i(x_1, .., x_n)$ is the n-ary projection function at the i-th parameter.
– $\mathsf{f}(\mathsf{f}_1(\widetilde{x}), .., \mathsf{f}_i(\widetilde{x}))$ is the function composed of $\mathsf{f}(x_1, \ldots, x_i), \mathsf{f}_1(\widetilde{x}), .., \mathsf{f}_i(\widetilde{x})$.
– $\mathsf{rec}\, z.[\mathsf{f}(\widetilde{x}, x', z), \mathsf{g}(\widetilde{x})]$ is the recursion function defined by $\mathsf{f}(\widetilde{x}, x', z), \mathsf{g}(\widetilde{x})$.
– $\mu z.\mathsf{f}(\widetilde{x}, z)$ is the minimization function over $\mathsf{f}(\widetilde{x}, z)$.

Theorem 3. $\mathbb{C} \sqsubseteq \pi^L$, $\mathbb{C} \sqsubseteq \pi^R$ and $\mathbb{C} \sqsubseteq \pi^S$.

The encoding of the natural numbers in π^L is as $[\![\underline{0}]\!]_c^{\pi^L} \stackrel{\text{def}}{=} c(z).\overline{z}$ and $[\![\underline{n+1}]\!]_c^{\pi^L} \stackrel{\text{def}}{=} (d)(c(z).\overline{z}d \mid [\![\underline{n}]\!]_d^{\pi^L})$. Every number is accessible at a global name. The encoding makes use of a special name $\bot$. This is harmless because $\bot$ never appears as a channel name in our encoding. Since an input number might be used several time when computing a recursive function, a persistent form of the above encoding is necessary. This is given by $[\![!\underline{0}]\!]_c^{\pi^L} \stackrel{\text{def}}{=} !c(z).\overline{z}$ and $[\![!\underline{n+1}]\!]_c^{\pi^L} \stackrel{\text{def}}{=} (d)(!c(z).\overline{z}d \mid [\![!\underline{n}]\!]_d^{\pi^L})$. In sequel we shall use Milner's encoding of the polyadic π-prefixes in terms of the monadic prefixes [Mil93]. This is given by

$$a(x_1, \cdots, x_k).T \stackrel{\text{def}}{=} a(z).\bar{z}(c).c(x_1).\cdots.c(x_k).T,$$
$$\bar{a}\langle n_1, \cdots, n_k \rangle.T \stackrel{\text{def}}{=} \bar{a}(d).d(y).\bar{y}n_1.\cdots.\bar{y}n_k.T.$$

It is not a very faithful translation [Fu16]. But this point should not concern us since the above encoding will only be used locally. Now for every π^L-term T, we would like to introduce a π^L-term $Rp(n, a).T$ that is capable of converting an input number to its persistent form. More specifically, it enables the following change-of-state internal action: $[\![\underline{n}]\!]_c^{\pi^L} \mid Rp(c, d).T \stackrel{\iota}{\longrightarrow} = [\![!\underline{n}]\!]_d^{\pi^L} \mid T$. The process $Rp(n, a).T$ is defined by

$$(f)(\bar{n}(c).c(y).([y{=}\bot](!a(z).\bar{z}\bot \mid T) \mid [y{\neq}\bot]\bar{y}(c).c(z).(d)\overline{f}\langle d, z\rangle.!d(z_1).\overline{z_1}\bot)$$
$$\mid !f(u, v).([v{=}\bot](!a(z).\bar{z}u \mid T) \mid [v{\neq}\bot]\bar{v}(c).c(z).(d)\overline{f}\langle d, z\rangle.!d(z_2).\overline{z_2}u)).$$

Once we have the process $[\![\underline{n+1}]\!]_c^{\pi^L}$ and $[\![!\underline{n+1}]\!]_c^{\pi^L}$, we might want to make a copy of them when necessary. This is achieved by $Cp(n, a).T$ defined by

$$(f)(\bar{n}(c).c(y).([y{=}\bot](a(z).\bar{z}\bot \mid T) \mid [y{\neq}\bot]\bar{y}(c).c(z).(d)\overline{f}\langle d, z\rangle.d(z_1).\overline{z_1}\bot)$$
$$\mid !f(u, v).([v{=}\bot](a(z).\bar{z}u \mid T) \mid [v{\neq}\bot]\bar{v}(c).c(z).(d)\overline{f}\langle d, z\rangle.d(z_2).\overline{z_2}u)),$$

which is very much similar to the previous process. Clearly the following interactions are admissible.

$$[\![\underline{n}]\!]_c^{\pi^L} \mid Cp(c, d).T \stackrel{\iota}{\longrightarrow} = [\![\underline{n}]\!]_d^{\pi^L} \mid T,$$
$$[\![!\underline{n}]\!]_c^{\pi^L} \mid Cp(c, d).T \stackrel{\iota}{\longrightarrow} = [\![!\underline{n}]\!]_c^{\pi^L} \mid [\![\underline{n}]\!]_d^{\pi^L} \mid T.$$

An n-ary function $f(x_1, \cdots, x_n)$ is translated to a process that picks up n inputs consecutively before outputting the result. The input and output actions must be carried out in particular channels. We write $[\![F_{a_1 \cdots a_n}^b(f(x_1, \cdots, x_n))]\!]^{\pi^L}$ for the translation of $f(x_1, \cdots, x_n)$ in π^L at the input channels $a_1, \ldots, a_n$ and the output channel b. The structural definition goes as follows:

- The successor, constant, projection and composition functions are defined as follows:

$$[\![F_{a_1}^b(s(x))]\!]^{\pi^L} \stackrel{\text{def}}{=} (d_1)Rp(a_1, d_1).(c)Cp(d_1, c).b(x).\bar{x}(c),$$
$$[\![F_{a_1 \cdots a_n}^b(in(x_1, \cdots, x_n))]\!]^{\pi^L} \stackrel{\text{def}}{=} (d_1)Rp(a_1, d_1).\cdots.(d_n)Rp(a_n, d_n).[\![\underline{i}]\!]_b^{\pi^L},$$
$$[\![F_{a_1 \cdots a_n}^b(p_n^i(x_1, \cdots, x_n))]\!]^{\pi^L} \stackrel{\text{def}}{=} (d_1)Rp(a_1, d_1).\cdots.(d_n)Rp(a_n, d_n).Cp(d_i, b),$$
$$[\![F_{a_1 \cdots a_n}^b(f(f_1(\tilde{x}), \cdots, f_i(\tilde{x})))]\!]^{\pi^L} \stackrel{\text{def}}{=} (d_1)Rp(a_1, d_1).\cdots.(d_n)Rp(a_n, d_n).(c_1 \cdots c_i)$$
$$([\![F_{c_1 \cdots c_i}^b(f(\tilde{x}))]\!]^{\pi^L} \mid [\![F_{d_1 \cdots d_n}^{c_1}(f_1(\tilde{x}))]\!]^{\pi^L}$$
$$\mid \cdots \mid [\![F_{d_1 \cdots d_n}^{c_i}(f_i(\tilde{x}))]\!]^{\pi^L})$$

– $[\![F^b_{a_1\cdots a_{n+1}}(\mathrm{rec}z.[\mathsf{f}(\widetilde{x},x',z),\mathsf{g}(\widetilde{x})])]\!]^{\pi^L}$ is the following process

$$(d_1)Rp(a_1,d_1).\cdots.(d_{n+1})Rp(a_{n+1},d_{n+1}).\overline{d_{n+1}}(c).c(y).$$
$$([y{=}\bot][\![F^b_{d_1,\cdots,d_n}(\mathsf{g}(\widetilde{x}))]\!]^{\pi^L}$$
$$\mid[y{\neq}\bot](f)(Rec\mid(g)\overline{f}\langle g,y\rangle.g(w).[\![F^b_{d_1,\cdots,d_n,w,y}\mathsf{f}(\widetilde{x},x',z)]\!]^{\pi^L}))$$

where Rec stands for

$$!f(u,v).\overline{u}(d).([v{=}\bot][\![F^d_{d_1,\cdots,d_n}(\mathsf{g}(\widetilde{x}))]\!]^{\pi^L}$$
$$\mid[v{\neq}\bot]\overline{v}(c).c(y).(g)\overline{f}\langle g,y\rangle.(g(w).[\![F^d_{d_1,\cdots,d_n,w,v}(\mathsf{f}(\widetilde{x},x',x''))]\!]^{\pi^L})).$$

– $[\![F^b_{a_1\cdots a_n}(\mu z.[\mathsf{f}(\widetilde{x},z)])]\!]^{\pi^L}$ is the following process

$$(d_1)Rp(a_1,d_1).\cdots.(d_n)Rp(a_n,d_n).(f)(Mu\mid\overline{f}(c).[\![\underline{0}]\!]^{\pi^L}_c)$$

where Mu stands for

$$!f(v).(g)Rp(v,g).(d)([\![F^d_{d_1,\cdots,d_n,g}(\mathsf{f}(\widetilde{x},z))]\!]^{\pi^L}$$
$$\mid\overline{d}(e).e(z).([z{=}\bot]Cp(g,b).\mid[z{\neq}\bot]\overline{f}(c).(c')Cp(g,c').c(y).\overline{y}c')).$$

This completes the definition of $[\![\_]\!]^{\pi^L}$. The reader can work out the encoding $[\![\_]\!]^{\pi^R}$ symmetrically. The proof of the completeness of π^S is subsumed by that for π^P. See Sect. 7 for more details. We only have to remark that the parametric definitions can be implemented in π^S using the replication operator.

6 Proof System

Based on Theorem 1 one may talk about complete equational proof systems for the absolute equality of the π-variants. In view of Theorem 3 no decidable proof system is possible for all processes. However the finite fragment consisting of **0**, the choice operator, the match/mismatch operator and the localization operator is decidable. Equational systems for various congruence relations on the finite π-processes are well-known. The paper by Parrow and Sangiorgi [PS95] deserves particular attention. A complete equational system AS for the absolute equality on the finite π-terms is studied in [FZ15]. In this section we shall briefly explain how to construct complete systems for π^L and π^R by extending AS.

| L | $\overline{n}(c).C[\sum_{i\in I}\overline{c}m_i.T_i]=\overline{n}(c).C[\mathbf{0}]$ |
| R | $\overline{n}(c).C[\sum_{i\in I}c(x).T_i]=\overline{n}(c).C[\mathbf{0}]$ |

Fig. 1. Axioms for the variants.

In Fig. 1 two axioms are proposed. The law L is valid for $\simeq^{\pi^L}$ and the law R is valid for $\simeq^{\pi^R}$. Let AS_L be $AS \cup \{L\}$ and AS_R be $AS \cup \{R\}$. The first indication of the power of AS_L, as well as AS_R, is a normalization lemma stating that all finite terms can be converted to some normal forms. Due to the presence of mobility the normal forms for the π^L-terms are a bit involved. But the main definition is the following.

Definition 6. *Suppose* $\mathcal{F}, \mathcal{G} \subseteq_f \mathcal{N} \cup \mathcal{N}_v$. *The* π^L-*term* T *is a normal form on* $\mathcal{F}$ *and* $\mathcal{G}$ *if it is of the form*

$$\sum_{i \in I} \lambda_i.T_i$$

such that for each $i \in I$ *one of the followings holds.*

1. *If* $\lambda_i = \tau$ *then* T_i *is a normal form on* $\mathcal{F}$ *and* $\mathcal{G}$.
2. *If* $\lambda_i = \bar{n}m$ *and* $n \notin \mathcal{G}$ *then* T_i *is a normal form on* $\mathcal{F}$ *and* $\mathcal{G}$.
3. *If* $\lambda_i = \bar{n}(c)$ *and* $n \notin \mathcal{G}$ *then* $T_i \equiv (\bigwedge_{n \in \mathcal{F}} c{\neq}n)T_i^c$ *for some normal form* T_i^c *on* $\mathcal{F} \cup \{c\}$ *and* $\mathcal{G} \cup \{c\}$.
4. *If* $\lambda_i = c(x)$ *then* T_i *is of the form*

$$(\bigwedge_{n \in \mathcal{F}} x{\neq}n)T_i^{\neq} + \sum_{m \in \mathcal{F}} [x{=}m]T_i^m$$

such that $T_i^{\neq}$ *is a normal form on* $\mathcal{F} \cup \{x\}$ *and* $\mathcal{G}$, *and, for each* $m \in \mathcal{F}$, $x \notin fv(T_i^m)$ *and* T_i^m *is a normal form on* $\mathcal{F}$ *and* $\mathcal{G}$.

The reader can easily work out the definition of the normal forms for the π^R-processes from Definitions 3, 4 and 6. The rest of the arguments and proofs are almost an reiteration of corresponding arguments and proofs in [FZ15]. Without further ado, let's state the main result.

Theorem 4. *The following statements are valid:*

1. $S \simeq^{\pi^L} T$ *if and only if* $AS_L \vdash \tau.S = \tau.T$ *for all finite* π^L-*terms* S, T.
2. $S \simeq^{\pi^R} T$ *if and only if* $AS_R \vdash \tau.S = \tau.T$ *for all finite* π^R-*terms* S, T.

The above theorem is concerned with π^L-terms, from which we can easily derive that $P \simeq^{\pi^L} Q$ if and only if $AS_L \vdash P = Q$ for all finite π^L-processes P, Q and that $P \simeq^{\pi^R} Q$ if and only if $AS_R \vdash P = Q$ for all finite π^R-processes P, Q.

7   Private Pi

The private π-calculus of Sangiorgi [San96a], denoted by π^P, is interesting in that it is a nontrivial model that stays between CCS and the π-calculus. All mobility admitted in π^P is internal. It is shown in [FL10] that π^P fails to be complete if recursion is provided by the replication operator. So normally π^P comes with the parametric definition. Here is its grammar:

$$T := \mathbf{0} \mid \sum_{i \in I} n(x).T_i \mid \sum_{i \in I} \bar{n}(c).T_i \mid T \mid T' \mid (c)T \mid D(p_1, \ldots, p_n).$$

A parametric definition is given by

$$D(x_1, \ldots, x_n) = T, \tag{3}$$

where $\{x_1, \ldots, x_n\}$ is the set of all the name variables in T. An instantiation of the parametric definition at $p_1, \ldots, p_n$, denoted by $D(p_1, \ldots, p_n)$, is the term $T\{p_1/x_1, \ldots, p_n/x_n\}$. The match and mismatch operators are absent in π^P because they are useless in this particular model. The operational semantics of π^P can be read off from the semantics of the π-calculus. We will not repeat it here.

The Turing completeness of π^P [BGZ03] does not imply the completeness of π^P since the latter is a strictly stronger property. We still need be assured that π^P is a legitimate model according to the Axiom of Completeness. Unlike in π^L and π^R the natural numbers in π^P must be defined in prefix form.

$$[\![0]\!]_p^{\pi^P} \stackrel{\text{def}}{=} \overline{p}(b_0, b_1).\overline{b_0}, \tag{4}$$

$$[\![n+1]\!]_p^{\pi^P} \stackrel{\text{def}}{=} \overline{p}(b_0, b_1).[\![n]\!]_{b_1}^{\pi^P}. \tag{5}$$

To get a feeling of how (4,5) work, let's see how the successor function Sc and the predecessor function Sb are defined.

$$Sc(u, v).T \stackrel{\text{def}}{=} u(x_0, x_1).(\overline{v}(e_0, e_1).\overline{e_1}(f_0, f_1).Com(x_0, x_1, f_0, f_1) \,|\, T),$$

$$Sb(u, v).T \stackrel{\text{def}}{=} u(x_0, x_1).(x_0.\overline{v}(e_0, e_1).\overline{e_0}$$
$$|\, x_1(y_0, y_1).\overline{v}(e_0, e_1).Com(y_0, y_1, e_0, e_1) \,|\, T),$$

where Com is introduced by the following parametric definition:

$$Com(x_0, x_1, y_0, y_1) = x_0.\overline{y_0} \,|\, x_1(z_0, z_1).\overline{y_1}(c_0, c_1).Com(z_0, z_1, c_0, c_1).$$

The interesting thing about Com is that it works in a lazy fashion. It is only when a produced number is being used can the copy mechanism of Com be invoked. This feature is prominent in all the following encodings. For example the persistent form of the natural number must make use of Com:

$$[\![!0]\!]_p^{\pi^P} = [\![!0]\!]_p^{\pi^P} \,|\, [\![0]\!]_p^{\pi^P},$$

$$[\![!n+1]\!]_p^{\pi^P} \stackrel{\text{def}}{=} (c)([\![!n]\!]_c^{\pi^P} \,|\, !PSc(c, p)),$$

where

$$PSc(u, v).T \stackrel{\text{def}}{=} \overline{v}(e_0, e_1).(u(x_0, x_1).\overline{e_1}(f_0, f_1).Com(x_0, x_1, f_0, f_1) \,|\, T).$$

The copy term can be simply defined in terms of Com:

$$Cp(u, v).T \stackrel{\text{def}}{=} u(x_0, x_1).(\overline{v}(e_1, e_2).Com(x_0, x_1, e_1, e_2) \,|\, T).$$

It is not difficult to see that the following actions are admissible:

$$[\![n]\!]_a^{\pi^P} \mid Cp(a,b).T \stackrel{\iota}{\longrightarrow} = [\![n]\!]_b^{\pi^P} \mid T,$$

$$[\![!n]\!]_a^{\pi^P} \mid Cp(a,b).T \stackrel{\iota}{\longrightarrow} = [\![!n]\!]_a^{\pi^P} \mid [\![n]\!]_b^{\pi^P} \mid T.$$

The definition of the replication term is slightly more involved:

$$Rp(u,v).T \stackrel{\text{def}}{=} u(x_0,x_1).(x_0.[\![0]\!]_v^{\pi^P} \mid x_1(y_0,y_1).(c)(!PSc(c,v) \mid R(y_0,y_1,c)) \mid T),$$

$$R(y_0,y_1,w) = y_0.[\![0]\!]_w^{\pi^P} \mid y_1(z_0,z_1).(c)(!PSc(c,w) \mid R(z_0,z_1,c)).$$

The reader is advised to verify that the following action is admissible:

$$[\![n]\!]_a^{\pi^P} \mid Rp(a,b).T \stackrel{\iota}{\longrightarrow} = [\![!n]\!]_b^{\pi^P} \mid T.$$

Another process useful to the encoding is the following:

$$Eq(u,v).(T_0,T_1) \stackrel{\text{def}}{=} u(x_0,x_1).v(y_0,y_1).$$
$$(d_0d_1)(E(x_0,x_1,y_0,y_1,d_0,d_1) \mid \overline{d_0}.T_0 \mid \overline{d_1}.T_1),$$

$$E(x_0,x_1,y_0,y_1,u_0,u_1) \stackrel{\text{def}}{=} x_0.(y_0.u_0 \mid y_1(z_0,z_1).u_1)$$
$$\mid x_1(z_0,z_1).(y_0.u_1 \mid y_1(w_0,w_1).E(z_0,z_1,w_0,w_1,u_0,u_1)).$$

The encoding $[\![\_]\!]^{\pi^P}$ of the computable functions in π^P can now be given. It should be enough to explain how the recursion functions and the minimization functions are interpreted.

- $[\![F_{a_1\cdots a_{n+1}}^b(\mathrm{rec}z.[\mathrm{f}(\tilde{x},x',z),\mathrm{g}(\tilde{x})])]\!]^{\pi^P}$ is defined by the following process

$$(d_1)Rp(a_1,d_1).\cdots.(d_n)Rp(a_n,d_n).(d_{n+1})Rp(a_{n+1},d_{n+1}).$$
$$(d_0)([\![!\overline{d_0}(0)]\!]^{\pi^P} \mid Eq(d_0,d_{n+1}).([\![F_{\tilde{d}}^b(\mathrm{g}(\tilde{x}))]\!]^{\pi^P},$$
$$(e'eg)Sb(d_{n+1},e').Rp(e',e).([\![F_{\tilde{d}ge}^b(\mathrm{f}(\tilde{x},x',x''))]\!]^{\pi^P} \mid Rec(g,e,d_0,\tilde{d})))),$$

where

$$Rec(u,v,z_0,\tilde{z}) = Eq(z_0,v).([\![F_{\tilde{z}}^u(\mathrm{g}(\tilde{x}))]\!]^{\pi^P},$$
$$(e'eg)Sb(v,e').Rp(e',e).([\![F_{\tilde{z}ge}^u(\mathrm{f}(\tilde{x},x',x''))]\!]^{\pi^P} \mid Rec(g,e,z_0,\tilde{z}))).$$

- $[\![F_{a_1\cdots a_n}^b(\mu z.[\mathrm{f}(\tilde{x},z)])]\!]^{\pi^P}$ is the following process

$$(d_1)Rp(a_1,d_1).\cdots.(d_n)Rp(a_n,d_n).$$
$$(d_0d')([\![!\overline{d_0}(0)]\!]^{\pi^P} \mid [\![F_{\tilde{d},d_0}^{d'}(\mathrm{f}(\tilde{x},z))]\!]^{\pi^P} \mid Eq(d_0,d').(Cp(d_0,b),$$
$$(e'ef)Sc(d_0,e').Rp(e',e).([\![F_{\tilde{d},e}^f(\mathrm{f}(\tilde{x},z))]\!]^{\pi^P} \mid Mu(f,e,d_0,\tilde{d},b)))),$$

where

$$Mu(u,v,z_0,\tilde{z},w) = Eq(z_0,u).(Cp(v,w),$$
$$(e'ef)Sc(v,e').Rp(e',e).([\![F_{\tilde{d},e}^f(\mathrm{f}(\tilde{x},z))]\!]^{\pi^P} \mid Mu(f,e,z_0,\tilde{z},w))).$$

We have coded up all the computable functions. Hence the next result.

Theorem 5. π^P, *and consequently* π^S *as well, are complete.*

8 Remark

We have provided the observational theories for π^L and π^R. Our attempt to do the same thing for π^S, π^P has not been successful. What has stopped us from getting a similar picture for the latter is the absence of an external characterization of the absolute equality for either variant. It is easy to conceive a family of explicit bisimilarities for π^S. But we probably need a new technique to show that it gives rise to an alternative way of defining the absolute equality in π^S. The external characterization of $=$ for π^P appears more elusive. For one thing the match and the mismatch operators are not of any use in any proof since they are redundant in π^P. So π^P poses a bigger challenge.

The completeness of all the four calculi raises the following question: What kind of universal processes does each of them have? It has been shown in [Fu] that the π-calculus proper has very powerful universal processes. The situations in $\pi^L, \pi^R, \pi^S, \pi^P$ remain to be seen.

Acknowledgments. This work has been supported by National Natural Science Foundation of China (61502296, 61472239, 61261130589) and Natural Science Foundation of Shanghai (15ZR1417000).

References

[ACS96] Amadio, R.M., Castellani, I., Sangiorgi, D.: On bisimulations for the asynchronous π-calculus. In: Montanari, U., Sassone, V. (eds.) CONCUR 1996. LNCS, vol. 1119, pp. 147–162. Springer, Heidelberg (1996). doi:10.1007/3-540-61604-7_53

[Bae96] Baeten, J.: Branching bisimilarity is an equivalence indeed. Inform. Process. Lett. **58**, 141–147 (1996)

[BGZ03] Busi, N., Gabbrielli, M., Zavattaro, G.: Replication vs. recursive definitions in channel based calculi. In: Baeten, J.C.M., Lenstra, J.K., Parrow, J., Woeginger, G.J. (eds.) ICALP 2003. LNCS, vol. 2719, pp. 133–144. Springer, Heidelberg (2003). doi:10.1007/3-540-45061-0_12

[BGZ04] Busi, N., Gabbrielli, M., Zavattaro, G.: Comparing recursion, replication, and iteration in process calculi. In: Díaz, J., Karhumäki, J., Lepistö, A., Sannella, D. (eds.) ICALP 2004. LNCS, vol. 3142, pp. 307–319. Springer, Heidelberg (2004). doi:10.1007/978-3-540-27836-8_28

[Bou92] Boudol, G.: Asynchrony and the π-calculus. Technical report RR-1702, INRIA Sophia-Antipolis (1992)

[CF11] Cai, X., Fu, Y.: The λ-calculus in the π-calculus. Math. Struct. Comput. Sci. **21**, 943–996 (2011)

[DNMV90] Nicola, R., Montanari, U., Vaandrager, F.: Back and forth bisimulations. In: Baeten, J.C.M., Klop, J.W. (eds.) CONCUR 1990. LNCS, vol. 458, pp. 152–165. Springer, Heidelberg (1990). doi:10.1007/BFb0039058

[FL10] Fu, Y., Lu, H.: On the expressiveness of interaction. Theoret. Comput. Sci. **411**, 1387–1451 (2010)

[Fu] Fu, Y.: The universal process. In: Logical Methods in Computer Science (to appear)

[Fu05]   Fu, Y.: On quasi open bisimulation. Theoret. Comput. Sci. **338**, 96–126 (2005)

[Fu10]   Fu, Y.: Theory by process. In: Gastin, P., Laroussinie, F. (eds.) CONCUR 2010. LNCS, vol. 6269, pp. 403–416. Springer, Heidelberg (2010). doi:10.1007/978-3-642-15375-4_28

[Fu13]   Fu, Y.: The Value-Passing Calculus. In: Liu, Z., Woodcock, J., Zhu, H. (eds.) Theories of Programming and Formal Methods. LNCS, vol. 8051, pp. 166–195. Springer, Heidelberg (2013). doi:10.1007/978-3-642-39698-4_11

[Fu16]   Fu, Y.: Theory of interaction. Theoret. Comput. Sci. **611**, 1–49 (2016)

[Fu17]   Fu, Y.: On the expressive power of name-passing communication. In: CONCUR 2017 (2017)

[FZ15]   Fu, Y., Zhu, H.: The name-passing calculus. arXiv:1508.00093 (2015)

[Gor08]   Gorla, D.: Comparing communication primitives via their relative expressive power. Inf. Comput. **206**, 931–952 (2008)

[GSV04]   Giambiagi, P., Schneider, G., Valencia, F.D.: On the expressiveness of infinite behavior and name scoping in process calculi. In: Walukiewicz, I. (ed.) FoSSaCS 2004. LNCS, vol. 2987, pp. 226–240. Springer, Heidelberg (2004). doi:10.1007/978-3-540-24727-2_17

[Hoa78]   Hoare, C.: Communicating sequential processes. Commun. ACM **21**, 666–677 (1978)

[HT91a]   Honda, K., Tokoro, M.: An object calculus for asynchronous communication. In: America, P. (ed.) ECOOP 1991. LNCS, vol. 512, pp. 133–147. Springer, Heidelberg (1991). doi:10.1007/BFb0057019

[HT91b]   Honda, K., Tokoro, M.: On asynchronous communication semantics. In: Tokoro, M., Nierstrasz, O., Wegner, P. (eds.) ECOOP 1991. LNCS, vol. 612, pp. 21–51. Springer, Heidelberg (1992). doi:10.1007/3-540-55613-3_2

[LPSS08] Lanese, I., Perez, J., Sangiorgi, D., Schmitt, A.: On the expressiveness and decidability of higher-order process calculi. In: Proceedings of LICS 2008, pp. 145–155 (2008)

[Mer00]   Merro, M.: Locality in the π-calculus and applications to object-oriented languages. PhD thesis, Ecole des Mines de Paris (2000)

[Mil89]   Milner, R.: Communication and Concurrency. Prentice Hall, Upper Saddle River (1989)

[Mil92]   Milner, R.: Functions as processes. Math. Struct. Comput. Sci. **2**, 119–146 (1992)

[Mil93]   Milner, R.: The polyadic π-calculus: a tutorial. In: Bauer, F.L., Brauer, W., Schwichtenberg, H. (eds.) Logic and Algebra of Specification. NATO ASI Series (Series F: Computer & Systems Sciences), vol. 94, pp. 203–246. Springer, Heidelberg (1993). doi:10.1007/978-3-642-58041-3_6

[MPW92]   Milner, R., Parrow, J., Walker, D.: A calculus of mobile processes. Inform. Comput. **100**, 1–40 (Part I), 41–77 (Part II) (1992)

[MS92]   Milner, R., Sangiorgi, D.: Barbed bisimulation. In: Kuich, W. (ed.) ICALP 1992. LNCS, vol. 623, pp. 685–695. Springer, Heidelberg (1992). doi:10.1007/3-540-55719-9_114

[MS04]   Merro, M., Sangiorgi, D.: On asynchrony in name-passing calculi. Math. Struct. Comput. Sci. **14**, 715–767 (2004)

[Pal03]   Palamidessi, C.: Comparing the expressive power of the synchronous and the asynchronous π-calculus. Math. Struct. Comput. Sci. **13**, 685–719 (2003)

[Par81] Park, D.: Concurrency and automata on infinite sequences. In: Deussen, P. (ed.) GI-TCS 1981. LNCS, vol. 104, pp. 167–183. Springer, Heidelberg (1981). doi:10.1007/BFb0017309

[Pri78] Priese, L.: On the concept of simulation in asynchronous, concurrent systems. Progress Cybern. Syst. Res. **7**, 85–92 (1978)

[PS95] Parrow, J., Sangiorgi, D.: Algebraic theories for name-passing calculi. Inf. Comput. **120**, 174–197 (1995)

[Rog87] Rogers, H.: Theory of Recursive Functions and Effective Computability. MIT Press, Cambridge (1987)

[San93] Sangiorgi, D.: From π-calculus to higher-order π-calculus – and back. In: Gaudel, M.-C., Jouannaud, J.-P. (eds.) CAAP 1993. LNCS, vol. 668, pp. 151–166. Springer, Heidelberg (1993). doi:10.1007/3-540-56610-4_62

[San96a] Sangiorgi, D.: π-calculus, internal mobility and agent-passing calculi. Theoret. Comput. Sci. **167**, 235–274 (1996)

[San96b] Sangiorgi, D.: A theory of bisimulation for π-calculus. Acta Informatica **3**, 69–97 (1996)

[SW01] Sangiorgi, D., Walker, D.: The π Calculus: A Theory of Mobile Processes. Cambridge University Press, Cambridge (2001)

[Tho95] Thomsen, B.: A theory of higher order communicating systems. Inf. Comput. **116**, 38–57 (1995)

[vGW89] van Glabbeek, R., Weijland, W.: Branching time and abstraction in bisimulation semantics. In: Information Processing 1989, North-Holland, pp. 613–618 (1989)

[Wal95] Walker, D.: Objects in the π-calculus. Inf. Comput. **116**, 253–271 (1995)

[XYL15] Xu, X., Yin, Q., Long, H.: On the computation power of name parameterization in higher-order processes. In: ICE 2015 (2015)

Reasoning About Periodicity on Infinite Words

Wanwei Liu[1(✉)], Fu Song[2], and Ge Zhou[1]

[1] College of Computer Science, National University of Defense Technology,
Changsha, China
wwliu@nudt.edu.cn
[2] School of Information Science and Technology,
ShanghaiTech University, Pudong, China

Abstract. Characterization of temporal properties is the original purpose of inventing of temporal logics. In this paper, we show that the property like "some event holds periodically" is not omega-regular. Such property is called "periodicity", which plays an important role in task scheduling and system design. To give a characterization of periodicity, we present the logic QPLTL, which is an extension of LTL via adding quantified step variables. Based on the decomposition theorem, we show that the SATISFIABILITY problem of QPLTL is **PSPACE**-complete.

1 Introduction

In some sense, concurrency and interleaving are the main course of complexity in distributed and parallel programs. For such a program, we are in general concerned about its interactive behaviors, rather than the final output. Hence, various temporal logics are designed to characterize such issues of parallel programs. Characterization of temporal properties is always one of the central topics in the research on temporal logics.

Linear-time temporal logic (LTL), is one of the most frequently used logics, which is obtained from propositional logic via merely adding two temporal connectives — X (next) and U (until). Simple as such logic is, LTL could express a majority of properties which are usually concerned about — for example, *responsibility* — such as the formula $G(req \rightarrow F\ ack)$, which depicts the assertion that "each *req*uest will be eventually *ack*nowledged".

In [Wol83], Wolper pointed out some properties, like "p holds at least in every even moment" (we refer to it as G^2p in this paper), cannot be expressed by any LTL formula. As a consequence, numerous extensions are presented to enhance the expressive power. To mention a few: In [VW94], ω-automata are employed as extended temporal connectives. In [BB87], Banieqbal and Barringer yield the linear-time version of modal μ-calculus. In [LS07], Leucker and Sánchez propose to use regular-expressions as prefixes of formulae. All of these logics are shown to be as expressive as the full ω-regular languages, or the whole set of (nondeterministic) Büchi automata [Büc62]. As a result, properties like G^2p can be described by these logics. In the view of formal languages, LTL formulae precisely correspond to *star-free* ω-regular languages [Tho79], in which

© Springer International Publishing AG 2017
K.G. Larsen et al. (Eds.): SETTA 2017, LNCS 10606, pp. 200–215, 2017.
https://doi.org/10.1007/978-3-319-69483-2_12

the Kleene-star ($*$) and the omega-power (ω) operators can only be applied to Σ (the alphabet). Meanwhile, in the automata-theoretic perspective, LTL formulae are inter-convertible with Büchi automata of special forms, e.g., *counter-free automata* or *aperiodic automata* (cf. [DG08] for a comprehensive survey).

We say that an event f happens *periodically*, if there exists some number $n > 0$ (called the *period*, which is not pre-given) such that f holds at least in every moment which is a multiple of n. Such kind of property is called *periodicity*, which plays an important role in task scheduling and system design. For example, when designing a synchronous circuit, we need to guarantee the clock interrupt generates infinitely and equidistantly. An aerospace control system has a diagnosis module which periodically checks whether the system works properly. If it was not, then the system immediately enters the recovery state.

Enlightened by the fact that "$G^2 p$ cannot be expressed by any LTL formula", we conjecture that "the periodicity property $\exists k. G^k p$ is not expressible by any ω-regular language". We give an affirmative proof of this conjecture in this paper, and hence deduce that ω-regular properties are not closed under infinite unions and/or intersections. To express periodicity within linear-time framework, we tentatively suggest to add quantified *step variables* into logics. For example, if we use μTL [BB87] as the base logic, the aforementioned property could be described by the formula $\exists k. \nu Z. (p \wedge X^k Z)$. However, such mechanism gives rise to an expressive power beyond expectations — we show that, if we allow nesting of step variables, all formulae of Peano arithmetic can be encoded — provided that the base logic involves the X-operator and two distinct propositions. Hence, it leads to undecidability for the SATISFIABILITY problem of the logic.

Thus, to obtain a decidable extension, we have to impose strong syntactic restriction to the logic. In this paper, using LTL as the base logic, we introduce the logic *Quantified Periodic LTL* (QPLTL, for short). In such logic, formulae are categorized into four groups, the first three are contained within ω-regular languages, and the last group is specially tailored for defining periodicity: Each of such formulae uses at most one step variable, and the occurrence of this variable must be of special form. We show that, for QPLTL, the SATISFIABILITY problem is also **PSPACE**-complete. Our proof approach is mainly based on the *decomposition theorem* [AS85], namely, each property could be decomposed into an intersection of a liveness property and a safety property, with which, we give a normal form of star-free liveness properties. Instead of deciding the satisfiability of the formulae, we give a decision procedure to decide their validity.

The rest part of this paper is organized as follows: Sect. 2 introduces some basic notations and definitions. In Sect. 3, we show that periodicity properties are not ω-regular. Thus, we suggest to add quantified step variables into linear-time logics to gain such an enhancement. However, we show that an unrestricted use of such extension will result in undecidability of SATISFIABILITY, provided that the base logic involves the X-operator and at least two propositions. We then present the logic QPLTL in Sect. 4 via imposing strict syntactic constraints to the use of step variables, by revealing a normal form of star-free liveness properties, we show that the SATISFIABILITY problem of the proposed logic is **PSPACE**-complete. We summarize this paper and discuss future work in Sect. 5.

2   Preliminaries

2.1   Finite and Infinite Words

Fix an alphabet Σ, whose elements are called *letters*. We call w an ω-*word* (or *infinite*-word) if $w \in \Sigma^\omega$, and we call w a *finite-word* if $w \in \Sigma^*$.

We use $|w|$ to denote the *length* of w, and definitely $|w| = \infty$ if w is an ω-word. For each $i < |w|$, let $w(i)$ be the ith letter of w. Remind the first letter should be $w(0)$.

Given $w_1 \in \Sigma^*$ and $w_2 \in \Sigma^* \cup \Sigma^\omega$, we denote by $w_1 \cdot w_2$ the *concatenation* of w_1 and w_2. Namely, we have: $(w_1 \cdot w_2)(i) = w_1(i)$ whenever $i < |w_1|$; and $(w_1 \cdot w_2)(i) = w_2(i - |w_1|)$ whenever $i \geq |w_1|$. In this case, we say that w_1 and w_2 are respectively a *prefix* and a *postfix* (or *suffix*) of $w_1 \cdot w_2$.

In the rest of this paper, we fix a (potentially infinite) set $\mathcal{P}$ of *propositions*, the elements range over p, p_1, p_2, etc., and when mentioning about "words", without explicit declaration, we let $\Sigma = 2^\mathcal{P}$.

2.2   Linear-Time Temporal Logic

Syntax. Formulae of linear-time temporal logic (LTL, [Pnu77]), ranging over $f, g, f_1, f_2, \ldots$, can be described by the following abstract grammar:

$$f ::= \bot \mid p \mid f \rightarrow f \mid \mathsf{X} f \mid f \, \mathsf{U} \, f.$$

We usually use the following derived operators as syntactic sugars:

$$\neg f \overset{\text{def}}{=} f \rightarrow \bot \qquad \top \overset{\text{def}}{=} \neg\bot \qquad f_1 \vee f_2 \overset{\text{def}}{=} \neg f_1 \rightarrow f_2$$
$$f_1 \wedge f_2 \overset{\text{def}}{=} \neg(\neg f_1 \vee \neg f_2) \qquad \mathsf{F} f \overset{\text{def}}{=} \top \, \mathsf{U} \, f \qquad f_1 \leftrightarrow f_2 \overset{\text{def}}{=} (f_1 \rightarrow f_2) \wedge (f_2 \rightarrow f_1)$$
$$f_1 \mathsf{R} f_2 \overset{\text{def}}{=} \neg(\neg f_1 \, \mathsf{U} \, \neg f_2) \qquad \mathsf{G} f \overset{\text{def}}{=} \neg\mathsf{F}\neg f \qquad f_1 \, \mathsf{W} \, f_2 \overset{\text{def}}{=} (f_1 \, \mathsf{U} \, f_2) \vee \mathsf{G} f_2$$

Semantics. The *satisfaction* relation ($\models$) can be defined w.r.t. an ω-word $w \in (2^\mathcal{P})^\omega$ and a position $i \in \mathbb{N}$. Inductively:

- $w, i \not\models \bot$ for each w and i.
- $w, i \models p$ iff $p \in w(i)$.
- $w, i \models f_1 \rightarrow f_2$ iff either $w, i \not\models f_1$ or $w, i \models f_2$.
- $w, i \models \mathsf{X} f$ iff $w, i + 1 \models f$.
- $w, i \models f_1 \, \mathsf{U} \, f_2$ iff $w, j \models f_2$ for some $j \geq i$ and $w, k \models f_1$ for each $k \in [i, j)$.

We may abbreviate $w, 0 \models f$ as $w \models f$. The *language* of f, denoted by $\mathscr{L}(f)$, consists of all ω-words initially satisfying f, i.e., $\mathscr{L}(f) = \{w \in (2^\mathcal{P})^\omega \mid w \models f\}$.

X, U and W are respectively the "next", "until" and "weak until" operators. According to the definition, $\mathsf{G} f$ holds if f holds at each position, and $\mathsf{F} f$ is true if f eventually holds at some position. In addition, $f_1 \, \mathsf{W} \, f_2$ holds if at every position, either f_1 holds, or at some previous position f_2 holds.

2.3   Automata on Finite/Infinite-Words

An *automaton* is a tuple $A = (Q, \Sigma, \delta, Q_0, F)$ where: Q is a finite set of *states*; Σ is a finite *alphabet*; $\delta : Q \times \Sigma \to 2^Q$, is the *transition function*; $Q_0 \subseteq Q$ is a set of *initial states*; and $F \subseteq Q$ is a set of *accepting states*.

The automaton A is *deterministic*, if $\#Q_0 = 1$, and $\#\delta(q, a) = 1$ for each $q \in Q$ and $a \in \Sigma$.

In this paper, we are both concerned about automata on finite and infinite words. For an automaton A on infinite (resp. finite) words, a *run* of A over a word $w \in \Sigma^\omega$ (resp. $w \in \Sigma^*$) is a sequence $q_0, q_1, \ldots \in Q^\omega$ (resp. $q_0, q_1, \ldots, q_{|w|} \in Q^*$), where $q_0 \in Q_0$ and $q_{i+1} \in \delta(q_i, w(i))$.

If A is an automaton on finite words, the run $q_0, q_1, \ldots, q_m$ is *accepting* if $q_m \in F$. Otherwise, if A is on infinite word, the run $q_0, q_1, \ldots$ is accepting if there are infinitely many i's having $q_i \in F$ — such kind of acceptance is called *Büchi* [Büc62] acceptance condition. For this reason, in this paper, automata on infinite words are also called Büchi automata.

For convenience, we in what follows use a three-letter-acronym to designate the type of an automaton: The first letter can be either "N" (non-deterministic) or "D" (deterministic). The second letter can either be "B" or "F", referring to Büchi and finite acceptance condition, respectively. The last letter is always "A", the acronym of automaton. For example, NBA stands for nondeterministic Büchi automaton, and DFA means deterministic automaton on finite words.

A word w is *accepted* by A, if A has an accepting run on it. We denote by $\mathscr{L}(A)$ the set of words accepted by A, call it the *language* of A. In the case of $L = \mathscr{L}(A)$, we say that L is *recognized* by A.

It is well known that NFAs and DFAs are of equivalent expressive power, because each NFA can be transformed into a DFA via the power-set construction. However, it is not the case for Büchi automata. NBAs are strictly more expressive than DBAs. As an example, let $\Sigma = \{a, b\}$, then the language consisting of "ω-words involving finitely many a's" can only be recognized by NBAs. For Büchi automata, determinization requires more complex acceptance conditions, such as *Rabin* or *parity* (cf. [Rab69, McN66]).

Languages recognized by NBAs are called ω-*regular* ones. Meanwhile, transformation from LTL formulae to Büchi automata has been well studied ever since the logic was presented (cf. [TH02, ST03] etc.).

Theorem 1. *Given an LTL formula f, there is an NBA A_f such that $\mathscr{L}(A_f) = \mathscr{L}(f)$.*

Theorem 2. *Given an automaton $A = (Q, \Sigma, \delta, Q_0, F)$, then there exists an automaton $A' = (Q', \Sigma, \delta', Q_0', F')$ such that $\mathscr{L}(A') = \mathscr{L}(A)$ and $\#Q_0' = 1$.*

Proof. Let q be some new state not belonging to Q, then: let $Q' = Q \cup \{q\}$; let the function δ' be the extension of δ by defining $\delta'(q, a) = \bigcup_{q_0 \in Q_0} \delta(q_0, a)$ for each $a \in \Sigma$; let $Q_0' = \{q\}$; and, if A is an NFA/DFA and $Q_0 \cap F \neq \emptyset$ then we let $F' = F \cup \{q\}$, otherwise, let $F' = F$.   □

2.4 Safety and Liveness

Generally speaking, each set $L \subseteq (2^\mathcal{P})^\omega$ defines a *property*, and a formula f *corresponds* to L if $\mathscr{L}(f) = L$. Remind that in this definition, f is not necessary to be an LTL formula. Actually, it could be a formula of more (or less) powerful logic within linear-time framework.

There are two kinds of fundamental properties relating to temporal logics, called *safety* and *liveness*. Informally, safety and liveness are described as "the bad thing never happens" and "the good thing eventually happens", respectively. Below we give a formal characterization, which is introduced in [AS85].

Safety: A formula f corresponds to a safety property if: For every ω-word w, $w \not\models f$ implies there is some finite prefix (called "*bad-prefix*") w' of w, such that $w' \cdot w'' \not\models f$ for each ω-word w''.
Liveness: A formula f corresponds to a liveness property if: For every finite-word w, there is some ω-word w' having $w \cdot w' \models f$.

Theorem 3 ([AS85, CS01]). *For safety and liveness properties, we have:*

1. *Safety properties are closed under finite unions and arbitrary intersections.*
2. *Liveness properties are closed under arbitrary unions, but not under intersections.*
3. $\top$ *is the only property which is both a safety and a liveness property.*
4. *For any property f, there exists a liveness property g and a safety property h such that $f = g \wedge h$.*

The last proposition is the so-called *decomposition theorem*. As an example, the LTL formula $f \, \mathsf{U} \, g$ can be decomposed as $(f \mathsf{W} g) \wedge \mathsf{F} g$, where $f \, \mathsf{W} \, g$ corresponds to a safety property, whereas $\mathsf{F} g$ corresponds to a liveness property.

Lemma 1. *If $f = \bigwedge_i f_i$ corresponds to a liveness property, then so does each f_i.*

Proof. Otherwise, according to the decomposition theorem, f_i can be written as the conjunction of some g_i and h_i, which respectively correspond to a liveness property and a safety property. If $h_i \not\leftrightarrow \top$, there is some ω-word w violating h_i. Just let $w' \in (2^\mathcal{P})^*$ be the bad-prefix of w w.r.t. h_i, we thus have $w' \cdot w'' \not\models f$ for every $w'' \in (2^\mathcal{P})^\omega$, contradiction! □

3 Periodicity and Step Variables

3.1 Periodicity: Beyond Omega-Regular

In [Wol83], Wolper pointed out that: The property $\mathsf{G}^2 p$, namely "p holds at every even moment", is not expressible in LTL. Indeed, $\mathscr{L}(\mathsf{G}^2 p) = \bigcap_{k \in \mathbb{N}} \mathscr{L}(\mathsf{X}^{2k} p)$, where X^n is the shorthand of n successive X-operators. This implies that languages captured by LTL formulae are not closed under infinite intersections.

This naturally enlightens us to make one step ahead — we would like to know: "Are ω-regular languages closed under infinite intersections/unions?" Now, let us consider the language $\bigcup_{k>0} \mathscr{L}(\mathsf{G}^k p)$, which consists of all ω-words along which p holds periodically. Remind that $w \models \mathsf{G}^k p$ if $w, i \times k \models p$ for each $i \in \mathbb{N}$.

Theorem 4. *The language $\bigcup_{k>0} \mathscr{L}(\mathsf{G}^k p)$ is not ω-regular.*

Proof. Assume by contradiction that this language is ω-regular, then there is an NBA A precisely recognizing it. Therefore, each ω-word being of the form $(\{p\} \cdot \emptyset^k)^\omega$ is accepted by A.

Suppose that A has n states, and let us fix some $k > n$. Suppose that the corresponding accepting run of A on $w_0 = (\{p\} \cdot \emptyset^k)^\omega$ is $\sigma_0 = s_0, s_1, s_2, \ldots$, and we denote s_i by $\sigma_0(i)$. Since we have totally n states, for each $t \in \mathbb{N}$, there exists a pair (i_t, j_t), s.t. $0 < i_t < j_t \le k$ and $\sigma_0(t \times (k+1) + i_t) = \sigma_0(t \times (k+1) + j_t)$.

Since $\sigma_0(i_0) = \sigma_0(j_0)$ in the case of $t = 0$, for any $\ell_0 > 0$, from Pumping lemma, A has the run

$$\sigma_1 = \sigma_0(0), \sigma_0(1), \ldots, \sigma_0(i_0), [\sigma_0(i_0+1), \ldots, \sigma_0(j_0)]^{\ell_0}, \sigma_0(j_0+1), \sigma_0(j_0+2) \ldots$$

on the word

$$w_1 = (\{p\} \cdot \emptyset^{i_0-1} \cdot \emptyset^{(j_0-i_0)\times\ell_0} \cdot (\emptyset)^{k-j_0+1}) \cdot (\{p\} \cdot \emptyset^k)^\omega.$$

σ_1 is definitely an accepting run because states occurring infinitely often in σ_0 are the same as that in σ_1. Let $L_0 = k + (j_0 - i_0) \times (\ell_0 - 1)$, then we have $w_1 = (\{p\} \cdot \emptyset^{L_0}) \cdot (\{p\} \cdot \emptyset^k)^\omega$. The above process is depicted by Fig. 1.

Also note that $\sigma_0(\ell) = \sigma_1(\ell+L_0-k)$ for each $\ell > k$, and hence $\sigma_0(k+1+i_1) = \sigma_0(k+1+j_1)$ implies $\sigma_1(L_0+1+i_1) = \sigma_1(L_0+1+j_1)$. Then, for any $\ell_1 > 0$, using Pumping lemma again, A also has an accepting run on the word

$$w_2 = (\{p\} \cdot \emptyset^{L_0}) \cdot (\{p\} \cdot \emptyset^{i_1-1} \cdot \emptyset^{(j_1-i_1)\times\ell_1} \cdot (\emptyset)^{k-j_1+1}) \cdot (\{p\} \cdot \emptyset^k)^\omega.$$

Now, let $L_1 = k + (j_1 - i_1) \times (\ell_1 - 1)$, then $w_2 = (\{p\} \cdot \emptyset^{L_0}) \cdot (\{p\} \cdot \emptyset^{L_1}) \cdot (\{p\} \cdot \emptyset^k)^\omega$.

Likewise and stepwise, we may obtain a sequence of ω-words accepted by A:

- $w_0 = (\{p\} \cdot \emptyset^k)^\omega$.
- $w_1 = (\{p\} \cdot \emptyset^{L_0}) \cdot (\{p\} \cdot \emptyset^k)^\omega$.
- $w_2 = (\{p\} \cdot \emptyset^{L_0}) \cdot (\{p\} \cdot \emptyset^{L_1}) \cdot (\{p\} \cdot \emptyset^k)^\omega$.
- $\ldots$

Since that w_{i+1} is constructed based on w_i, we can always choose a proper ℓ_{i+1} to guarantee that $L_{i+1} > L_i$ for each i. Then, consider the limit

$$w_\infty = (\{p\} \cdot \emptyset^{L_0}) \cdot (\{p\} \cdot \emptyset^{L_1}) \cdot (\{p\} \cdot \emptyset^{L_2}) \cdot \ldots$$

of all such w_is: On one hand, we have $w_\infty \in \mathscr{L}(A)$ because the corresponding run σ_∞ is accepting — observe that each occurrence of an accepting state in σ_0 will be "postponed" to at most finitely many steps in σ_∞. On the other hand, the "distance" between two adjacent "occurrences" of p monotonically increases, and it would be eventually larger than any fixed number — this implies that p cannot have a period w.r.t. w_∞. We thus get a contradiction because A accepts some word not belonging to $\bigcup_{k>0} \mathscr{L}(\mathsf{G}^k p)$. $\square$

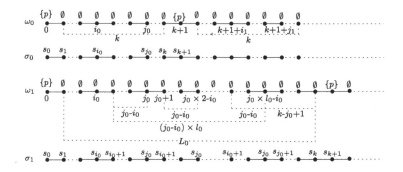

Fig. 1. The construction of w_1 from w_0.

Note that for each fixed number $k > 0$, the language $\mathscr{L}(\mathsf{G}^k p)$ is ω-regular, and it is not the case for $\bigcup_{k>0} \mathscr{L}(\mathsf{G}^k p)$, and we thus have the following corollary.

Corollary 1. *ω-regular languages are not closed under infinite unions and/or intersections.*

Remark 1. We say that ω-regular languages are not closed under infinite intersections because such languages are closed under complement.   □

3.2 Logics with Step Variables

To express periodicity, we tentatively propose to add (quantified) *step variables* into logics. As an example, if we choose μTL as the base logic, then the afore-mentioned property can be described as[1] $\exists k.\nu Z.(p \wedge \mathsf{X}^k \mathsf{X} Z)$ and here k is a step variable. As another example, one can see that $\forall k.\mathsf{X}^k \mathsf{X}^k p$ is precisely the description of $\mathsf{G}^2 p$.

From now on, we fix a set $\mathcal{K}$ of step variables, whose elements range over k, k_1, k_2, etc., and each step variable is interpreted as a natural number. In addition, we require that the base logic must involve the X-operator to designate "distance between events". Therefore, at first thought, each such extension involves the following logic (called *core logic*)

$$f ::= \bot \mid p \mid f \to f \mid \mathsf{X}f \mid \mathsf{X}^k f \mid \exists k.f$$

as fragment[2]. For convenience, we define $\forall k.f$ as the shorthand of $\neg\exists k.\neg f$ (recall that $\neg g$ stands for $g \to \bot$).

Since we have step variables, when giving semantics of a formula, besides an ω-word and a position, an evaluation $v : \mathcal{K} \to \mathbb{N}$ is also required. For the core logic, we define the semantics as follows.

---

[1] Note that we here have an extra X in the formula, because k can be assigned to 0.

[2] Remind that step variables cannot be instantiated as concrete numbers in such logic, hence we have both $\mathsf{X}f$ and $\mathsf{X}^k f$ in the grammar.

- $w, i \not\models_v \bot$ for every $w \in (2^P)^\omega$, $i \in \mathbb{N}$.
- $w, i \models_v p$ iff $p \in w(i)$.
- $w, i \models_v f_1 \to f_2$ iff either $w, i \not\models_v f_1$ or $w, i \models_v f_2$.
- $w, i \models_v Xf$ iff $w, i + 1 \models_v f$.
- $w, i \models_v X^k f$ iff $w, i + v(k) \models_v f$.
- $w, i \models_v \exists k.f$ iff there is some $n \in \mathbb{N}$ such that $w, i \models_{v[k \leftarrow n]} f$.

Here, $v[k \leftarrow n]$ is also an evaluation which is almost identical to v, except that it assigns n to k. To simplify notations, when f is a closed formula[3], we often omit v from the subscript; meanwhile, we can also omit i whenever $i = 0$.

Though such kind of extensions seems to be natural and succinct, however, we will show that the SATISFIABILITY problem, even if for the core logic, is not decidable. But before that, we first introduce the following notations:

- We abbreviate $\underbrace{X \ldots X}_{n \text{ times}} f$ and $\underbrace{X^k \ldots X^k}_{n \text{ times}} f$ as $X^n f$ and $X^{n \times k} f$, respectively, where $n \in \mathbb{N}$ and $k \in \mathcal{K}$.
- We sometimes directly write $X^{t_1} X^{t_2} f$ as $X^{t_1 + t_2} f$.

Note that in this setting, both the addition (+) and the multiplication ($\times$) are communicative and associative. Meanwhile, "$\times$" is distributive w.r.t. "+", namely, $t_1 \times t_2 + t_1 \times t_3$ can be rewritten as $t_1 \times (t_2 + t_3)$.

Theorem 5. *The* SATISFIABILITY *problem of the core logic is undecidable.*

Proof. Our goal is to show that "each formula of Peano arithmeticcan be encoded with the core logic". To this end, we need to build the following predicates:

1. Fix a proposition $p \in \mathcal{P}$, let f_p be $\forall k_1.\forall k_2.\exists k_3.(X^{k_1+k_3} p \not\leftrightarrow X^{k_1+k_2+k_3+1} p)$. Actually, f_p just depicts the *non-shifting* property of p. i.e., $w \models f_p$ only if for each $i, j \in \mathbb{N}$ with $i < j$, there is some t having: either $(w, i + t \models p$ and $w, j + t \not\models p)$ or $(w, i + t \not\models p$ and $w, j + t \models p)$ — indeed, one can just view k_1 as i and view $k_1 + k_2 + 1$ as j. Note that f_p is satisfied by any ω-word along which p occurs infinitely often and the distance between two adjacent occurrences of p monotonically increases.

2. Let $P_=(k_1, k_2) \stackrel{\text{def}}{=} f_p \wedge \forall k.(X^{k_1+k} p \leftrightarrow X^{k_2+k} p)$. Hence, $w, i \models_v P_=(k_1, k_2)$ iff $v(k_1) = v(k_2)$. Because, if $v(k_1) \neq v(k_2)$, according to the definition of f_p, there must exist some $n \in \mathbb{N}$, such that p differs from $w(v(k_1) + n)$ and $w(v(k_2) + n)$.

3. Let $P_<(k_1, k_2) \stackrel{\text{def}}{=} \exists k.P_=(k_1 + k + 1, k_2)$. Then, $w, i \models_v P_<(k_1, k_2)$ iff $v(k_1) < v(k_2)$.

4. Subsequently, we use $P_+(k_1, k_2, k_3)$ to denote $P_=(k_1 + k_2, k_3)$. According to the definition, $w, i \models_v P_+(k_1, k_2, k_3)$ iff $v(k_3) = v(k_1 + k_2) = v(k_1) + v(k_2)$.

---

[3] That is, f involves no free variable.

5. Now, let us fix another proposition $q \in \mathcal{P}$ and define

$$f_q \overset{\text{def}}{=} q \ \wedge \ \mathsf{X}q \ \wedge \ \forall k_1.\exists k_2.\mathsf{X}^{k_1+k_2}q \ \wedge$$
$$\forall k_1.\forall k_2.\forall k_3.(\mathsf{X}^{k_1}q \wedge \mathsf{X}^{k_2}q \wedge \mathsf{X}^{k_3}q \wedge$$
$$P_<(k_1,k_2) \wedge P_<(k_2,k_3) \wedge$$
$$\forall k_4.((P_<(k_1,k_4) \wedge P_<(k_4,k_2)) \vee (P_<(k_2,k_4) \wedge P_<(k_4,k_3)) \rightarrow \neg\mathsf{X}^{k_4}q)$$
$$\rightarrow \exists k_5.\exists k_6.(P_+(k_5,k_1,k_2) \wedge P_+(k_6,k_2,k_3) \wedge P_+(2,k_5,k_6)))$$

We may assert that $w, i \models f_q$ iff i is a complete square number (i.e., $i = j^2$ for some j). Let us explain: The first line indicates that q holds infinitely often, and it holds at the positions of 0 and 1. For every three adjacent positions k_1, k_2, k_3 at which q holds (hence, q does not hold between k_1 and k_2, nor between k_2 and k_3), we have $(k_3 - k_2) = 2 + (k_2 - k_1)$. Inductively, we can show that q becomes true precisely at positions 0, 1, 4, ..., $(n-1)^2$, n^2, $(n+1)^2$, ... [4].

6. We let $P_2(k_1, k_2)$ be

$$f_q \wedge \mathsf{X}^{k_2}q \wedge \mathsf{X}^{k_2 + 2 \times k_1 + 1}q \wedge \neg\exists k_3.(P_<(k_2,k_3) \wedge P_<(k_3, 2 \times k_1 + k_2 + 1) \wedge \mathsf{X}^{k_3}q)$$

then $w \models_v P_2(k_1,k_2)$ iff $v(k_2) = (v(k_1))^2$.

7. Lastly, we define

$$P_\times(k_1,k_2,k_3) \overset{\text{def}}{=} \exists k_4.\exists k_5.\exists k_6.(P_2(k_1,k_4) \wedge P_2(k_2,k_5)$$
$$\wedge P_2(k_1+k_2,k_6) \wedge P_=(2 \times k_3 + k_4 + k_5, k_6))$$

Then, once $w \models_v P_\times(k_1,k_2,k_3)$ holds, we can infer that there is some evaluation v' which agrees with v on k_1, k_2 and k_3 (i.e., $v'(k_i) = v(k_i)$ for $i = 1, 2, 3$) having

$$\begin{cases} v'(k_4) = (v'(k_1))^2 \\ v'(k_5) = (v'(k_2))^2 \\ v'(k_6) = (v'(k_1) + v'(k_2))^2 \\ v'(k_6) = v'(k_4) + v'(k_5) + 2 \times v'(k_3) \end{cases}$$

and we thus subsequently have $v(k_3) = v(k_1) \times v(k_2)$.

Therefore, "addition", "multiplication", and the "less than" relation over natural numbers can be encoded in terms of the core logic. Since quantifiers are also involved here, the SATISFIABILITY problem of Peano arithmetic, which is known to be undecidable (cf. [Göd31, Chu36]), is now reduced to that of the core logic. □

4   The Logic QPLTL

Theorem 5 indicates that the unrestricted use of step variables leads to undecidability for the SATISFIABILITY problem. To obtain a decidable extension, we need to impose strong syntactic constraints to the logic. In this paper, we investigate the logic *Quantified Periodic LTL* (QPLTL for short), based on LTL.

---

[4] The encoding of f_q is enlightened by [Sch10].

4.1 Syntax and Semantics

A QPLTL formula can be one of the following forms:

(I)   An LTL formula.
(II) A formula like $f_1 \, U^n \, f_2$ or $f_1 \, W^n \, f_2$, where f_1 and f_2 are LTL formulae, n is a positive natural number.
(III) Boolean or temporal combinations of (I) and (II).
(IV) A formula being of the form $\exists\!\!\!\forall k.(f_1 \, U^k \, f_2)$ or $\exists\!\!\!\forall k.(f_1 \, W^k \, f_2)$, where $\exists\!\!\!\forall \in \{\exists, \forall\}$, $k \in \mathcal{K}$, f_1 and f_2 are two LTL formulae.

Given an ω-word $w \in (2^{\mathcal{P}})^\omega$, and a position $i \in \mathbb{N}$, we define the *satisfaction* relation as follows.

- Satisfaction of an LTL formula is defined in the way same as before.
- $w, i \models f_1 \, U^n \, f_2$ iff there is some $t \in \mathbb{N}$ such that $w, i + t \times n \models f_2$ and $w, i + j \times n \models f_1$ for every $j < t$.
- $w, i \models f_1 \, W^n \, f_2$ iff either $w, i \models f_1 \, U^n \, f_2$ or $w, i + t \times n \models f_1$ for each $t \in \mathbb{N}$.
- Boolean and temporal combinations are defined accordingly with the corresponding operators.
- $w, i \models \exists k.(f_1 \, U^k \, f_2)$ (resp. $w, i \models \exists k.(f_1 \, W^k \, f_2)$) iff there is some $n > 0$ such that $w, i \models f_1 \, U^n \, f_2$ (resp. $w, i \models f_1 \, W^n \, f_2$).
- $w, i \models \forall k.(f_1 \, U^k \, f_2)$ (resp. $w, i \models \forall k.(f_1 \, W^k \, f_2)$) iff for each $n > 0$ we have $w, i \models f_1 \, U^n \, f_2$ (resp. $w, i \models f_1 \, W^n \, f_2$).

As usual, we directly write $w, 0 \models f$ as $w \models f$, and we also define the following derived notations[5].

$$F^n f \stackrel{\text{def}}{=} \top \, U^n \, f \qquad\qquad G^n f \stackrel{\text{def}}{=} f \, W^n \, \bot$$
$$\exists\!\!\!\forall k.F^k f \stackrel{\text{def}}{=} \exists\!\!\!\forall k.\top \, U^k \, f \qquad \exists\!\!\!\forall k.G^k f \stackrel{\text{def}}{=} \exists\!\!\!\forall k.f \, W^k \, \bot$$

Indeed, from the proof of Theorem 5, one can see that nested use of quantifiers leads to undecidability of satisfaction decision. Thus, in QPLTL, we use at most one step variable in a formula.

Remark 2. $f_1 \, U^1 \, f_2$ and $f_1 \, W^1 \, f_2$ can be just written as $f_1 \, U \, f_2$ and $f_1 \, W \, f_2$, which coincide with the definitions given in LTL.

Also remind that a step variable must be interpreted as a positive number — beware that this is different from that in Sect. 3.2. This is just for the following consideration: If we allow assigning 0 to a step variable, the formula $\exists k.G^k p$ is no longer the description of "p holds periodically", because this formula is weaker than $G^0 p$ (which is equivalent to p). Note that in such setting, the "core logic" remains undecidable — the proof can be obtained via doing a simple adaptation from that of Theorem 5.

---

[5] We do not consider the "releases" (R) operator here, which is the duality of U. Indeed, $f_1 R f_2$ is equivalent to $f_2 \, W(f_1 \wedge f_2)$.

To make things compatible, for the formulae like $f_1 \mathsf{U}^n f_2$ or $f_1 \mathsf{W}^n f_2$, we don't allow $n = 0$. Actually, according to the definition, we can see that $f_1 \mathsf{U}^0 f_2$ and $f_1 \mathsf{W}^0 f_2$ are just merely f_2 and $f_1 \vee f_2$, and such operators are redundant.

Note that X^n is still the abbreviation of n successive Xs. We don't explicitly add the formulae like $\exists k.\mathsf{X}^k f$ and $\forall k.\mathsf{X}^k f$ into the logic, because they are essentially $\mathsf{XF}f$ and $\mathsf{XG}f$, respectively.   □

4.2 The Decision Problem

In this section, we will show that the SATISFIABILITY problem of QPLTL is decidable. Indeed, this proof also reveals the close connection among liveness, safety and periodicity.

We can equivalently transform each LTL formula f to make it involve only literals (i.e., $\top$, $\bot$, formulae like p or $\neg p$), $\vee$, $\wedge$, X, U and W. We in what follows call it the *positive normal form* (PNF, for short) of f. An LTL formula is U-*free* if its PNF involves no U-operator.

Lemma 2 ([CS01]). *Each U-free LTL formula corresponds to a safety property.*

Below we give a characterization of LTL formulae corresponding to liveness properties. In some sense, it could be considered as a normal form of such kind of star-free properties.

Theorem 6. *If the LTL formula f corresponds to a liveness property, then it can be equivalently written as $\bigwedge_i (f_i' \vee \mathsf{F}f_i'')$, where each f_i' corresponds to a safety property.*

Proof. Suppose that f is already in its PNF, then we conduct a series of transformations on f.

First of all, we use the pattern $f_1 \mathsf{U} f_2 \leftrightarrow (f_1 \mathsf{W} f_2) \wedge \mathsf{F}f_2$ to replace each occurrence of U operator with that of W and F.

Then, use the following rules

$$\mathsf{X}(f_1 \wedge f_2) \leftrightarrow \mathsf{X}f_1 \wedge \mathsf{X}f_2 \qquad \mathsf{X}(f_1 \vee f_2) \leftrightarrow \mathsf{X}f_1 \vee \mathsf{X}f_2$$
$$\mathsf{X}(f_1 \mathsf{W} f_2) \leftrightarrow (\mathsf{X}f_1)\mathsf{W}(\mathsf{X}f_2) \qquad \mathsf{XF}f' \leftrightarrow \mathsf{FX}f'$$

to push X inward, until X occurrs only before a literal or another X.

Subsequently, repeatedly use the following schemas[6]

$$(f_1 \wedge f_2)\mathsf{W} f_3 \leftrightarrow (f_1 \mathsf{W} f_3) \wedge (f_2 \mathsf{W} f_3)$$
$$f_1 \mathsf{W}(f_2 \vee f_3) \leftrightarrow (f_1 \mathsf{W} f_2) \vee (f_1 \mathsf{W} f_3)$$
$$(f_1 \vee \mathsf{F}f_2)\mathsf{W} f_3 \leftrightarrow \mathsf{F}(f_2 \wedge \mathsf{X}(f_1 \mathsf{W} f_3)) \vee (f_1 \mathsf{W} f_3) \vee \mathsf{F}(f_3 \wedge \mathsf{F}f_2) \vee \mathsf{FGF}f_2$$
$$f_1 \mathsf{W}(f_2 \wedge \mathsf{F}f_3) \leftrightarrow (f_1 \mathsf{W} f_2) \wedge (\mathsf{F}(f_2 \wedge \mathsf{F}f_3) \vee \mathsf{G}f_1)$$
$$(\mathsf{F}f_1)\mathsf{W} f_2 \leftrightarrow f_2 \vee \mathsf{F}(f_1 \wedge \mathsf{X}f_2) \vee \mathsf{F}(f_2 \wedge \mathsf{F}f_1) \vee \mathsf{FGF}f_1$$
$$f_1 \mathsf{W}(\mathsf{F}f_2) \leftrightarrow \mathsf{G}f_1 \vee \mathsf{F}f_2$$

---

[6] For example, when dealing with $(p_1 \vee p_2 \wedge \mathsf{F}p_3)\mathsf{W} p_4$, we need first transform the first operand into disjunctive normal form — i.e., rewrite it as $((p_1 \vee \mathsf{F}p_3) \wedge (p_2 \vee \mathsf{F}p_3))\mathsf{W} p_4$, and then conduct the transformation with the first rule and the third rule. Similarly, for the second operand, we need first transform it into conjunctive normal form.

until the following holds: For each subformula $f' = f_1 \mathsf{W} f_2$, if f_1 or f_2 involves F, then f' must be contained in the scope of some other F. Remind that we equivalently write $\mathsf{GF}f_i$ as $\mathsf{FGF}f_i$ in the second rule and the fifth rule to fulfill such requirement[7]. Note that when the third or the fifth schema is applied, one need further push Xs inward using the previous group of rules.

Lastly, we write the resulting formula into the conjunctive normal form $\bigwedge_i f_i$ where each $f_i = \bigvee_j f_{i,j}$. We now show that it must be of the desired form.

- First of all, since f corresponds to a liveness property, so does each f_i (cf. Lemma 1).
- Second, for each $f_i = \bigvee_j f_{i,j}$, if there is no such $f_{i,j}$ whose outermost operator is F, then f_i is U-free[8]. From Lemma 2, f_i also corresponds to a safety property. Hence, such f_i is essentially equivalent to $\top$ (cf. Theorem 3), and we can simply remove this conjunct.
- Then, for the other f_is: By applying the scheme $\mathsf{F}g_1 \vee \ldots \vee \mathsf{F}g_n \leftrightarrow \mathsf{F}(g_1 \vee \ldots \vee g_n)$, we can just preserve one disjunct having F as the outermost operator, denote it by $\mathsf{F}f_i''$. Since other disjuncts are U-free, the disjunction of them, denoted by f_i', corresponds to a safety property (cf. Theorem 3).

Then, the above discussion concludes the proof.   □

Lemma 3 ([AS87]). *Given an LTL formula f, the question "whether f corresponds to a liveness property" is decidable. In addition, given a safety LTL formula, then there is an automaton on finite words recognizing all its bad-prefixes.*

Theorem 7. *Given two LTL formulae f_1 and f_2, where f_2 corresponds to a safety property, then the question "whether exists some ω-word w such that $w \models \mathsf{G}f_1$ and $w \models \exists k.\mathsf{G}^k \neg f_2$" is decidable.*

Proof. Let $A_1 = (Q_1, 2^\mathcal{P}, \delta_1, \{q_1\}, F_1)$ be the NBA recognizing $\mathscr{L}(\mathsf{G}f_1)$, and $A_2 = (Q_2, 2^\mathcal{P}, \delta_2, \{q_2\}, F_2)$ be the NFA[9] recognizing the bad-prefixes of f_2.

For each $q \in Q_1$, we can define a *product* $A_1^q \otimes A_2 \overset{\text{def}}{=} (Q_1 \times Q_2, 2^\mathcal{P}, \delta', \{(q, q_2)\})$, where

$$\delta'((q_1', q_2'), a) = \begin{cases} \{(q_1'', q_2'') \mid q_1'' \in \delta_1(q_1', a), q_2'' \in \delta_2(q_2', a)\} & q_2' \notin F \\ \{(q_1'', q_2'') \mid q_1'' \in \delta_1(q_1', a), q_2'' \in Q_2\} & q_2' \in F \end{cases}.$$

The product is almost an automaton, but we are not concerned about its run over words, hence the accepting state set is not given here. Instead, we will define the notion of *accepting loops*: An accepting loop is a finite sequence $(q_{1,0}, q_{2,0}), (q_{1,1}, q_{2,1}), \ldots, (q_{1,m}, q_{2,m}) \in (Q_1 \times Q_2)^*$ such that:

---

[7] Note that the operator G is derived from W.

[8] Because, the previous transformations could guarantee that: If the outermost operator of $f_{i,j}$ is not F, then no F occurs in $f_{i,j}$.

[9] Remind that each automaton can be equivalently transformed into another one having a unique initial state, and the transformation is linear (cf. Theorem 2). Indeed, to obtain a finite alphabet, here we may temporarily take $\mathcal{P}$ as the set constituted with propositions occurring in f_1 or f_2.

1. $q_{1,0} = q_{1,m}$ and $q_{2,0} = q_{2,m}$.
2. For each $0 \leq i < m$, there is some a_i having $(q_{1,i+1}, q_{2,i+1}) \in \delta'((q_{1,i}, q_{2,i}), a_i)$.
3. There exists some $0 \leq i < m$ such that $q_{1,i} \in F_1$.
4. There also exists some $0 \leq j < m$ such that $q_{2,j} \in F_2$.

We will then show the following claim:

There is some ω-word $w \in (2^{\mathcal{P}})^{\omega}$ making $w \models \mathsf{G}f_1$ and $w \models \exists k.\mathsf{G}^k \neg f_2$ iff there is some $q \in Q_1$ such that $A_1^q \otimes A_2$ involves an accepting loop starting from (q, q_2).

$\Longrightarrow$: Suppose that $w \models \mathsf{G}f_1$ and $w \models \exists k.\mathsf{G}^k \neg f_2$ with the period n, namely $w \models \mathsf{G}^n \neg f_2$ — which implies $w, i \times n \not\models f_2$ for every $i \in \mathbb{N}$.

Let $\sigma = \sigma(0), \sigma(1), \sigma(2), \ldots \in Q_1^{\omega}$ be an accepting run of A_1 on w. Then, there exists some $q_f \in F_1$ such that there are infinitely many i's having $\sigma(i) = q_f$. Because the run is infinite, there must exist some $q \in Q_1$ fulfilling: there are infinitely many i's having $\sigma(i \times n) = q$. W.l.o.g., suppose that $\sigma(i_0 \times n) = q$.

Because $w, i_0 \times n \not\models f_2$, there exists some bad-prefix w' of f_2, which starts from $w(i_0 \times n)$. Let i_1 be the number fulfilling that: $\sigma(i_1 \times n) = q$, and $(i_1 - i_0) \times n > |w'|$, and there exists some $\ell \in [i_0 \times n, i_1 \times n)$ having $\sigma(\ell) = q_f \in F_1$.

Since w' is a bad-prefix of f_2, there is a finite accepting run $\sigma' = \sigma'(0), \sigma'(1), \ldots, \sigma'(|w'|)$ of A_2 on w', where $\sigma'(0) = q_2$ and $\sigma'(|w'|) \in F_2$.

Let $t = (i_1 - i_0) \times n$, since $t > |w'|$, according to the construction, we can prolong σ' by defining $\sigma'(j) = \sigma'(|w'|)$ for each $j \in (|w'|, t)$ and $\sigma'(t) = q_2$.

For each $j \in [0, t]$, we let $\sigma''(j) = \sigma(i_0 \times n + j)$. Thus, we get the accepting loop $(\sigma''(0), \sigma'(0)), \ldots, (\sigma''(t), \sigma'(t))$ in the product. Indeed, according to the construction, we can see that both $(\sigma''(0), \sigma'(0))$ and $(\sigma''(t), \sigma'(t))$ are (q, q_2), and the loop is accepting.

$\Longleftarrow$: Conversely, suppose $(q_{1,0}, q_{2,0}), \ldots, (q_{1,n}, q_{2,n})$ to be an accepting loop of $A_1^q \otimes A_2$ where $q_{1,0} = q_{1,n} = q$ and $q_{2,0} = q_{2,n} = q_2$. We can, of course, assume that q is reachable from q_1 in A_1 — if not so, such state can be safely removed.

Suppose that $(q_{1,i+1}, q_{2,i+1}) \in \delta'((q_{1,i}, q_{2,i}), a_i)$, we let $w = (a_0 \cdot a_1 \cdot \ldots \cdot a_{n-1})^{\omega}$. Also let $m \in [0, n)$ be the minimal index having $q_{2,m} \in F_2$, then $a_0 \cdot a_1 \cdot \ldots \cdot a_{m-1}$ is a bad-prefix of f_2, hence w violates f_2 with the period n, and thus $w \models \exists k.\mathsf{G}^k \neg f_2$.

What left is to ensure that $w \models \mathsf{G}f_1$ also holds. Suppose that q_1 reaches q via reading the finite word w_0, then $w_0 \cdot w \in \mathscr{L}(A_1)$, and thus $w_0 \cdot w \models \mathsf{G}f_1$. Consequently, we have $w \models \mathsf{G}f_1$.   □

Theorem 8. *Given an LTL formula f, the question "whether $\exists k.\mathsf{G}^k \neg f$ is satisfiable" is decidable.*

Proof. First of all, we may decompose f as $g \wedge h$ where g corresponds to a liveness property and h corresponds to a safety property.

If $h \not\leftrightarrow \top$, then there is a finite-word w_0 acting as the bad-prefix of h, and hence a bad-prefix of f. Therefore, the ω-word $w_0^{\omega} = w_0 \cdot w_0 \cdot w_0 \cdot \ldots$ violates f periodically, and thus $\exists k.\mathsf{G}^k \neg f$ must be satisfiable.

In what follows, we just consider the case that $h \leftrightarrow \top$, and hence f corresponds to a liveness property. From Theorem 6, let us further assume the normal form of f is $\bigwedge_{i=1}^{m} f_i$, where $f_i = f_i' \vee \mathsf{F} f_i''$, and f_i' corresponds to a safety property.

Thus, f is periodically violated, if and only if: there is some ω-word w, a set of indices $J \subseteq \{1, 2, \ldots, m\}$, and a positive number n, such that for each $i \in \mathbb{N}$ we have $w, i \times n \not\models f_j$ for some $j \in J$.

Then, the obligation is to detect the existence of such word w, index set J and period n. First, we may choose the set J, and the number of such choices are finite. We can further assume that $w \models \mathsf{G}\neg f_j''$ for each $j \in J$, because:

- If there is some $j \in J$ having $w \models \mathsf{GF} f_j''$, then the disjunct f_j is never violated by w. In this case, we may choose some $J' \subseteq J \setminus \{j\}$ to be the index set.
- If there are finitely many i's having $w, i \models f_j''$, then we may choose some postfix of w to be the word.

Now, given J, the problem becomes: Find some ω-word w, which fulfills $w \models \bigwedge_{j \in J} \mathsf{G}\neg f_j''$ and $w \models \exists k.\mathsf{G}^k \neg \bigwedge_{j \in J} f_j'$. Since that $\bigwedge_{j \in J} \mathsf{G}\neg f_j''$ is equivalent to $\mathsf{G} \bigwedge_{j \in J} \neg f_j''$, and $\bigwedge_{j \in J} f_j'$ corresponds to a safety property, from Theorem 7, we know that this problem is decidable.   □

Theorem 9. *The* SATISFIABILITY *problem of QPLTL is decidable.*

Proof. Since formulae of Type (I)–(III) can be expressed by some logics equal to ω-regular languages, such as μTL or ETL, we here just consider formulae of Type (IV).

First, we lift the negation operator ($\neg$) to that group of formulae by defining

$$\neg \exists\!\!\!/ k.(f_1 \mathsf{U}^k f_2) \stackrel{\text{def}}{=} \overline{\exists\!\!\!/} k.(\neg f_2 \mathsf{W}^k (\neg f_1 \wedge \neg f_2))$$
$$\neg \exists\!\!\!/ k.(f_1 \mathsf{W}^k f_2) \stackrel{\text{def}}{=} \overline{\exists\!\!\!/} k.(\neg f_2 \mathsf{U}^k (\neg f_1 \wedge \neg f_2))$$

where $\overline{\exists\!\!\!/}$ is $\forall$(resp. $\exists$) if $\exists\!\!\!/$ is $\exists$(resp. $\forall$). Indeed, for each such formula f, we can examine that $w \models f$ iff $w \not\models \neg f$ for every ω-word w. Hence, such lifting is admissible. This also implies that formulae of Type (IV) are closed under negation.

Then, for such a formula, instead of deciding its *satisfiability*, we would rather decide its *validity*, because f is satisfiable iff $\neg f$ is not valid. Below gives the decision approach.

(1) $\exists k.(f_1 \mathsf{W}^k f_2)$ is valid iff $f_1 \vee f_2$ is valid.
(2) $\forall k.(f_1 \mathsf{W}^k f_2)$ is valid iff $f_1 \vee f_2$ is valid.
(3) $\exists k.(f_1 \mathsf{U}^k f_2)$ is valid iff $f_2 \vee (f_1 \wedge \mathsf{F} f_2)$ is valid.
(4) $\forall k.(f_1 \mathsf{U}^k f_2)$ is valid iff $f_1 \vee f_2$ is valid and $\exists k.\mathsf{G}^k \neg f_2$ is not satisfiable.

For (1) and (2): If $\exists\!\!\!/ k.(f_1 \mathsf{W}^k f_2)$ is valid, then $w \not\models f_2$ implies $w \models f_1$ for each ω-word w, hence $f_1 \vee f_2$ must be valid. Conversely, if $f_1 \vee f_2$ is valid, then $w \models f_1 \mathsf{W}^n f_2$ holds for any positive natural number n.

For (3): The "only if" direction is also trivial, we just show the "if" direction. Indeed, if $f_2 \vee (f_1 \wedge \mathsf{F} f_2)$ is valid, then for each ω-word w, according to the definition, $w \models f_2$ implies $w \models \exists k.(f_1 \mathsf{U}^k f_2)$ holds; otherwise, if $w \models f_1 \wedge \mathsf{F} f_2$, w.l.o.g., assume that $w, 0 \models f_1$ and $w, n \models f_2$, this also guarantees $w \models \exists k.(f_1 \mathsf{U}^k f_2)$ because $w \models f_1 \mathsf{U}^n f_2$.

As for the "only if" direction of (4), in the same way as before, we can infer that $f_1 \vee f_2$ should be valid if $\forall k.(f_1 \mathsf{U}^k f_2)$ is. Meanwhile, since for every ω-word w and every $n > 0$ there exists some $i \in \mathbb{N}$ having $w, i \times n \models f_2$, hence $\exists k.\mathsf{G}^k \neg f_2$ should not be satisfiable. Conversely, $\exists k.\mathsf{G}^k \neg f_2$ is not satisfiable implies that $\forall k.\mathsf{F}^k f_2$ is valid. Then, for each ω-word w and each $n > 0$, there exists a minimal number i making $w, i \times n \models f_2$, and since $f_1 \vee f_2$ is valid, we have $w, j \times n \models f_1$ for every $j < i$, hence $w \models f_1 \mathsf{U}^n f_2$. $\qquad \square$

From the decision procedure, one can examine that each step could be accomplished with polynomial (in the size of the formula) space. Since **PSPACE** is closed under relativization, namely, $\textbf{PSPACE}^{\text{PSPACE}} = \textbf{PSPACE}$, we can thus conclude that SATISFIABILITY of QPLTL is in **PSPACE**. On the other hand, the SATISFIABILITY problem of LTL is also **PSPACE**-hard. Thus, we have the following conclusion.

Corollary 2. *The* SATISFIABILITY *problem of QPLTL is **PSPACE**-complete.*

5   Discussion and Future Work

In this paper, we suggest to use quantified step variables to describe periodicity. As an attempt, using LTL as the base logic, we can obtain one of its decidable periodic extension — QPLTL.

Indeed, formulae of Type (I)–(III) constitutes a proper super logic of LTL, whereas a subset of whole ω-regular properties. It is interesting to study the relation between this set and ω-regular languages.

For formulae of Type (IV), actually, we may make a bit relaxation on that part. For example, consider the formula $f = \exists k.\mathsf{F}\mathsf{G}^k p$, which gives the assertion "p eventually holds periodically", and call such property *soft periodicity*. We can see that f is satisfiable if and only if $\exists k.\mathsf{G}^k p$ is satisfiable.

To make the logic more flexible in syntax and more expressive, as a future work, we need carefully study some extensions of QPLTL. For example, Boolean combinations of formulae of Type (IV), or combinations of Type (III) and (IV). Further, we are also wonder about the decidability of the extension built up from more expressive logics, such as linear-time μTL. The key issue is also to establish the corresponding normal form of general liveness properties.

Acknowledgement. The first author would thank Normann Decker, Daniel Thoma and Martin Leucker for the fruitful discussion on this problem. We would also thank the anonymous reviewers for their valuable comments on an earlier version of this paper. Wanwei Liu is supported by Natural Science Foundation of China (No. 61103012, No. 61379054, and No. 61532007). Fu Song is supported by Natural Science Foundation of China (No. 61402179 and No. 61532019).

References

[AS85]   Alpern, B., Schneider, F.B.: Defining liveness. Inf. Process. Lett. **21**, 181–185 (1985)

[AS87]   Alpern, B., Schneider, F.B.: Recognizing safety and liveness. Distrib. Comput. **2**(3), 117–126 (1987)

[BB87]   Banieqbal, B., Barringer, H.: Temporal logic with fixed points. In: Banieqbal, B., Barringer, H., Pnueli, A. (eds.) Temporal Logic in Specification. LNCS, vol. 398, pp. 62–74. Springer, Heidelberg (1989). doi:10.1007/3-540-51803-7_22

[Büc62] Büchi, J.R.: On a decision method in restricted second order arithmetic. In: Proceedings of the International Congress on Logic, Method and Philosophy of Science 1960, pp. 1–12, Palo Alto. Stanford University Press (1962)

[Chu36] Church, A.: A note on the Entscheidungsproblem. J. Symb. Log. **1**, 101–102 (1936)

[CS01]   Clarke, E.M., Schlingloff, B.: Model checking. In: Robinson, A., Voronkov, A. (eds.) Handbook of Automated Reasoning. vol. 2, Chapt. 24, pp. 1369–1522. MIT and Elsevier Science Publishers (2001)

[DG08]   Diekert, V., Gastin, P.: First-order definable languages. In: Flum, J., Grädel, E., Wilke, T. (eds.) Logic and Automata: History and Perspectives [in Honor of Wolfgang Thomas], vol. 2, pp. 261–306. Amsterdam University Press (2008)

[Göd31] Gödel, K.: Über formal unentscheidbare sätze der principia mathematica und verwandter system I. Monatshefte für Methematik und Physik **38**, 173–198 (1931)

[LS07]   Leucker, M., Sánchez, C.: Regular linear temporal logic. In: Jones, C.B., Liu, Z., Woodcock, J. (eds.) ICTAC 2007. LNCS, vol. 4711, pp. 291–305. Springer, Heidelberg (2007). doi:10.1007/978-3-540-75292-9_20

[McN66] McNaughton, R.: Testing and generating infinite sequences by a finite automaton. Inf. Comput. **9**, 521–530 (1966)

[Pnu77] Pnueli, A.: The temporal logic of programs. In: Proceedings of 18th IEEE Symposium on Foundation of Computer Science (FOCS 1977), pp. 46–57. IEEE Computer Society (1977)

[Rab69] Rabin, M.O.: Decidability of second order theories and automata on infinite trees. Trans. AMS **141**, 1–35 (1969)

[Sch10] Schweikardt, N.: On the expressive power of monadic least fixed point logic. Theor. Comput. Sci. **350**(2–3), 1123–1135 (2010)

[ST03]   Sebastiani, R., Tonetta, S.: "More deterministic" vs. "smaller" Büchi automata for efficient LTL model checking. In: Geist, D., Tronci, E. (eds.) CHARME 2003. LNCS, vol. 2860, pp. 126–140. Springer, Heidelberg (2003). doi:10.1007/978-3-540-39724-3_12

[TH02]   Taurainen, H., Heljanko, K.: Testing LTL formula translation into Büchi automata. STTT **4**, 57–70 (2002)

[Tho79] Thomas, W.: Star-free regular sets of omega-sequences. Inf. Control **42**, 148–156 (1979)

[VW94]   Vardi, M.Y., Wolper, P.: Reasoning about infinite computations. Inf. Comput. **115**(1), 1–37 (1994)

[Wol83] Wolper, P.: Temporal logic can be more expressive. Inf. Control **56**(1–2), 72–99 (1983)

On Equivalence Checking of Nondeterministic Finite Automata

Chen Fu[1,3]([✉]), Yuxin Deng[2], David N. Jansen[1], and Lijun Zhang[1,3]

[1] State Key Laboratory of Computer Science,
Institute of Software, Chinese Academy of Sciences, Beijing, China
fchen@ios.ac.cn
[2] Shanghai Key Laboratory of Trustworthy Computing,
East China Normal University, Shanghai, China
[3] University of Chinese Academy of Sciences, Beijing, China

Abstract. We provide a comparative study of some typical algorithms for language equivalence in nondeterministic finite automata and various combinations of optimization techniques. We find that their practical efficiency mostly depends on the density and the alphabet size of the automaton under consideration. Based on our experiments, we suggest to use HKC (Hopcroft and Karp's algorithm up to congruence) [4] if the density is large and the alphabet is small; otherwise, we recommend the antichain algorithm (Wulf, Doyen, Henzinger, Raskin) [6]. Bisimulation equivalence and memoisation both pay off in general. When comparing highly structured automata over a large alphabet, one should use symbolic algorithms.

1 Introduction

Checking whether two *nondeterministic finite automata (NFA)* accept the same language is important in many application domains such as compiler construction and model checking. Unfortunately, solving this problem is costly: it is PSPACE-complete [16].

However, the problem is much easier when restricted to *deterministic finite automata (DFA)* where nondeterminism is ruled out. Checking language equivalence for DFA can be done using either minimisation [10,12,18] or Hopcroft and Karp's algorithm (HK algorithm) [9]. The former searches for equivalent states in the whole state space of a finite automaton or the disjoint union of two automata. This works well in practice because for DFA, bisimulation equivalence, simulation equivalence and language equivalence coincide, and both bisimulation and simulation can be computed in polynomial time. The HK algorithm is more appropriate in the case where one only wants to know if two particular states are language equivalent because it is an "on-the-fly" algorithm that explores merely the part of state space that is really needed. It should be mentioned that the HK algorithm exploits a technique nowadays called *coinduction* [15].

A straightforward idea for checking the language equivalence of two NFA is to convert them into DFA through a standard powerset construction, and then

© Springer International Publishing AG 2017
K.G. Larsen et al. (Eds.): SETTA 2017, LNCS 10606, pp. 216–231, 2017.
https://doi.org/10.1007/978-3-319-69483-2_13

execute an equivalence checking algorithm for DFA. Since there are exponentially many state sets in the powerset, one would like to avoid constructing them as much as possible. In particular, if one only needs to decide if two specific sets of states in a nondeterministic finite automaton are equivalent, one can construct the state sets on-the-fly, and simultaneously try to build a bisimulation that relates these sets. With this approach, the number of constructed sets is usually much smaller than the exponential worst-case bound. In terms of implementation, it is easy to adapt a naive version of the HK algorithm from DFA to NFA: The algorithm maintains two sets *todo* and R. The set *todo* contains the pairs (X, Y) to be checked, where X and Y are two sets of NFA states. The set R contains the checked (equivalent) pairs. It is a bisimulation relation when the algorithm terminates successfully. There are several optimizations for this algorithm:

1. Reduce the set R by constructing a relation that is not a bisimuation but is a bisimulation up to equivalence, e.g. the HK algorithm [9], or up to congruence, e.g. the HKC algorithm [4].
2. Often *todo* is implemented as a list that may contain repeated elements. Avoid these repetitions by some memoisation techniques [13].
3. Represent the automata symbolically rather than explicitly by using *binary decision diagrams (BDD)* [13,19].
4. Saturate the given automata with respect to bisimulation equivalence or simulation preorder.

An alternative approach to checking NFA equivalence is to use *antichain* algorithms [2,6]. The basic idea is to check language inclusion in both directions. This approach also exploits the *coinduction* technique: in order to check whether the language of a set of NFA states X is a subset of the language of a set of NFA states Y, it simultaneously tries to build a simulation relation, relating each state $x \in X$ (as a state in the NFA) to Y (as a state in the DFA). This algorithm can also be optimized by reducing the list *todo* by memoisation or by reducing the list R with antichains. The antichain algorithm can be enhanced by exploiting any preorder contained in language inclusion [2]. For example, the simulation preorder can be used for this purpose.

In this paper we investigate the mentioned algorithms and their combinations with optimizations to achieve the best time efficiency. We find that in most cases the antichain algorithm is stable and often outperforms other algorithms. In contrast, the performance of HK and HKC algorithms may vary a lot, depending on the size of the alphabet and the density of transitions in the automata under consideration. When the size of the alphabet is small (e.g. 2) and the density is large (e.g. 1.25 or 1.5), HKC is the best choice. Otherwise, computing congruence closures is very costly and renders HKC impractical to use. One should use memoisation because it mostly accelerates the algorithms. Further, if the considered automata are highly structured and over a large alphabet, one should try the symbolic algorithms because the BDDs are usually small in such situations. Finally we suggest to minimize the automata by bisimilarity instead of saturating the automata by similarity before performing the algorithms. Although the latter is more powerful, the time efficiency of computing the bisimilarity makes the total time shorter.

The rest of this paper is organized as follows. In Sect. 2 we recall some basic concepts. In Sect. 3 we introduce the HK, HKC and antichain algorithms and relevant optimizations. In Sect. 4 we assess those techniques introduced previously by comparing their running times experimentally. We discuss related work in Sect. 5 and summarize our recommendations in the concluding Sect. 6.

2   Preliminaries

Finite Automata. A Nondeterministic Finite Automaton (NFA) A is a tuple $(\Sigma, Q, I, F, \delta)$ where: Σ is an alphabet, Q is a finite set of states, $I \subseteq Q$ is a non-empty set of initial states, $F \subseteq Q$ is a set of accepting states, and $\delta \subseteq Q \times \Sigma \times Q$ is the transition relation. A word $u = u_1 u_2 \ldots u_n$ is accepted by $q \in Q$ if there exists a sequence $q_0 u_1 q_1 u_2 \ldots u_n q_n$ such that $q_0 = q$, $q_j \in \delta(q_{j-1}, u_j)$ for all $0 < j \leq n$ and $q_n \in F$. Define $L(q) = \{u \mid u \text{ is accepted by } q\}$ as the language of q and $L(A) = \bigcup_{q \in I} L(q)$ as the language of A. Two NFA A, B are said to be language equivalent iff $L(A) = L(B)$.

An NFA is called deterministic if $|I| = 1$ and $|\delta(q, a)| \leq 1$ for all $q \in Q$ and $a \in \Sigma$. For each NFA $A = (\Sigma, Q, I, F, \delta)$, we can use the standard powerset construction [3, Sect. 4.1] to transform it to a DFA $A^\sharp = (\Sigma, Q^\sharp, I^\sharp, F^\sharp, \delta^\sharp)$ with the same language.

The NFA equivalence checking problem is to decide whether two given NFA accept the same language.

Simulation and Bisimulation

Definition 1. *Let R, $R' \subseteq Q \times Q$ be two binary relations on states, we say that R s-progresses to R', denoted $R \rightarrowtail_s R'$, if $x \, R \, y$ implies:*

- *if x is accepting, then y is accepting;*
- *for any $a \in \Sigma$ and $x' \in \delta(x, a)$, there exists some $y' \in \delta(y, a)$ such that $x' \, R' \, y'$.*

A simulation is a relation R such that $R \rightarrowtail_s R$ and a bisimulation is a relation R such that $R \rightarrowtail_s R$ and $R^{-1} \rightarrowtail_s R^{-1}$, where R^{-1} is the inverse relation of R.

The largest simulation and bisimulation are called *similarity* and *bisimilarity,* denoted by $\precsim$ and $\sim$, respectively. For any NFA, if a bisimulation between two states can be found, then they are language equivalent. Similarly, for any NFA, if a simulation between two states can be found, for example $x \precsim y$, then $L(x) \subseteq L(y)$. The reverse direction of these two implications holds in general only for DFA. Computing similarity needs $O(|\delta| \cdot |Q|)$ time [1,7,8,14], while computing bisimilarity is faster, as it is in $O(|\delta| \cdot \log |Q|)$ [18]. Bisimulation is a sound proof technique for checking language equivalence of NFA and it is also complete for DFA. Simulation is a sound proof technique for checking language inclusion of NFA and it is also complete for DFA.

Binary Decision Diagrams. A standard technique [13,19] for working with automata over a large alphabet consists in using BDDs to represent the automata. A Binary Decision Diagram (BDD) over a set of variables $X_n = \{x_1, x_2, \ldots, x_n\}$ is a directed, acyclic graph having leaf nodes and internal nodes. There is exactly one root node in a BDD; each internal node is labelled with a variable and has two outgoing edges whose ends are other nodes. The leaf nodes are labelled with 0 or 1. After fixing the order of variables, any BDD can be transformed into a reduced one which has the fewest nodes [5]. In the sequel, we only work with reduced ordered BDDs, which we simply call BDDs.

BDDs can be used to represent functions of type $2^{X_n} \rightarrow \{0,1\}$. Here, we use BDDs to represent NFA. The advantage is that one often does not need many variables. For example, if there are 2^k letters, one only needs k variables to encode (the characteristic function of) a set of letters.

3   Algorithms and Optimizations

3.1   A Naive Algorithm for Language Equivalence Checking

A naive adaptation of Hopcroft and Karp's algorithm from DFA to NFA is shown in Algorithm 1. Starting with the two sets of initial states, we do the powerset construction on-the-fly for both NFA and simultaneously try to build a bisimulation relating these two sets. The sets of states of the NFA become the states of the DFA constructed by powerset construction. We use two sets: *todo* and R. We call a pair (X, Y) a bad pair if one of X and Y is accepting but the other is not. Whenever we pick a pair from the set *todo*, we check if it is a bad pair; if it isn't, we generate their successors and insert these successor pairs into *todo*. The set R is used to store the processed pairs: if a pair is in R, the

---

Algorithm 1. The `Naive Eq` algorithm for checking NFA equivalence

Input: two NFA $A = (\Sigma, Q_A, I_A, F_A, \delta_A)$ and $B = (\Sigma, Q_B, I_B, F_B, \delta_B)$
Output: "Yes" if $L(A) = L(B)$, otherwise "No"
 1: $R := \emptyset$, $todo := \{(I_A, I_B)\}$
 2: **while** $todo \neq \emptyset$ **do**
 3:     Pick $(X, Y) \in todo$ and remove it
 4:     **if** $(X, Y) \notin R$ **then**
 5:        **if** (X, Y) is a bad pair **then**
 6:           **return** "No, $L(A) \neq L(B)$."
 7:        **end if**
 8:        **for all** $a \in \Sigma$ **do**
 9:           $todo := todo \cup \{(\delta_A^\sharp(X, a), \delta_B^\sharp(Y, a))\}$
10:        **end for**
11:        $R := R \cup \{(X, Y)\}$
12:     **end if**
13: **end while**
14: **return** "Yes, $L(A) = L(B)$."

---

states are language equivalent if the pairs in *todo* are language equivalent. In a formula, $R \rightarrowtail_s R \cup todo$ and $R^{-1} \rightarrowtail_s R^{-1} \cup todo^{-1}$. If the algorithm terminates with the return value "yes", R is a bisimulation between $A^{\sharp}$ and $B^{\sharp}$.

When checking (X, Y), the algorithm eventually determinizes both parts corresponding to X and Y, that is, it compares X (regarded as state of $A^{\sharp}$) with Y (regarded as state of $B^{\sharp}$). Based on the naive algorithm, one can imagine several optimizations: One idea is to try to reduce the number of pairs in R; the algorithm proposed in [9] by Hopcroft and Karp (called the HK algorithm) does so. In [4] Bonchi and Pous extend the HK algorithm by exploiting the technique of bisimulation up to congruence and obtain the HKC algorithm, in which R contains even fewer pairs. Another idea is to reduce the number of pairs in *todo* by so-called memoisation. The observation is very simple: one does not need to insert the same pair into *todo* more than once. (In practice, the set *todo* is often implemented as a list, so it actually can "contain" an element multiple times.) Besides these, one can do some preprocessing: One can use bisimilarity or similarity to saturate the NFA before running the algorithms, in the hope to accelerate the main algorithm. In addition, we also test whether it is a good idea to use BDDs to represent transition functions and then perform symbolic algorithms.

3.2   Reducing R

We only need to know whether there *exists* a bisimulation relating two sets, and it is unnecessary to build the whole bisimulation. So we can reduce R to a relation that is contained in – and sufficient to infer – a bisimulation.

HK Algorithm. Hopcroft and Karp [9] propose that if an encountered pair is not in R but in its reflexive, symmetric and transitive closure, we can also skip this pair. Ignoring the concrete data structure to store equivalence classes, the HK algorithm consists in simply replacing Line 4 in Algorithm 1 with

4:      **if** $(X, Y) \notin rst(R)$ **then**

where rst is the function mapping each relation $R \subseteq \mathcal{P}(Q) \times \mathcal{P}(Q)$ into its reflexive, symmetric, and transitive closure. With this optimization, the number of pairs in R will be reduced. When the algorithm returns "yes", R is no longer a bisimulation, but is contained in a bisimulation [4] and one can infer a bisimulation from R.

HKC Algorithm. Based on the simple observation that if $L(X_1) = L(Y_1)$ and $L(X_2) = L(Y_2)$ then $L(X_1 \cup X_2) = L(Y_1 \cup Y_2)$, Bonchi and Pous [4] improve the HK algorithm with congruence closure. One gets the HKC algorithm just by replacing Line 4 in Algorithm 1 with

4:      **if** $(X, Y) \notin rstu(R \cup todo)$ **then**

where $rstu$ is the reflexive, symmetric, and transitive closure extended with the following union of relations: $u(R)$ is the smallest relation containing R satisfying:

if $X_1 \mathrel{R} Y_1$ and $X_2 \mathrel{R} Y_2$, then $(X_1 \cup X_2) \mathrel{u(R)} (Y_1 \cup Y_2)$. Note that Bonchi and Pous use $R \cup todo$ rather than R because this helps to skip more pairs, and this is safe since all pairs in $todo$ will eventually be processed [4].

When the HKC algorithm returns "yes", R is also contained in a bisimulation [4] and sufficient to infer one.

In order to check whether a pair is in the equivalence closure of a relation, Hopcroft and Karp use disjoint sets forests to represent equivalence classes, which allow to check $(X, Y) \notin rst(R)$ in almost constant amortised time. Unfortunately, this data structure cannot help one to do the checking for congruence closure $rstu$. Bonchi and Pous use a set rewriting approach to do this. However, this requires to scan the pairs in R one by one, which makes it slow. As we shall show in the experiments, this has a great impact on the performance of HKC.

3.3 Reducing *todo*

The pairs in $todo$ are those to be processed. However, if this set is implemented as a list or similar data structure, there are often redundancies. We can remember that some element has already been inserted into $todo$ earlier; this is called *memoisation*. In Line 9, we check whether we have inserted the same pair into $todo$ earlier and only insert the pair if it never has been in $todo$. (Note that this also skips pairs that have in the meantime moved from $todo$ to R.) We can use hash sets to check this condition in constant time.

3.4 BDD Representation

Pous [13] proposed a symbolic algorithm for checking language equivalence of finite automata over large input alphabets. By processing internal nodes, the symbolic algorithm may insert fewer pairs into $todo$, which makes us want to know whether it can save time if we perform the symbolic algorithm instead of the explicit one.

The symbolic version of the HK and HKC algorithms and of memoisation can be easily constructed from the explicit ones. The only difference is that the pairs of sets of states become pairs of BDD nodes, including leaf nodes and internal nodes. If a pair of internal nodes is skipped, then all its successors are also skipped. This is why the symbolic algorithm may have fewer pairs in $todo$.

3.5 Preprocessing Operations

Bonchi and Pous [4] extend the HKC algorithm to exploit the similarity preorder. It suffices to notice that for any similarity pair $x \precsim y$ (in the NFA), we have $\{x, y\} \sim \{y\}$ (in the DFA). So to check whether $(X, Y) \in rstu(R \cup todo)$, it suffices to compute the congruence closure of X and Y w.r.t. the pairs from $R \cup todo \cup \{(\{x, y\}, \{y\}) | x \precsim y\}$. This may allow to skip more pairs. However, the time required to compute similarity may be expensive.

Since bisimilarity can be computed in less time than similarity, it may be advantageous to replace similarity with bisimilarity. So we can replace the similarity with bisimilarity to get another algorithm, i.e. computing the congruence closure of X and Y w.r.t. the pairs from $R \cup todo \cup \{(\{x\}, \{y\}) | x \sim y\}$.

As a matter of fact, we can use this technique as a preprocessing operation. For similarity, we saturate the original NFA w.r.t. the similarity preorder before running the algorithms. For bisimilarity, we choose another approach, that is taking a quotient according to bisimilarity. This amounts to saturating the NFA w.r.t. bisimilarity, but it is more efficient. Note that we do not take a quotient according to simulation equivalence, because it is less powerful than saturating the NFA w.r.t. similarity. Although taking a quotient according to bisimilarity may be less powerful than saturating by similarity, using bisimilarity makes the total time shorter in many cases, which is shown in Sect. 4.

3.6   Algorithms for Language Inclusion Checking

Instead of directly checking language equivalence for NFA, it is possible to check two underlying language inclusions: for any pair (X, Y), we check whether $L(X) \subseteq L(Y)$ and $L(Y) \subseteq L(X)$. If both of them hold, we have $L(X) = L(Y)$. So it is enough to solve the problem of checking $L(X) \subseteq L(Y)$. The naive algorithm is quite similar to the one for checking equivalence. Here, we call a pair (x, Y) a bad pair if x is accepting but Y not. The idea of the algorithm is still: Whenever we pick a pair from $todo$, we check whether it is a bad pair; if not, we insert the successor pairs into $todo$.

When checking (X, Y), the algorithm eventually determinizes the part corresponding to Y and remains nondeterministic for X, that is, it compares $x \in X$ from A and Y from $B^\sharp$. The sets $todo$ and R will therefore be subsets of $Q_A \times \mathcal{P}(Q_B)$. Again, the naive algorithm allows for several optimizations. Memoisation can be used without modifications. The antichain algorithm proposed in [6] aims to reduce the number of pairs in R, and it can be enhanced by exploiting similarity [2].

Given a partial order $(X, \sqsubseteq)$, an *antichain* is a subset $Y \subseteq X$ containing only incomparable elements. The antichain algorithm exploits antichains over the set $Q_A \times \mathcal{P}(Q_B)$, equipped with the partial order $(x_1, Y_1) \sqsubseteq (x_2, Y_2)$ iff $x_1 = x_2$ and $Y_1 \subseteq Y_2$.

In order to check $L(X) \subseteq L(Y)$ for two sets of states X, Y, the antichain algorithm ensures that R is an antichain of pairs (x', Y'). If one of these pairs p is larger than a previously encountered pair $p' \in R$ (i.e. $p' \sqsubseteq p$) then the language inclusion corresponding to p is subsumed by p' so that p can be skipped. Conversely, if there are some pairs $p_1, \ldots, p_n \in R$ which are all larger than p (i.e. $p \sqsubseteq p_i$ for all $1 \leq i \leq n$), one can safely remove them: they are subsumed by p and, by doing so, the set R remains an antichain. We denote the antichain algorithm as "AC".

Abdulla et al. [2] propose to accelerate the antichain algorithm by exploiting similarity. The idea is that when processing a pair (x, Y), if there is a previously encountered pair (x', Y') such that $x \precsim x'$ and $Y' \precsim Y$ (which means $\forall y' \in Y', \exists y \in Y : y' \precsim y$), then (x, Y) can be skipped because it is subsumed by (x', Y'). For the same reason as in Sect. 3.5, we can also use bisimilarity instead of similarity.

Here, we can still take a quotient according to bisimilarity before performing the algorithms. However, we can not saturate the NFA like in Sect. 3.5 because the algorithms need to maintain an antichain and if we saturate the NFA, there would be lots of pairs which can be actually skipped. So we can only exploit similarity while running the algorithms, which sometimes slows them down.

4   Experiments

In this section, we describe some experiments to compare the performance of the algorithms mentioned above. We implemented all these algorithms in Java. For the symbolic one, we use the JavaBDD library [20]. We compute bisimilarity according to the algorithm in [18] and similarity according to [8], which, as far as we know, are the two fastest algorithms to compute bisimilarity and similarity, respectively. All the implementation details are available at https://github.com/fuchen1991/EBEC.

We conducted the experiments on random automata and automata obtained from model-checking problems. All the experiments were performed on a machine with an Intel i7-2600 3.40 GHz CPU and 8 GB RAM.

Random automata: We generate different random NFA by changing three parameters: the number of states ($|Q|$), the number of letters ($|\Sigma|$), and the density (d), which is the average out-degree of each state with respect to each letter. Although Tabakov and Vardi [17] empirically showed that one statistically gets the most challenging NFA with $d = 1.25$, we find that the algorithms are quite sensitive to changes in density and the densities of many NFA from model checking are much smaller than this value, so we test more values for this parameter. For each setting, we generated 100 NFA. To make sure that all the algorithms meet their worst cases, there are no accepting states in the NFA (So we have to skip the operation of removing non-coaccessible states, otherwise this reduces each NFA to the empty one). The two initial state sets are two distinct singleton sets.

Automata from model checking: Bonchi and Pous [4] use the same automata as Abdulla, Chen, Holík et al. [2], which come from the model checking of various programs (the bakery algorithm, bubble sort, and a producer-consumer system) and are available from L. Holík's website. We also use these automata. The difference between our work and Bonchi and Pous's is that they only show the performance of HKC and AC after preprocessing with similarity, while we compare more algorithms.

We record the running time of each algorithm on the above NFAs, measured as the average over four executions. We depict all the data by boxplots – a method for graphically depicting groups of numerical data through their quartiles. In a boxplot, the box denotes the values between the lower quartile and the upper quartile, and the horizontal line in the box denotes the median value. Numbers which are outside 1.5 times the interquartile range above the upper quartile or below the lower quartile are regarded as outliers and shown as individual points. Finally, the two end points of the vertical line outside the box denote the minimal and maximal values that are not considered to be outliers.

4.1   Memoisation Accelerates the Algorithms

In most situations, memoisation saves time because it reduces the number of pairs in *todo*. But the effects on the three algorithms are different, that is, it saves more time to optimize HKC with memoisation than HK and AC. For HKC, repeated pairs are skipped because they are in the congruence closure of R (Algorithm 1, Line 4). This check costs much more time than the corresponding checks of HK and AC. Besides, memoisation needs less time than all the three methods to remove repeated pairs. So as we can see in Fig. 1, the huge difference of the number of pairs in *todo* leads to the huge difference of time for HKC, but leads to small difference for HK and AC.

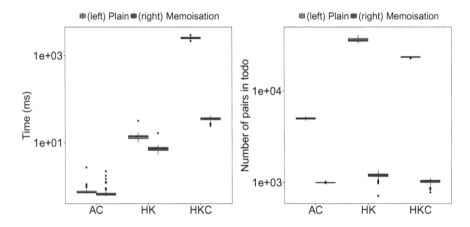

Fig. 1. Comparison of algorithms without and with memoisation ($|Q| = 500$, $|\Sigma| = 50$, $d = 0.1$). The y-axis is logarithmic.

When memoisation only reduces a few pairs in *todo*, it just costs a little more time because memoisation requires only constant time. So it always pays off to use memoisation. In the following, we always use memoisation for HK, HKC, and AC.

4.2   BDDs Are only Suitable for Highly Structured NFA

As discussed in Sect. 3.4, storing NFA with BDDs and performing symbolic algorithms can reduce the pairs in *todo*. However, we find that this variant slows down the algorithms on random NFA for most settings, as shown in Fig. 2(a). This is because it is hard to generate highly structured random NFA.

The size of the BDD highly depends on the structure of the automaton. If an automaton has many symmetries, there are fewer nodes in the BDD; this makes the symbolic algorithms run faster. Also, the performance of explicit algorithms on automata over large alphabet is bad, while BDDs can represent large alphabets with few variables, so symbolic algorithms are preferable for this kind of automata.

In order to show this, we let $|Q| = 4$, $|\Sigma| = 2^{17} = 131072$, and $d = 4.0$, which may be impractical, but a clear example to exhibit the advantage of symbolic algorithms. In this NFA, each state has a transition to all states on all input symbols, so the algorithms only need to process two pairs, namely the pair of initial state sets and the pair of sets containing all states. An explicit algorithm needs to scan every symbol in Σ, while there is only one node in the BDD, which makes symbolic algorithm faster. The result is shown in Fig. 2(b).

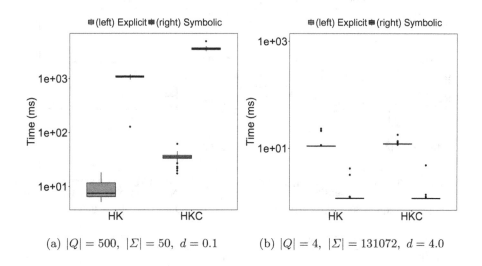

(a) $|Q| = 500$, $|\Sigma| = 50$, $d = 0.1$ (b) $|Q| = 4$, $|\Sigma| = 131072$, $d = 4.0$

Fig. 2. Comparison of explicit and symbolic algorithms. The time-axis is logarithmic.

4.3   Comparison of HK, HKC and AC

In addition to $|Q|$, $|\Sigma|$ and d, there is another parameter that should be considered: the number of transitions, $|\delta| = |Q| \cdot |\Sigma| \cdot d$. Here, we compare the performance of HK, HKC and AC under different settings of these parameters.

First, let us fix $|\Sigma| = 2$ and $d = 1.25$ and vary the state space size $|Q|$. The result is shown in Fig. 3(a). When increasing $|Q|$ (and also $|\delta|$), the time required by all the three algorithms increases, but HK's time increases much stronger than the others'. HKC performs best, and AC has some bad outliers, which can be more than 100 times slower than HKC.

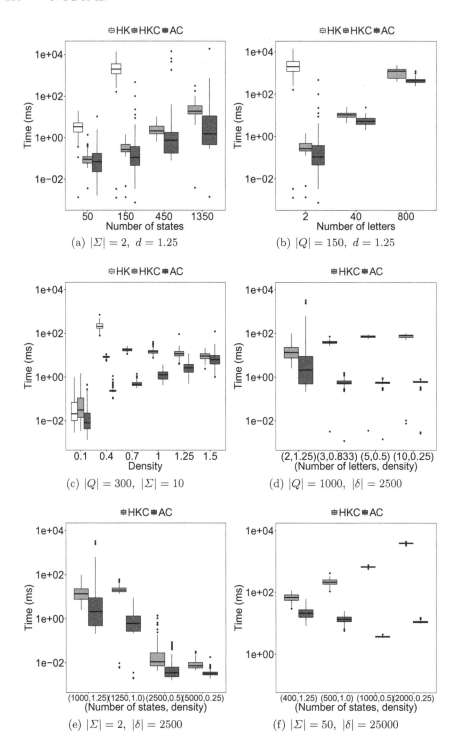

Fig. 3. Comparison of HK, HKC and AC. The time-axis is logarithmic.

Then, we fix $|Q| = 150$ and $d = 1.25$ and vary the alphabet size $|\Sigma|$. The result is shown in Fig. 3(b): Upon increasing $|\Sigma|$ (and also $|\delta|$), the time required by all the three algorithms increases, and again HK's time increases the most. The average performance of AC is best overall; however, if $|\Sigma| = 2$, there are some very slow outliers, so HKC may be preferable, as its performance is not much worse than the average of AC.

Next, we fix $|Q| = 300$ and $|\Sigma| = 10$ and vary the density d. The measurements are shown in Fig. 3(c). Basically, the time first increases then decreases as the density grows. But the peak values appear at different densities for the three algorithms. The peak value of HKC appears to be near $d = 1$, but for HK and AC at a much larger density. We do not expect very large densities in practice; if the density is maximal $(d = |Q|)$, then it can be found very quickly that all states are language equivalent. Still, HK always performs worst. When the density is between 0.1 and 1.25, AC performs 10 to 100 times faster than HKC, but when the density is 1.5, HKC can be several times faster than AC.

Until now, we have seen that HK always performs worse than at least one of the other two algorithms, and no one of HKC and AC always performs best. In the following, we only compare the performance of HKC and AC to find out in which situation HKC performs better and in which situation AC performs better.

Let us fix $|Q| = 500$ and $|\delta| = 2500$ and let $|\Sigma|$ and d vary. The result is shown in Fig. 3(d). The running time remains approximately the same as we fix $|Q|$ and $|\delta|$, but HKC performs better (does not have slow outliers) at $|\Sigma| = 2, d = 1.25$ and worse in other settings. Then we fix $|\Sigma| = 2$ and $|\delta| = 2500$. The result is shown in Fig. 3(e). We can see that HKC is always worse except $d = 1.25$. Finally, we fix $|\Sigma| = 50$ and $|\delta| = 25000$. The result is shown in Fig. 3(f). We can see that HKC is always slower than AC when $|\Sigma| = 50$. The smaller the density is, the larger difference between the performance of HKC and AC is.

In conclusion, HKC performs better when d is large and $|\Sigma|$ is small. In this setting, the maximal value of AC is always very large, which can be 10 to 100 times slower than HKC. Moreover, there are always some bad outliers for AC. So HKC is a better choice in this setting. But If d is small or $|\Sigma|$ is large, AC can even be more than 100 times faster than HKC.

4.4   Automata from Model Checking

Now we compare the performance of HK, HKC, AC and all these three algorithms with preprocessing with similarity and bisimilarity on the automata from model checking. Bonchi and Pous state that HKC can mimic AC even for language inclusion problem. Here, we also use these algorithms to check language inclusion for the automata.

We perform all the three algorithms without any preprocessing ("Plain"), with saturating w.r.t. similarity ("Sim"), and minimizing w.r.t. bisimilarity ("Bisim"), respectively. We separate the results into those for which the inclusion holds and those for which the inclusion does not hold. Figure 4 shows the total running time of each algorithm. First, we find that the densities of these automata are all between 0.05 and 0.49, and over half of them are smaller than 0.19. Their alphabet sizes are between 7 and 36, and over half of them are smaller than 20. We also observe that the performance difference of the algorithms is similar to Fig. 3(c) with $d = 0.1$. This is approximately in agreement with our conclusion for random automata. Second, as we can see the total running time of "Bisim" is much shorter than "Sim" because computing bisimilarity is often much faster than similarity. Preprocessing with similarity may even be slower than no preprocessing at all, but this does not happen for bisimilarity.

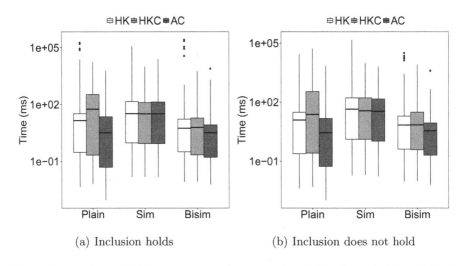

(a) Inclusion holds   (b) Inclusion does not hold

Fig. 4. Comparison of different preprocessing operations. The time-axis is logarithmic.

4.5 Tools: EBEC, hknt, and VATA

We also conducted experiments with other tools. Abdulla et al. [2] implemented their algorithm in their tool "VATA" [11] in C++, and Bonchi and Pous [4] implemented the HKC algorithm in their tool "hknt" in OCaml.

We again run experiments on the same automata sets used in Sect. 4.4. The result is shown in Fig. 5. In this figure, "EBEC" denotes our tool in which we choose the antichain algorithm with memoisation and preprocessing with bisimilarity, since it is the optimal combination according our previous experiments. "hknt" denotes Bonchi and Pous's tool running their HKC algorithm and "VATA, VATA_sim" the tool of Abdulla et al. running the basic antichain algorithm and antichain algorithm with preprocessing with similarity respectively.

We choose 10 min as the timeout and find that for a few tests, hknt and VATA_sim does not terminate in this time, while EBEC and VATA both terminate. We also see that our tool performs about 10 times faster than hknt in both situations. When the inclusion holds, EBEC is 2–3 times faster than VATA and VATA_sim. When the inclusion does not hold, our tool EBEC has the same

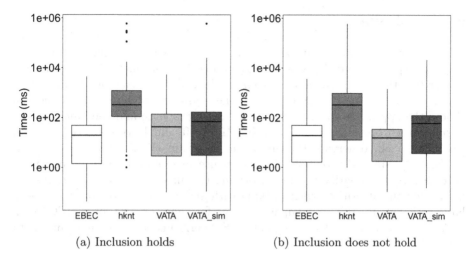

(a) Inclusion holds   (b) Inclusion does not hold

Fig. 5. Comparison of the tools. The time-axis is logarithmic.

performance as VATA on most automata and is up to 2.5 times slower on some automata. However, one should note that if the inclusion does not hold, the performance highly depends on the order of visiting the states.

5   Related Work

To our knowledge, VATA implemented by Abdulla et al. [2] is the most efficient implementation currently available for the antichain algorithms. They compare their algorithm with the naive algorithm and the basic antichain algorithm [6], but not with other algorithms.

Bonchi and Pous propose that HKC can be optimized by similarity [4], but their implementation of the algorithm to compute similarity is slow, because it is based on the algorithm proposed by Henzinger et al. [7], which is no longer the fastest known one. Bonchi and Pous compare HKC with HK and AC only on random automata, and they also compare HKC and AC after preprocessing with similarity. However, they do not show the total time of these two algorithms. In our work, we find that preprocessing with similarity is hardly beneficial, it even makes the total time longer sometimes. So we propose preprocessing with bisimilarity, and the experiments show that it is preferable.

Pous [13] proposes a symbolic algorithm for checking language equivalence of finite automata over large alphabets and applies it to Kleene algebra with tests. We have implemented the symbolic version of HK and HKC, and show that the symbolic algorithm is suitable for highly structured automata, especially those over large alphabets.

6   Conclusion

We have reviewed various algorithms and optimization techniques for checking language equivalence of NFA, and compared their performance by experiments.

We find that their practical efficiencies depend very much on the automata under consideration. But according to the automata to be checked, one can choose the appropriate algorithm: If the density is large and the alphabet size is small, then one should choose HKC (Hopcroft and Karp's algorithm up to congruence) [4], otherwise the antichain algorithm (Wulf, Doyen, Henzinger, Raskin) [6]. Moreover, one should always use memoisation and minimize the automata by bisimilarity before performing the algorithm. One may choose to use a symbolic algorithm when working with highly structured automata over a large alphabet. Finally, we compared the performance of our tool "EBEC" with "VATA" and "hknt" and showed that EBEC performs better on most automata we tested.

Acknowledgements. This work has been supported by the National Natural Science Foundation of China (Grants 61532019, 61472473, 61672229), the CAS/SAFEA International Partnership Program for Creative Research Teams, the Sino-German CDZ project CAP (GZ 1023) and Shanghai Municipal Natural Science Foundation (16ZR1409100).

References

1. Abdulla, P.A., Bouajjani, A., Holík, L., Kaati, L., Vojnar, T.: Computing Simulations over Tree Automata. In: Ramakrishnan, C.R., Rehof, J. (eds.) TACAS 2008. LNCS, vol. 4963, pp. 93–108. Springer, Heidelberg (2008). doi:10.1007/978-3-540-78800-3_8
2. Abdulla, P.A., Chen, Y.-F., Holík, L., Mayr, R., Vojnar, T.: When simulation meets antichains. In: Esparza, J., Majumdar, R. (eds.) TACAS 2010. LNCS, vol. 6015, pp. 158–174. Springer, Heidelberg (2010). doi:10.1007/978-3-642-12002-2_14
3. Baier, C., Katoen, J.-P.: Principles of Model Checking. MIT Press, Cambridge (2008)
4. Bonchi, F., Pous, D.: Checking NFA equivalence with bisimulations up to congruence. In: Principles of Programming Languages (POPL 2013), pp. 457–468. ACM, New York (2013)
5. Bryant, R.E.: Graph-based algorithms for boolean function manipulation. IEEE Trans. Comput. **35**(8), 677–691 (1986). https://doi.org/10.1109/TC.1986.1676819
6. De Wulf, M., Doyen, L., Henzinger, T.A., Raskin, J.-F.: Antichains: a new algorithm for checking universality of finite automata. In: Ball, T., Jones, R.B. (eds.) CAV 2006. LNCS, vol. 4144, pp. 17–30. Springer, Heidelberg (2006). doi:10.1007/11817963_5
7. Henzinger, M.R., Henzinger, T.A., Kopke, P.W.: Computing simulations on finite and infinite graphs. In: Symposium on Foundations of Computer Science (FOCS), Milwaukee, pp. 453–462. IEEE Computer Society (1995)
8. Holík, L., Šimáček, J.: Optimizing an LTS-simulation algorithm. Comput. Inform. **29**(6+), 1337–1348 (2010). http://www.cai.sk/ojs/index.php/cai/article/view/147

9. Hopcroft, J.E., Karp, R.M.: A linear algorithm for testing equivalence of finite automata. Technical report, 71–114, Cornell University, Ithaca (1971). http://hdl.handle.net/1813/5958

10. Hopcroft, J.: An $n \log n$ algorithm for minimizing states in a finite automaton. In: Kohavi, Z., Paz, A. (eds.) Theory of Machines and Computations, pp. 189–196. Academic Press, New York (1971)

11. Lengál, O., Šimáček, J., Vojnar, T.: VATA: a library for efficient manipulation of non-deterministic tree automata. In: Flanagan, C., König, B. (eds.) TACAS 2012. LNCS, vol. 7214, pp. 79–94. Springer, Heidelberg (2012). doi:10.1007/978-3-642-28756-5_7

12. Paige, R., Tarjan, R.E.: Three partition refinement algorithms. SIAM J. Comput. **16**(6), 973–989 (1987)

13. Pous, D.: Symbolic algorithms for language equivalence and Kleene algebra with tests. In: Principles of Programming Languages (POPL 2015), pp. 357–368. ACM, New York (2015)

14. Ranzato, F., Tapparo, F.: A new efficient simulation equivalence algorithm. In: Symposium on Logic in Computer Science (LICS), pp. 171–180. IEEE Computer Society (2007)

15. Rutten, J.J.M.M.: Automata and coinduction (an exercise in coalgebra). In: Sangiorgi, D., de Simone, R. (eds.) CONCUR 1998. LNCS, vol. 1466, pp. 194–218. Springer, Heidelberg (1998). doi:10.1007/BFb0055624

16. Stockmeyer, L.J., Meyer, A.R.: Word problems requiring exponential time. In: Proceedings of the 5th Annual ACM Symposium on Theory of Computing (STOC), pp. 1–9 (1973)

17. Tabakov, D., Vardi, M.Y.: Experimental evaluation of classical automata constructions. In: Sutcliffe, G., Voronkov, A. (eds.) LPAR 2005. LNCS, vol. 3835, pp. 396–411. Springer, Heidelberg (2005). doi:10.1007/11591191_28

18. Valmari, A.: Bisimilarity minimization in $O(m \log n)$ time. In: Franceschinis, G., Wolf, K. (eds.) PETRI NETS 2009. LNCS, vol. 5606, pp. 123–142. Springer, Heidelberg (2009). doi:10.1007/978-3-642-02424-5_9

19. Veanes, M.: Applications of symbolic finite automata. In: Konstantinidis, S. (ed.) CIAA 2013. LNCS, vol. 7982, pp. 16–23. Springer, Heidelberg (2013). doi:10.1007/978-3-642-39274-0_3

20. Whaley, J.: JavaBDD. http://javabdd.sourceforge.net/. Accessed 13 June 2017

A New Decomposition Method for Attractor Detection in Large Synchronous Boolean Networks

Andrzej Mizera[1], Jun Pang[1,2], Hongyang Qu[3], and Qixia Yuan[1(✉)]

[1] Faculty of Science, Technology and Communication, University of Luxembourg,
Luxembourg, Luxembourg
{andrzej.mizera,jun.pang,qixia.yuan}@uni.lu
[2] Interdisciplinary Centre for Security, Reliability and Trust,
University of Luxembourg, Luxembourg, Luxembourg
[3] Department of Automatic Control and Systems Engineering,
University of Sheffield, Sheffield, UK
h.qu@sheffield.ac.uk

Abstract. Boolean networks is a well-established formalism for modelling biological systems. An important challenge for analysing a Boolean network is to identify all its attractors. This becomes challenging for large Boolean networks due to the well-known state-space explosion problem. In this paper, we propose a new SCC-based decomposition method for attractor detection in large synchronous Boolean networks. Experimental results show that our proposed method is significantly better in terms of performance when compared to existing methods in the literature.

1 Introduction

Boolean networks (BN) is a well-established framework widely used for modelling biological systems, such as gene regulatory networks (GRNs). Although it is simple by mainly focusing on the wiring of a system, BN can still capture the important dynamic property of the modelled system, e.g., the *attractors*. In the literature, attractors are hypothesised to characterise cellular phenotypes [1] or to correspond to functional cellular states such as proliferation, apoptosis, or differentiation [2]. Hence, attractor identification is of vital importance to the study of biological systems modelled as BNs.

Attractors are defined based on the BN's state space (often represented as a transition system or graph), the size of which is exponential in the number of nodes in the network. Therefore, attractor detection becomes non-trivial when it comes to a large network. In the BN framework, algorithms for detecting attractors have been extensively studied in the literature. The first study dates back to the early 2000s when Somogyi and Greller proposed an enumeration and simulation method [3]. The idea is to enumerate all the possible states and to run simulation from each of them until an attractor is found. This method is

Q. Yuan—Supported by the National Research Fund, Luxembourg (grant 7814267).

© Springer International Publishing AG 2017
K.G. Larsen et al. (Eds.): SETTA 2017, LNCS 10606, pp. 232–249, 2017.
https://doi.org/10.1007/978-3-319-69483-2_14

largely limited by the network size as its running time grows exponentially with the number of nodes in the BN. Later on, the performance of attractor detection has been greatly improved with the use of two techniques. The first technique is binary decision diagrams (BDDs). This type of methods [4,5] encodes Boolean functions of BNs with BDDs and represents the network's corresponding transition system with BDD structures. Using the BDD operations, the forward and backward reachable states can be often efficiently computed. Detecting attractors is then reduced to finding fix point sets of states in the corresponding transition system. The other technique makes use of satisfiability (SAT) solvers. It transforms attractor detection in BNs into a SAT problem [6]. An unfolding of the transition relation of the BN for a bounded number of steps is represented as a propositional formula. The formula is then solved by a SAT solver to identify a valid path in the state transition system of the BN. The process is repeated iteratively for larger and larger bounded numbers of steps until all attractors are identified. The efficiency of the algorithm largely relies on the number of unfolding steps required and the number of nodes in the BN. Recently, a few decomposition methods [7–9] were proposed to deal with large BNs. The main idea is to decompose a large BN into small components based on its structure, to detect attractors of the small components, and then recover the attractors of the original BN.

In this paper, we propose a new decomposition method for attractor detection in BNs, in particular, in large synchronous BNs, where all the nodes are updated synchronously at each time point. Considering the fact that a few decomposition methods have already been introduced, we explain our new method by showing its main differences from the existing ones. First, our method carefully considers the semantics of synchronous BNs and thus it does not encounter a problem that the method proposed in [7] does. We explain this in more details in Sect. 3. Second, our new method considers the dependency relation among different subnetworks when detecting attractors of them while our previous method [8] does not require this. We show with experimental results that this consideration can significantly improve the performance of attractor detection in large BNs. Further, the decomposition method in [9] is designed for asynchronous networks while here we extend it for synchronous networks. As a consequence, the key operation *realisation* for the synchronous BNs is completely re-designed with respect to the one for asynchronous BNs in [9].

2   Preliminaries

2.1   Boolean Networks

A Boolean network (BN) is composed of two elements: binary-valued nodes, which represents elements of a biological system, and Boolean functions, which represent interactions between the elements. The concept of BNs was first introduced in 1969 by S. Kauffman for analysing the dynamical properties of GRNs [10], where each gene was assumed to be in only one of two possible states: ON/OFF.

Definition 1 (Boolean network). *A Boolean network $G(V, \boldsymbol{f})$ consists of a set of nodes $V = \{v_1, v_2, \ldots, v_n\}$, also referred to as genes, and a vector of Boolean functions $\boldsymbol{f} = (f_1, f_2, \ldots, f_n)$, where f_i is a predictor function associated with node v_i $(i = 1, 2, \ldots, n)$. A state of the network is given by a vector $\boldsymbol{x} = (x_1, x_2, \ldots, x_n) \in \{0, 1\}^n$, where $x_i \in \{0, 1\}$ is a value assigned to node v_i.*

Since the nodes are binary, the state space of a BN is exponential in the number of nodes. Each node $v_i \in V$ has an associated subset of nodes $\{v_{i_1}, v_{i_2}, \ldots, v_{i_{k(i)}}\}$, referred to as the set of *parent nodes* of v_i, where $k(i)$ is the number of parent nodes and $1 \leq i_1 < i_2 < \cdots < i_{k(i)} \leq n$. Starting from an initial state, the BN evolves in time by transiting from one state to another. The state of the network at a discrete time point t $(t = 0, 1, 2, \ldots)$ is given by a vector $\boldsymbol{x}(t) = (x_1(t), x_2(t), \ldots, x_n(t))$, where $x_i(t)$ is a binary-valued variable that determines the value of node v_i at time point t. The value of node v_i at time point $t+1$ is given by the predictor function f_i applied to the values of the parent nodes of v_i at time t, i.e., $x_i(t+1) = f_i(x_{i_1}(t), x_{i_2}(t), \ldots, x_{i_{k(i)}}(t))$. For simplicity, with slight abuse of notation, we use $f_i(x_{i_1}, x_{i_2}, \ldots, x_{i_{k(i)}})$ to denote the value of node v_i at the next time step. For any $j \in [1, k(i)]$, node v_{i_j} is called a *parent node* of v_i and v_i is called a *child node* of v_{i_j}.

In general, the Boolean predictor functions can be formed by combinations of any logical operators, e.g., logical AND $\wedge$, OR $\vee$, and NEGATION $\neg$, applied to variables associated with the respective parent nodes. The BNs are divided into two types based on the time evolution of their states, i.e., *synchronous* and *asynchronous*. In synchronous BNs, values of all the variables are updated simultaneously; while in asynchronous BNs, one variable at a time is randomly selected for update.

In this paper, we focus on synchronous BNs. The transition relation of a synchronous BN is given by

$$T\left(\boldsymbol{x}(t), \boldsymbol{x}(t+1)\right) = \bigwedge_{i=1}^{n} \left(x_i(t+1) \leftrightarrow f_i(x_{i_1}(t), x_{i_2}(t), \cdots, x_{i_{k_i}}(t)) \right). \quad (1)$$

It states that in every step, all the nodes are updated synchronously according to their Boolean functions.

In many cases, a BN $G(V, \boldsymbol{f})$ is studied as a state transition system. Formally, the definition of state transition system is given as follows.

Definition 2 (State transition system). *A state transition system $\mathcal{T}$ is a 3-tuple $\langle S, S_0, T \rangle$ where S is a finite set of states, $S_0 \subseteq S$ is the initial set of states and $T \subseteq S \times S$ is the transition relation. When $S = S_0$, we write $\langle S, T \rangle$.*

A BN can be easily modelled as a state transition system: the set S is just the state space of the BN, so there are 2^n states for a BN with n nodes; the initial set of states S_0 is the same as S; finally, the transition relation T is given by Eq. 1.

Example 1. A BN with 3 nodes is shown in Fig. 1a. Its Boolean functions are given as: $f_1 = \neg(x_1 \wedge x_2)$, $f_2 = x_1 \wedge \neg x_2$, and $f_3 = \neg x_2$. In Fig. 1a, the three circles v_1, v_2 and v_3 represent the three nodes of the BN. The edges between nodes represent the interactions between nodes. Applying the transition relation to each of the states, we can get the corresponding state transition system. For better understanding, we demonstrate the state transition system as a state transition graph in this paper. The corresponding state transition graph of this example is shown in Fig. 1b.

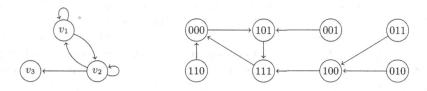

(a) A BN with 3 nodes.   (b) Transition graph of the BN in Example 1.

Fig. 1. The Boolean network in Example 1 and its state transition graph.

In the transition graph of Fig. 1b, the three states $(000), (1 * 1)^1$ can be reached from each other but no other state can be reached from any of them. This forms an *attractor* of the BN. The formal definition of an *attractor* is given as follows.

Definition 3 (Attractor of a BN). *An* attractor *of a BN is a set of states satisfying that any state in this set can be reached from any other state in this set and no state in this set can reach any other state that is not in this set.*

When analysing an attractor, we often need to identify transition relations between the attractor states. We call an attractor together with its state transition relation an *attractor system* (AS). The states constituting an attractor are called *attractor states*. The attractors of a BN characterise its long-run behaviour [11] and are of particular interest due to their biological interpretation.

For synchronous BNs, each state of the network can only have at most one outgoing transition. Therefore, the transition graph of an attractor in a synchronous BN is simply a loop. By detecting all the loops in a synchronous BN, one can identify all its attractors.

Example 2. The BN given in Example 1 has one attractor, i.e., $\{(000), (1 * 1)\}$.

---

[1] We use $*$ to denote that the bit can have value either 1 or 0, so $(1 * 1)$ actually denotes two states: 101 and 111.

2.2   Encoding BNs in BDDs

Binary decision diagrams (BDDs) were introduced by Lee in [12] and Akers in [13] to represent Boolean functions [12,13]. BDDs have the advantage of memory efficiency and have been applied in many model checking algorithms to alleviate the state space explosion problem. A BN $G(V, \boldsymbol{f})$ can be modelled as a state transition system, which can then be encoded in a BDD.

Each variable in V can be represented by a binary BDD variable. By slight abuse of notation, we also use V to denote the set of BDD variables. In order to encode the transition relation, another set V' of BDD variables, which is a copy of V, is introduced: V encodes the possible current states, i.e., $\boldsymbol{x}(t)$, and V' encodes the possible next states, i.e., $\boldsymbol{x}(t+1)$. Hence, the transition relation can be viewed as a Boolean function $T : 2^{|V|+|V'|} \rightarrow \{0, 1\}$, where values 1 and 0 indicate a valid and an invalid transition, respectively. Our attractor detection algorithm, which will be discussed in the next section, also utilizes two basis functions: $Image(X, T) = \{s' \in S \mid \exists s \in X \text{ such that } (s, s') \in T\}$, which returns the set of target states that can be reached from any state in $X \subseteq S$ with a single transition in T; $Preimage(X, T) = \{s' \in S \mid \exists s \in X \text{ such that } (s', s) \in T\}$, which returns the set of predecessor states that can reach a state in X with a single transition. To simplify the presentation, we define $Preimage^i(X, T) = \underbrace{Preimage(...(Preimage(X, T)))}_{i}$ with $Preimage^0(X, T) = X$. In this way, the set of all states that can reach a state in X via transitions in T is defined as an iterative procedure $Predecessors(X, T) = \bigcup\limits_{i=0}^{n} Preimage^n(X, T)$ such that $Preimage^n(X, T) = Preimage^{n+1}(X, T)$. Given a set of states $X \subseteq S$, the projection $T|_X$ of T on X is defined as $T|_X = \{(s, s') \in T \mid s \in X \wedge s' \in X\}$.

3   The New Method

In this section, we describe in details the new SCC-based decomposition method for detecting attractors of large synchronous BNs and we prove its correctness. The method consists of three steps. First, we divide a BN into sub-networks called *blocks* and this step is performed on the *network structure*, instead of the state transition system of the network. Second, we detect attractors in blocks. Last, we recover attractors of the original BN based on attractors of the blocks.

3.1   Decompose a BN

We start by giving the formal definition of a block.

Definition 4 (Block). *Given a BN $G(V, \boldsymbol{f})$ with $V = \{v_1, v_2, \ldots, v_n\}$ and $\boldsymbol{f} = \{f_1, f_2, \ldots, f_n\}$, a block $B(V^B, \boldsymbol{f}^B)$ is a subset of the network, where $V^B \subseteq V$. For any node $v_i \in V^B$, if B contains all the parent nodes of v_i, its Boolean function in B remains the same as in G, i.e., f_i; otherwise, the Boolean function is undetermined, meaning that additional information is required to determine* *the value of v_i in B. We call the nodes with undetermined Boolean functions as* undetermined nodes. *We refer to a block as an* elementary block *if it contains no undetermined nodes.*

We consider synchronous networks in this paper and therefore a block is also under the synchronous updating scheme, i.e., all the nodes in the block will be updated at each given time point no matter this node is undetermined or not.

We now introduce a method to construct blocks, using SCC-based decomposition. Formally, the standard graph-theoretical definition of an SCC is as follows.

Definition 5 (SCC). *Let $\mathcal{G}$ be a directed graph and $\mathcal{V}$ be its vertices. A strongly connected component (SCC) of $\mathcal{G}$ is a maximal set of vertices $C \subseteq \mathcal{V}$ such that for every pair of vertices u and v, there is a directed path from u to v and vice versa.*

We first decompose a given BN into SCCs. Figure 2a shows the decomposition of a BN into four SCCs: Σ_1, Σ_2, Σ_3, and Σ_4. A node outside an SCC that is a parent to a node in the SCC is referred to as a *control node* of this SCC. In Fig. 2a, node v_1 is a control node of Σ_2 and Σ_4; node v_2 is a control node of Σ_3; and node v_6 is a control node of Σ_4. The SCC Σ_1 does not have any control node. An SCC together with its control nodes forms a *block*. For example, in Fig. 2a, Σ_2 and its control node v_1 form one block B_2. Σ_1 itself is a block, denoted as B_1, since the SCC it contains does not have any control node. If a control node in a block B_i is a determined node in another block B_j, block B_j is called a *parent* of block B_i and B_i is a child of B_j. By adding directed edges from all parent blocks to all their child blocks, we form a directed acyclic graph (DAG) of the blocks as the blocks are formed from SCCs. As long as the block graph is guaranteed to be a DAG, other strategies to form blocks can be used. Two blocks can be merged into one larger block. For example, blocks B_1 and B_2 can be merged together to form a larger block $B_{1,2}$.

A state of a block is a binary vector of length equal to the size of the block which determines the values of all the nodes in the block. In this paper, we use a number of operations on the states of a BN and its blocks. Their definitions are given below.

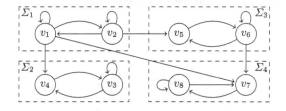
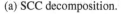
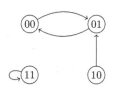

(a) SCC decomposition.   (b) Transition graph of block B_1.

Fig. 2. SCC decomposition and the transition graph of block B_1.

Definition 6 (Projection map, Compressed state, Mirror states). *For a BN G and its block B, where the set of nodes in B is $V^B = \{v_1, v_2, \ldots, v_m\}$ and the set of nodes in G is $V = \{v_1, v_2, \ldots, v_m, v_{m+1}, \ldots, v_n\}$, the projection map $\pi_B : X \to X^B$ is given by $x = (x_1, x_2, \ldots, x_m, x_{m+1}, \ldots, x_n) \mapsto \pi_B(x) = (x_1, x_2, \ldots, x_m)$. For any set of states $S \subseteq X$, we define $\pi_B(S) = \{\pi_B(x) : x \in S\}$. The projected state $\pi_B(x)$ is called a* compressed state *of x. For any state $x^B \in X^B$, we define its set of* mirror states *in G as $\mathcal{M}_G(x^B) = \{x \mid \pi_B(x) = x^B\}$. For any set of states $S^B \subseteq X^B$, its set of mirror states is $\mathcal{M}_G(S^B) = \{x \mid \pi_B(x) \in S^B\}$.*

The concept of projection map can be naturally extended to blocks. Given a block with nodes $V^B = \{v_1, v_2, \ldots, v_m\}$, let $V^{B'} = \{v_1, v_2, \ldots, v_j\} \subseteq V^B$. We can define $\pi_{B'} : X^B \to X^{B'}$ as $x = (x_1, x_2, \ldots, x_m) \mapsto \pi_{B'}(x) = (x_1, x_2, \ldots, x_j)$, and for a set of states $S^B \subseteq X^B$, we define $\pi_{B'}(S^B) = \{\pi_{B'}(x) : x \in S^B\}$.

3.2 Detect Attractors in a Block

An elementary block does not depend on any other block while a non-elementary block does. Therefore, they can be treated separately. We first consider the case of elementary blocks. An elementary block is in fact a BN; therefore, the definition of attractors in a BN can be directly taken to the concept of an elementary block.

Definition 7 (Preservation of attractors). *Given a BN G and an elementary block B in G, let $\mathcal{A} = \{A_1, A_2, \ldots, A_m\}$ be the set of attractors of G and $\mathcal{A}^B = \{A_1^B, A_2^B, \ldots, A_{m'}^B\}$ be the set of attractors of B. We say that B preserves the attractors of G if for any $k \in [1, m]$, there is an attractor $A_{k'}^B \in \mathcal{A}^B$ such that $\pi_B(A_k) \subseteq A_{k'}^B$.*

Example 3. Consider the BN G_1 in Example 1. Its set of attractors is $\mathcal{A} = \{\{(000), (1 * 1)\}\}$. Nodes v_1 and v_2 form an elementary block B_1. Since B_1 is an elementary block, it can be viewed as a BN. The transition graph of B_1 is shown in Fig. 3a. Its set of attractors is $\mathcal{A}^{B_1} = \{\{(00), (1*)\}\}$ (nodes are arranged as v_1, v_2). We have $\pi_{B_1}(\{(000), (1 * 1)\}) = \{(00), (1*)\} \in \mathcal{A}^{B_1}$, i.e., block B_1 preserves the attractors of G_1.

With Definition 7, we have the following lemma and theorem.

(a) Transition graph of Block B_1 in G_1.   (b) Transition graph of the "realisation".

Fig. 3. Two transition graphs used in Examples 3 and 4.

Lemma 1. *Given a BN G and an elementary block B in G, let Φ be the set of attractor states of G and Φ^B be the set of attractor states of B. If B preserves the attractors of G, then $\Phi \subseteq \mathcal{M}_G(\Phi^B)$.*

Theorem 1. *Given a BN G, let B be an elementary block in G. B preserves the attractors of G.*

For an elementary block B, the mirror states of its attractor states cover all G's attractor states according to Lemma 1 and Theorem 1. Therefore, by *searching from the mirror states only instead of the whole state space*, we can detect all the attractor states.

We now consider the case of non-elementary blocks.

Definition 8 (Parent SCC, Ancestor SCC). *An SCC Σ_i is called a* parent SCC *(or* parent *for short) of another SCC Σ_j if Σ_i contains at least one control node of Σ_j. Denote $P(\Sigma_i)$ the set of parent SCCs of Σ_i. An SCC Σ_k is called an* ancestor SCC *(or* ancestor *for short) of an SCC Σ_j if and only if either (1) Σ_k is a parent of Σ_j or (2) Σ_k is a parent of $\Sigma_{k'}$, where $\Sigma_{k'}$ is an ancestor of Σ_j. Denote $\Omega(\Sigma_j)$ the set of ancestor SCCs of Σ_i.*

For an SCC Σ_j, if it has no parent SCC, then this SCC can form an elementary block; if it has at least one parent, then it must have an ancestor that has no parent, and all its ancestors $\Omega(\Sigma_j)$ together can form an elementary block, which is also a BN. The SCC-based decomposition will usually result in one or more non-elementary blocks.

Definition 9 (Crossability, Cross operations). *Let G be a BN and let B_i be a non-elementary block in G with the set of nodes $V^{B_i} = \{v_{p_1}, v_{p_2}, \ldots, v_{p_s}, v_{q_1}, v_{q_2}, \ldots, v_{q_t}\}$, where q_k ($k \in [1,t]$) are the indices of the control nodes also contained by B_i's parent block B_j and p_k ($k \in [1,s]$) are the indices of the remaining nodes. We denote the set of nodes in B_j as $V^{B_j} = \{v_{q_1}, v_{q_2}, \ldots, v_{q_t}, v_{r_1}, v_{r_2}, \ldots, v_{r_u}\}$, where r_k ($k \in [1,u]$) are the indices of the non-control nodes in B_j. Let further $\boldsymbol{x}^{B_i} = (x_1, x_2, \ldots, x_s, y_1^i, y_2^i, \ldots, y_t^i)$ be a state of B_i and $\boldsymbol{x}^{B_j} = (y_1^j, y_2^j, \ldots, y_t^j, z_1, z_2, \ldots, z_u)$ be a state of B_j. States $\boldsymbol{x}^{B_i}$ and $\boldsymbol{x}^{B_j}$ are said to be* crossable, *denoted as $\boldsymbol{x}^{B_i}\ C\ \boldsymbol{x}^{B_j}$, if the values of their common nodes are the same, i.e., $y_k^i = y_k^j$ for all $k \in [1,t]$. The cross operation of two crossable states $\boldsymbol{x}^{B_i}$ and $\boldsymbol{x}^{B_j}$ is defined as $\Pi(\boldsymbol{x}^{B_i}, \boldsymbol{x}^{B_j}) = (x_1, x_2, \ldots, x_s, y_1^i, y_2^i, \ldots, y_t^i, z_1, z_2, \ldots, z_u)$. The notion of crossability naturally extends to two blocks without common nodes: any two states of these blocks are crossable.*

We say $S^{B_i} \subseteq X^{B_i}$ and $S^{B_j} \subseteq X^{B_j}$ are crossable, denoted as $S^{B_i}\ C\ S^{B_j}$, if at least one of the sets is empty or the following two conditions hold: (1) for any state $\boldsymbol{x}^{B_i} \in S^{B_i}$, there exists a state $\boldsymbol{x}^{B_j} \in S^{B_j}$ such that $\boldsymbol{x}^{B_i}$ and $\boldsymbol{x}^{B_j}$ are crossable; (2) vice versa. The cross operation on two crossable non-empty sets of states S^{B_i} and S^{B_j} is defined as $\Pi(S^{B_i}, S^{B_j}) = \{\Pi(\boldsymbol{x}^{B_i}, \boldsymbol{x}^{B_j}) \mid \boldsymbol{x}^{B_i} \in S^{B_i}, \boldsymbol{x}^{B_j} \in S^{B_j}$ and $\boldsymbol{x}^{B_i}\ C\ \boldsymbol{x}^{B_j}\}$. When one of the two sets is empty, the cross operation simply returns the other set, i.e., $\Pi(S^{B_i}, S^{B_j}) = S^{B_i}$ if $S^{B_j} = \emptyset$ and $\Pi(S^{B_i}, S^{B_j}) = S^{B_j}$ if $S^{B_i} = \emptyset$.

Let $\mathcal{S}^{B_i} = \{S^{B_i} \mid S^{B_i} \subseteq X^{B_i}\}$ be a family of sets of states in B_i and $\mathcal{S}^{B_j} = \{S^{B_j} \mid S^{B_j} \subseteq X^{B_j}\}$ be a family of sets of states in B_j. We say $\mathcal{S}^{B_i}$ and $\mathcal{S}^{B_j}$ are crossable, denoted as $\mathcal{S}^{B_i} \mathcal{C} \mathcal{S}^{B_j}$ if (1) for any set $S^{B_i} \in \mathcal{S}^{B_i}$, there exists a set $S^{B_j} \in \mathcal{S}^{B_j}$ such that S^{B_i} and S^{B_j} are crossable; (2) vice versa. The cross operation on two crossable families of sets $\mathcal{S}^{B_i}$ and $\mathcal{S}^{B_j}$ is defined as $\Pi(\mathcal{S}^{B_i}, \mathcal{S}^{B_j}) = \{\Pi(S_i, S_j) \mid S_i \in \mathcal{S}^{B_i}, S_j \in \mathcal{S}^{B_j} \text{ and } S_i \mathcal{C} S_j\}$.

Proposition 1. *Let V^C be the set of control nodes shared by two blocks B_i and B_j, i.e., $V^C = V^{B_i} \cap V^{B_j}$ and let $S^{B_i} \subseteq X^{B_i}$ and $S^{B_j} \subseteq X^{B_j}$. Then $S^{B_i} \mathcal{C} S^{B_j}$ is equivalent to $\pi_C(S^{B_i}) = \pi_C(S^{B_j})$.*

After decomposing a BN into SCCs, there is at least one SCC with no control nodes. Hence, there is at least one elementary block in every BN. Moreover, for each non-elementary block we can construct, by merging all its predecessor blocks, a single parent elementary block. We detect the attractors of the elementary blocks and use the detected attractors to guide the values of the control nodes of their child blocks. The guidance is achieved by considering *realisations of the dynamics of a non-elementary block with respect to the attractors of its parent elementary block*, shortly referred to as *realisations of a non-elementary block*. In some cases, a realisation of a non-elementary block is simply obtained by assigning new Boolean functions to the control nodes of the block. However, in many cases, it is not this simple and a realisation of a non-elementary block is obtained by explicitly constructing a transition system of this block corresponding to the considered attractor of the elementary parent block. Since the parent block of a non-elementary block may have more than one attractor, a non-elementary block may have more than one realisation.

Definition 10 (Realisation of a non-elementary block). *Let B_i be a non-elementary block formed by merging an SCC with its control nodes. Let nodes $u_1, u_2, \ldots, u_r$ be all the control nodes of B_i which are also contained in its elementary parent block B_j (we can merge B_i's ancestor blocks to form B_j if B_i has more than one parent block or has a non-elementary parent block). Let $A_1^{B_j}, A_2^{B_j}, \ldots, A_t^{B_j}$ be the attractor systems of B_j. For any $k \in [1, t]$, a realisation of block B_i with respect to $A_k^{B_j}$ is a state transition system such that*

1. *the transitions are as follows: for any transition $\boldsymbol{x}^{B_j} \rightarrow \tilde{\boldsymbol{x}}^{B_j}$ in the attractor system of $A_k^{B_j}$, there is a transition $\boldsymbol{x}^{B_{i,j}} \rightarrow \tilde{\boldsymbol{x}}^{B_{i,j}}$ in the realisation such that $\boldsymbol{x}^{B_{i,j}} \mathcal{C} \boldsymbol{x}^{B_j}$ and $\tilde{\boldsymbol{x}}^{B_{i,j}} \mathcal{C} \tilde{\boldsymbol{x}}^{B_j}$; each transition in the realisation is caused by the update of all nodes synchronously: the update of non-control nodes of B_i is regulated by the Boolean functions of the nodes and the update of nodes in its parent block B_j is regulated by the transitions of the attractor system of $A_k^{B_j}$;*

In the realisation of a non-elementary block all the nodes of its single elementary parent block are considered and not only the control nodes of the parent block. This allows to distinguish the potentially different states in which the values of control nodes are the same. Without this, a state in the state transition graph of the realisation may have more than one out-going transition, which is contrary to the fact that the out-going transition for a state in a synchronous network is always determined. Although the definition of attractors can still be applied to such a transition graph, the attractor detection algorithms for synchronous networks, e.g., SAT-based algorithms, may not work any more. Moreover, the meaning of attractors in such a graph are not consistent with the synchronous semantics and therefore the detected "attractors" may not be attractors of the synchronous BN. Note that the decomposition method mentioned in [7] did not take care of this and therefore produces incorrect results in certain cases. We now give an example to illustrate one of such cases.

Example 4. Consider the BN in Example 1, which can be divided into two blocks: block B_1 with nodes v_1, v_2 and block B_2 with nodes v_2, v_3. The transition graph of B_1 is shown in Fig. 3a and its attractor is $(00) \rightarrow (10) \rightarrow (11)$. If we do not include the node v_1 when forming the realisation of B_2, we will get a transition graph as shown in Fig. 3b, which contains two states with two out-going transitions. This is contrary to the synchronous semantics. Moreover, recovering attractors with the attractors in this graph will lead to a non-attractor state of the original BN, i.e., (001).

For asynchronous networks, however, such a distinction is not necessary since the situation of multiple out-going transitions is in consistent with the asynchronous updating semantics. Definition 10 forms the basis for a key difference between this decomposition method for synchronous BNS and the one for asynchronous BNs proposed in [9].

Constructing realisations for a non-elementary block is a key process for obtaining its attractors. For each realisation, the construction process requires the knowledge of all the transitions in the corresponding attractor of its elementary parent block. In Sect. 4, we explain in details how to implement it with BDDs. We now give some examples.

Example 5. Consider the BN in Fig. 2a. Its Boolean functions are given as follows:

$$\begin{cases} f_1 = x_1 \wedge x_2, & f_2 = x_1 \vee \neg x_2, & f_3 = \neg x_4 \wedge x_3, & f_4 = x_1 \vee x_3, \\ f_5 = x_2 \wedge x_6, & f_6 = x_5 \wedge x_6, & f_7 = (x_1 \vee x_6) \wedge x_8, & f_8 = x_7 \vee x_8. \end{cases} \quad (2)$$

The network contains four SCCs $\Sigma_1, \Sigma_2, \Sigma_3$ and Σ_4. For any Σ_i ($i \in [1, 4]$), we form a block B_i by merging Σ_i with its control nodes. Block B_1 is an elementary block and its transition graph is shown in Fig. 2b. Block B_1 has two attractors, i.e., $\{(0*)\}$ and $\{(11)\}$. Regarding the first attractor, block B_3 has a realisation by setting the nodes v_1 and v_2 (nodes from its parent block B_1) to contain transitions $\{(00) \rightarrow (01), (01) \rightarrow (00)\}$. The transition graph of this realisation is shown in Fig. 4a. Regarding the second attractor, block B_3 has a realisation by setting nodes v_1 and v_2 to contain only the transition $\{(11) \rightarrow (11)\}$. Its transition graph is shown in Fig. 4b.

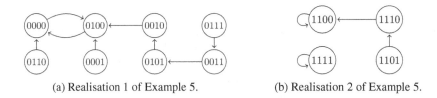

(a) Realisation 1 of Example 5.   (b) Realisation 2 of Example 5.

Fig. 4. Transition graphs of two realisations in Example 5.

A realisation of a non-elementary block takes care of the dynamics of the undetermined nodes, providing a transition system of the block. Therefore, we can extend the attractor definition to realisations of non-elementary blocks as follows.

Definition 11 (Attractors of a non-elementary block). *An attractor of a realisation of a non-elementary block is a set of states satisfying that any state in this set can be reached from any other state in this set and no state in this set can reach any other state that is not in this set. The attractors of a non-elementary block is the set of the attractors of all realisations of the block.*

With the above definition, we can extend Definition 10 by allowing B_j to be a non-elementary block as well. As long as B_i's parent block B_j contains all the control nodes of block B_i, the attractors of B_j can be used to form the realisations of B_i, no matter B_j is elementary or not. Observe that using a non-elementary block as a parent block does not change the fact that the attractor states of the parent block contain the values of all the nodes in the current block and all its ancestor blocks.

Computing attractors for non-elementary blocks requires the knowledge of the attractors of its parent blocks. Therefore, we need to order the blocks so that for any block B_i, the attractors of its parent blocks are always detected before it. The blocks are ordered topologically. For easier explanation of the order, we introduce the concept of topological credit as follows. For simplification, we refer topological credit as credit in the remaining part of the paper.

Definition 12 (Topological credit). *Given a BN G, the an elementary block B_i of G has a topological credit of 0, denoted as $\mathcal{P}(B_i) = 0$. Let B_j be a non-elementary block and $B_{j_1}, \ldots, B_{j_{p(j)}}$ be all its parent blocks. The topological credit of B_j is defined as $\mathcal{P}(B_j) = max_{k=1}^{p(j)}(\mathcal{P}(B_{j_k})) + 1$, where $p(j)$ is the number of parent blocks of B_j.*

3.3   Recovering Attractors for the Original BN

After computing attractors for all the blocks, we need to recover attractors for the original BN, with the help of the following theorem.

Theorem 2. *Let G be a BN and let B_i be one of its blocks. Denote $\Omega(B_i)$ as the block formed by all B_i's ancestor blocks and denote $\mathcal{X}(B_i)$ as the block formed by merging B_i with $\Omega(B_i)$. $\mathcal{X}(B_i)$ is in fact an elementary block, which is also a BN. The attractors of block B_i are in fact the attractors of $\mathcal{X}(B_i)$.*

Theorem 3. *Given a BN G, where B_i and B_j are its two blocks, let $\mathcal{A}^{B_i}$ and $\mathcal{A}^{B_j}$ be the set of attractors for B_i and B_j, respectively. Let $B_{i,j}$ be the block got by merging the nodes in B_i and B_j. Denote the set of all attractor states of $B_{i,j}$ as $S^{B_{i,j}}$. If B_i and B_j are both elementary blocks, $\mathcal{A}^{B_i} \subset \mathcal{A}^{B_j}$ and $\cup_{A \in \Pi(\mathcal{A}^{B_i}, \mathcal{A}^{B_j})} A = S^{B_{i,j}}$.*

The above developed theoretical background with Theorems 2 and 3 being its core result, allows us to design a new decomposition-based approach towards detection of attractors in large synchronous BNs. The idea is as follows. We divide a BN into blocks according to the detected SCCs. We order the blocks in the ascending order based on their credits and detect attractors of the ordered blocks one by one in an iterative way. We start from detecting attractors of elementary blocks (credit 0), and continue to detect blocks with higher credits after constructing their realisations. According to Theorem 2, by detecting the attractors of a block, we in fact obtain the attractors of the block formed by the current block and all its ancestor blocks. Hence, after the attractors of all the blocks have been detected, either we have obtained the attractors of the original BN or we have obtained the attractors of several elementary blocks of this BN. According to Theorem 3, we can perform a cross operation for any two elementary blocks (credits 0) to recover the attractor states of the two merged blocks. The resulting merged block will form a new elementary block, i.e., one with credit 0. The attractors can be easily identified from the set of attractor states. By iteratively performing the cross operation until a single elementary block containing all the nodes of the BN is obtained, we can recover the attractors of the original BN. The details of this new algorithm are discussed in the next section. In addition, we have the following corollary that extends Theorem 3 by allowing B_i and B_j to be non-elementary blocks. This corollary will be used in the next section.

Corollary 1. *Given a BN G, where B_i and B_j are its two blocks, let $\mathcal{A}^{B_i}$ and $\mathcal{A}^{B_j}$ be the set of attractors for B_i and B_j, respectively. Let $B_{i,j}$ be the block got by merging the nodes in B_i and B_j. Denote the set of attractor states of $B_{i,j}$ as $S^{B_{i,j}}$. It holds that $\mathcal{A}^{B_i} \subset \mathcal{A}^{B_j}$ and $\cup_{S \in \Pi(\mathcal{A}^{B_i}, \mathcal{A}^{B_j})} S = S^{B_{i,j}}$.*

4   A BDD-Based Implementation

We describe the SCC-based attractor detection method in Algorithm 1. As mentioned in Sect. 2.2, we encode BNs in BDDs; hence most operations in this algorithm is performed with BDDs. Algorithm 1 takes a BN G and its corresponding transition system $\mathcal{T}$ as inputs, and outputs the set of attractors of G. In this

---

Algorithm 1. SCC-based decomposition algorithm

1: **procedure** SCC_DETECT(G, $\mathcal{T}$)
2:     $B :=$ FORM_BLOCK(G); $\mathcal{A} := \emptyset$; $B_a := \emptyset$; $k :=$ size of B;
3:     initialise dictionary $\mathcal{A}^\ell$; $//\mathcal{A}^\ell$ is a dictionary storing the set of attractors for each block
4:     **for** $i := 1; i <= k; i + +$ **do**
5:         **if** B_i is an elementary block **then**
6:             $\mathcal{T}^{B_i} :=$ transition system converted from B_i;   //see Sect. 2.2 for more details
7:             $\mathcal{A}_i :=$ DETECT($\mathcal{T}^{B_i}$); $\mathcal{A}^\ell.add((B_i, \mathcal{A}_i))$;
8:         **else** $\mathcal{A}_i := \emptyset$;
9:             **if** B_i^p is the only parent block of B_i **then**
10:                 $\mathcal{A}_i^p := \mathcal{A}^\ell.getAtt(B_i^p)$;   //obtain attractors of B_i^p
11:             **else** $B^p := \{B_1^p, B_2^p, \ldots, B_m^p\}$ be the parent blocks of B_i (ascending ordered);
12:                 $B_c := B_1^p$;   //B^p is ordered based on credits
13:                 **for** $j := 2; j <= m; j + +$ **do**
14:                     $B_{c,j} :=$ a new block containing nodes in B_c and B_j^p;
15:                     **if** ($\mathcal{A}_i^p := \mathcal{A}^\ell.getAtt(B_{c,j})$) $== \emptyset$ **then**
16:                         $A := \Pi(\mathcal{A}^\ell.getAtt(B_c), \mathcal{A}^\ell.getAtt(B_j))$; $\mathcal{A}_i^p := D(A)$;
17:                         //$D(A)$ returns all the attractors from attractor states sets A
18:                         $\mathcal{A}^\ell.add(B_{c,j}, \mathcal{A}_i^p)$;
19:                     **end if**
20:                     $B_c := B_{c,j}$;
21:                 **end for**
22:             **end if**
23:             **for** $A \in \mathcal{A}_i^p$ **do**
24:                 $\mathcal{T}^{B_i}(A) := \langle S^{B_i}(A), T^{B_i}(A) \rangle$;   //obtain the realisation of B_i with A
25:                 $\mathcal{A}_i := \mathcal{A}_i \cup$ DETECT($\mathcal{T}^{B_i}(A)$);
26:             **end for**
27:             $\mathcal{A}^\ell.add((B_i, \mathcal{A}_i))$;   //the add operation will not add duplicated elements
28:             $\mathcal{A}^\ell.add((B_{i,ancestors}, \mathcal{A}_i))$;   //$B_{i,ancestors}$ means B_i and all its ancestor blocks
29:             **for** any $B^p \in \{B_1^p, B_2^p, \ldots, B_m^p\}$ **do**   //$B_1^p, B_2^p, \ldots, B_m^p$ are parent blocks of B_i
30:                 $\mathcal{A}^\ell.add((B_{i,p}, \mathcal{A}_i))$;
31:             **end for**
32:         **end if**
33:     **end for**
34:     **for** $B_i \in B$ and B_i has no child block **do**
35:         $\mathcal{A} = D(\Pi(\mathcal{A}^\ell.get(B_i), \mathcal{A}))$;
36:     **end for**
37:     **return** $\mathcal{A}$.
38: **end procedure**

39: **procedure** FORM_BLOCK(G)
40:     decompose G into SCCs and form blocks with SCCs and their control nodes;
41:     order the blocks in an ascending order according to their credits; $B := (B_1, \ldots, B_k)$;
42:     **return** B.   //B is the list of blocks after ordering
43: **end procedure**

---

algorithm, we denote by DETECT($\mathcal{T}$) a basic function for detecting attractors of a given transition system $\mathcal{T}$. Lines 23–26 of this algorithm describe the process for detecting attractors of a non-elementary block. The algorithm detects the attractors of all the realisations of the non-elementary block and performs the union operation of the detected attractors. For this, if the non-elementary block has only one parent block, its attractors are already computed as the blocks are considered in the ascending order with respect to their credits by the main

for loop in line 4. Otherwise, all the parent blocks are considered in the **for** loop in lines 13–21. By iteratively applying the cross operation in line 16 to the attractor sets of the ancestor blocks in the ascending order, the attractor states of a new block formed by merging all the parent blocks are computed as assured by Corollary 1. The attractors are then identified from the attractor states with one more operation. The correctness of the algorithm is stated as Theorem 4.

Theorem 4. *Algorithm 1 correctly identifies the set of attractors of a given BN G.*

We continue to illustrate in Example 6 how Algorithm 1 detects attractors.

Example 6. Consider the BN shown in Example 5 and its four blocks. Block B_1 is an elementary block and it has two attractors, i.e., $\mathcal{A}_1 = \{\{(0*)\}, \{(11)\}\}$. To detect the attractors of block B_2, we first form realisations of B_2 with the attractors of its parent block B_1. B_1 has two attractors so there are two realisations for B_2. The transition graphs of the two realisations are shown in Figs. 5a and b. We get two attractors for block B_2, i.e., $\mathcal{A}_2 = \{\{(0 * 00)\}, \{(1101)\}\}$. Those two attractors are also attractors for the merged block $B_{1,2}$, i.e., $\mathcal{A}_{1,2} = \mathcal{A}_2$. In

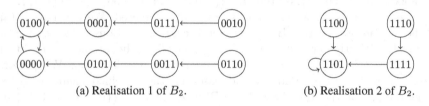

(a) Realisation 1 of B_2.   (b) Realisation 2 of B_2.

Fig. 5. Two realisations used in Example 6.

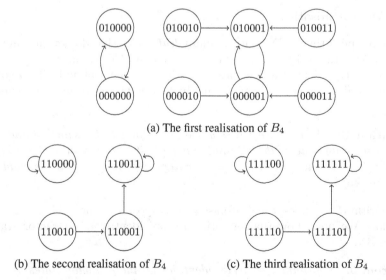

(a) The first realisation of B_4

(b) The second realisation of B_4 (c) The third realisation of B_4

Fig. 6. Transition graphs of the three realisations for block B_4.

Algorithm 2. Leaf-based optimisation

1: **procedure** LEAF_DETECT(G)
2:     form an elementary block B by removing all the leaves of G;
3:     $\mathcal{A}^B :=$ SCC_DETECT (B); $\varPhi^B := \cup_{A^B \in \mathcal{A}^B} A^B$;   //detect attractors of B
4:     $\mathcal{T} :=$ transition system of G with state space restricted to $\mathcal{M}_G(\varPhi^B)$;
5:     $\mathcal{A} :=$ DETECT $(\mathcal{T})$;
6:     **return** $\mathcal{A}$.
7: **end procedure**

Example 5, we have shown the two realisations of B_3 regarding the two attractors of B_1. Clearly, B_3 has two attractors, i.e., $\mathcal{A}_3 = \{\{(0*00)\}, \{(1100)\}, \{(1111)\}\}$. B_4 has two parent blocks. Therefore, we need to merge the two parent blocks to form a single parent block. Since the attractors of the merged block $B_{1,3}$ are the same as B_3, we directly obtain the attractors of $B_{1,3}$, i.e., $\mathcal{A}_{1,3} = \mathcal{A}_3 = \{\{(0*00)\}, \{(1100), (1111)\}\}$. There are three attractors so there will be three realisations for block B_4. The transition graph of the three realisations are shown in Fig. 6. From the transition graphs, we easily get the attractors of B_4, i.e., $\mathcal{A}_4 = \{\{(0*0000)\}, \{(0*0001)\}, \{(110000)\}, \{(110011)\}, \{(111100)\}, \{(111111)\}\}$.

Now the attractors for all the blocks have been detected. We can now obtain all the attractors of the BN by several cross operations. We start from the block with the largest credit, i.e., block B_4. The attractors of B_4 in fact cover blocks B_1, B_3 and B_4. The remaining block is B_2. We perform a cross operation on $\mathcal{A}_2$ and $\mathcal{A}_4$ and based on the obtained result we detect the attractors of the BN, i.e., $\mathcal{A} = D(\varPi(\mathcal{A}_2, \mathcal{A}_4) = \{\{(0*000000)\}, \{(0*000001)\}, \{(11010011)\}, \{(11010000)\}, \{(11011111)\}, \{(11011100)\}\}$.

4.1   An Optimisation

It often happens that a BN contains many *leaf* nodes that do not have any child node. Each of the leaf nodes will be treated as an SCC in our algorithm and it is not worth the effort to process an SCC with only one leaf node. Therefore, we treat leaf nodes in a special way. Formally, leaf nodes are recursively defined as follows.

Definition 13. *A node in a BN is a* leaf node *(or* leaf *for short) if and only if it is not the only node in the BN and either (1) it has no child nodes except for itself or (2) it has no other children after iteratively removing all its child nodes which are leaf nodes.*

Algorithm 2 outlines the leaf-based decomposition approach for attractor detection. We now show that Algorithm 2 can identify all attractor states of a given BN.

Theorem 5. *Algorithm 2 correctly identifies all the attractor states of a given BN G.*

5   Experimental Results

We have implemented the decomposition algorithm presented in Sect. 4 in the model checker MCMAS [14]. In this section, we demonstrate the efficiency of our method by comparing our method with the state-of-the-art decomposition method mentioned in [8] which is also based on BDD implementation. We generate 33 random BN models with different number of nodes using the tool ASSA-PBN [15,16] and compare the performance of the two methods on these 33 models. All the experiments are conducted on a computer with an Intel Xeon W3520@2.67 GHz CPU and 12 GB memory.

We name our proposed decomposition method M_1 and the one in [8] M_2. There are two possible implementations of the DETECT function used in Algorithm 1 as mentioned in [8]: monolithic and enumerative. We use the monolithic one which is shown to be more suitable for small networks as the decomposed sub-networks are relatively small. Since the method in [8] uses similar leaf reduction technique, we make comparisons on both the original models and the models whose leaves are removed in order to eliminate the influence of leaf nodes. We set the expiration time to 3 h. Before removing leaf nodes, there are 11 cases that both methods fail to process. Among the other 22 cases, our method is faster than M_2 in 16 cases, which is approximately 73%. After removing leaf nodes, there are 5 cases that both methods fail to process. Among the other 28 cases, our method is faster than M_2 in 25 cases, which is approximately 89%. We demonstrate the results for 7 models in Table 1 and the remaining result can be found in [17]. Since our method considers the dependency relation between different blocks, the attractors of all the blocks need to be computed; while method M_2 can ignore the blocks with only leaf nodes. Therefore, the performance of our method is more affected by the leaf nodes. This is why the percentage that our method is faster than M_2 is increased from 73% to 89% when leaf nodes are removed. Notably, after eliminating the influence of leaf nodes, our method is significantly faster than M_2. The "–" in Table 1 means the method fails to process the model within 3 h. The speedup is therefore not applicable (N/A) for

Table 1. Selected results for the performance comparison of methods M_1 and M_2.

Model ID	# nodes	# non-leaves	# attractors	Original models			Models with leaves removed		
				$t_{M_2}[s]$	$t_{M_1}[s]$	Speedup	$t_{M_2}[s]$	$t_{M_1}[s]$	Speedup
1	100	7	32	4.56	0.86	5.3	0.58	0.02	29.0
2	120	9	1	18.13	0.95	19.1	1.10	0.04	27.5
3	150	19	2	201.22	1.66	121.2	0.74	0.02	37.0
4	200	6	16	268.69	7.04	38.2	0.97	0.02	48.5
5	250	25	12	533.57	11.16	47.8	0.90	0.04	22.5
6	300	88	1	–	–	N/A	238.96	65.33	3.7
7	450	43	8	–	60.82	N/A	3704.33	0.17	21790.2

this result. The speedup is computed as t_{M_2}/t_{M_1}, where t_{M_1} is the time cost for M_1 and t_{M_2} is the time cost for M_2. All the time shown in Table 1 is in seconds. In general, we obtain a larger speedup when the number of attractors is relatively small. This is due to that our method takes the attractors of the parent block into account when forming a realisation of a non-elementary block and the number of realisations increases with the number of attractors. Summarising, our new method shows a significant improvement on the state-of-the-art decomposition method.

6 Conclusion and Future Work

We have introduced a new SCC-based decomposition method for attractor detection of large synchronous BNs. Although our decomposition method shares similar ideas on how to decompose a large network with existing decomposition methods, our method differs from them in the key process and has significant advantages.

First, our method is designed for synchronous BNs, as a consequence the key process for constructing realisations in our method is totally different from the one in [9], which is designed for asynchronous networks. Secondly, our method considers the dependency relation among the sub-networks. The method in [8] does not rely on this relation and only takes the detected attractors in sub-networks to restrict the initial states when recovering the attractors for the original network. In this way, the decomposition method in [8] potentially cannot scale up very well for large networks, as it still requires a BDD encoding of the transition relation of the whole network. This is our main motivation to extend our previous work [9] towards synchronous BNs. Experimental results show that our method is significantly faster than the one in [8]. Lastly, we have shown that the method proposed in [7] cannot compute correct results in certain cases.

Our current implementation is based on BDDs. One future work is to use SAT-solvers to implement the DETECT function as SAT-based methods are normally more efficient in terms of attractor detection for synchronous BNs [6].

References

1. Kauffman, S.: Homeostasis and differentiation in random genetic control networks. Nature **224**, 177–178 (1969)
2. Huang, S.: Genomics, complexity and drug discovery: insights from Boolean network models of cellular regulation. Pharmacogenomics **2**(3), 203–222 (2001)
3. Somogyi, R., Greller, L.D.: The dynamics of molecular networks: applications to therapeutic discovery. Drug Discov. Today **6**(24), 1267–1277 (2001)
4. Garg, A., Xenarios, I., Mendoza, L., De Micheli, G.: An efficient method for dynamic analysis of gene regulatory networks and *in silico* gene perturbation experiments. In: Speed, T., Huang, H. (eds.) RECOMB 2007. LNCS, vol. 4453, pp. 62–76. Springer, Heidelberg (2007). doi:10.1007/978-3-540-71681-5_5

5. Garg, A., Di Cara, A., Xenarios, I., Mendoza, L., De Micheli, G.: Synchronous versus asynchronous modeling of gene regulatory networks. Bioinformatics **24**(17), 1917–1925 (2008)
6. Dubrova, E., Teslenko, M.: A SAT-based algorithm for finding attractors in synchronous Boolean networks. IEEE/ACM Trans. Comput. Biol. Bioinform. **8**(5), 1393–1399 (2011)
7. Guo, W., Yang, G., Wu, W., He, L., Sun, M.: A parallel attractor finding algorithm based on Boolean satisfiability for genetic regulatory networks. PLOS ONE **9**(4), e94258 (2014)
8. Yuan, Q., Qu, H., Pang, J., Mizera, A.: Improving BDD-based attractor detection for synchronous Boolean networks. Sci. China Inf. Sci. **59**(8), 080101 (2016)
9. Mizera, A., Pang, J., Qu, H., Yuan, Q.: Taming asynchrony for attractor detection in large Boolean networks (Technical report) (2017). http://arxiv.org/abs/1704.06530
10. Kauffman, S.A.: Metabolic stability and epigenesis in randomly constructed genetic nets. J. Theor. Biol. **22**(3), 437–467 (1969)
11. Shmulevich, I., Dougherty, E.R.: Probabilistic Boolean Networks: The Modeling and Control of Gene Regulatory Networks. SIAM Press (2010)
12. Lee, C.Y.: Representation of switching circuits by binary-decision programs. Bell Syst. Tech. J. **38**(4), 985–999 (1959)
13. Akers, S.B.: Binary decision diagrams. IEEE Trans. Comput. **100**(6), 509–516 (1978)
14. Lomuscio, A., Qu, H., Raimondi, F.: MCMAS: An open-source model checker for the verification of multi-agent systems. Int. J. Softw. Tools Technol. Transf. (2015)
15. Mizera, A., Pang, J., Yuan, Q.: ASSA-PBN: an approximate steady-state analyser of probabilistic Boolean networks. In: Finkbeiner, B., Pu, G., Zhang, L. (eds.) ATVA 2015. LNCS, vol. 9364, pp. 214–220. Springer, Cham (2015). doi:10.1007/978-3-319-24953-7_16
16. Mizera, A., Pang, J., Yuan, Q.: ASSA-PBN 2.0: a software tool for probabilistic Boolean networks. In: Bartocci, E., Lio, P., Paoletti, N. (eds.) CMSB 2016. LNCS, vol. 9859, pp. 309–315. Springer, Cham (2016). doi:10.1007/978-3-319-45177-0_19
17. Mizera, A., Pang, J., Qu, H., Yuan, Q.: Benchmark Boolean networks. http://satoss.uni.lu/software/ASSA-PBN/benchmark/attractor_syn.xlsx

Construction of Abstract State Graphs for Understanding Event-B Models

Daichi Morita[1(✉)], Fuyuki Ishikawa[2], and Shinichi Honiden[1,2]

[1] The University of Tokyo, Tokyo, Japan
[2] National Institute of Informatics, Tokyo, Japan
{d-morita,f-ishikawa,honiden}@nii.ac.jp

Abstract. Event-B is a formal method that supports correctness by construction in system modeling using stepwise refinement. However, it is difficult to understand the rigorous behaviors of models from Event-B specifications, such as the reachable state space or the possible sequences of events. This is because the Event-B model is described in a style that lists events that have concurrently been enabled depending on their guard conditions. This paper proposes a method that helps in understanding the rigorous behaviors of an Event-B model by creating an abstract state graph. The core of our method involves dividing the concrete state space by using the guard conditions of individual events to extract states that are essential to enable possible transitions to be understood. Moreover, we further divided the state space by using the guard conditions of events in the models before refinement to support understanding of changes in behaviors between the models before and after refinement. Our unique approach facilitated finding of invariants that were not specified but held, which were useful for validation.

1 Introduction

Event-B [1] is a formal specification language based on first-order predicate logic and set theory. It adopts a refinement mechanism to allow developers to gradually build a model while ensuring its correctness by using these mathematical methods. Developers in Event-B modeling start from the most abstract machine to build the model and then refine it by building a more concrete machine that introduces new aspects so that it is closer to the comprehensive machine to be obtained. This refinement process is continued until the comprehensive machine is obtained that includes all the target aspects. This refinement mechanism reduces the difficulty in rigorously modeling and verifying a complicated model by enabling focus on individual small steps.

Event-B is a state-based formal method and the behavior of an Event-B model is expressed by states and transitions. Thus, it is important for developers to comprehend the reachable state space or the possible sequences of events in terms of validation. However, this is difficult because infinite sets, such as

This work is partially supported by JSPS KAKENHI Grant Number 17H01727.

© Springer International Publishing AG 2017
K.G. Larsen et al. (Eds.): SETTA 2017, LNCS 10606, pp. 250–265, 2017.
https://doi.org/10.1007/978-3-319-69483-2_15

integers, can be used as types of variables and thus the size of the state space can be infinite or too large to comprehend.

ProB [12] is one of the standard tools to check Event-B models. Although ProB provides several methods of visualizing the state space [9,11], there have been problems with these methods. They have required developers to specify that the range of infinite sets be finite to generate a graph because they constructed the state space by exhaustive simulation. Thus, the state space was restricted and developers could miss unexpected behaviors outside the space. Moreover, there are no methods of graph visualization that takes refinement into consideration, even though it would be useful to know how a concrete machine can refine an abstract one.

This paper proposes two methods of graph visualization of an Event-B model from the specifications without simulation to enable behaviors to be rigorously understood. The first method involves constructing an abstract state graph using predicate abstraction [8], which is useful to enable developers to explore the full state space and not to overlook the differences in behaviors, according to the range of infinite sets. Our key idea was to use guard conditions of events for predicate abstraction, which allowed us to extract essential insights into possible transitions (event occurrences) in each state. The second method was graph visualization that took refinement into account. It is useful for developers to validate behaviors by checking the correspondence between the states and transitions of abstract and concrete machines. Moreover, these graphs are useful for developers to find stronger invariants than those described in the specifications and help them to validate the state space. The unique feature in our approach is that we first constructed apparently reachable states from the specified predicates, such as invariants, and we then examined actual (un)reachability. This approach could expose unexpectedly unreachable states, which represented implicit expectations or faults.

We have organized the rest of this paper as follows. Section 2 provides the necessary background on Event-B and Sect. 3 describes our methods of generating graphs. Section 4 explains how we evaluated our methods by providing various applications. Section 5 relates our study to other studies and Sect. 6 concludes the paper.

2   Preliminaries

2.1   Event-B Models

An Event-B model consists of two modules, which are called machine and context. A context contains a set of constants and a set of axioms. Axioms are predicates that denote the constraints that the constants must satisfy. A machine contains a set of variables, a set of invariants and a set of events. Invariants are predicates that denote the safety properties that developers require variables and constants to satisfy. That they are actually invariants is verified by proving some statements, which are called "proof obligations". An event mainly consists of some guards and actions. Guards are predicates to denote the conditions under which an event is to be enabled. Actions are called before-after predicates that denote the relationships between the values of variables just before and after an event. The values of the constants cannot be changed by the events. Thus, the states of the model consist of the dynamic values of the variables and the static values of the constants. The transitions between them are triggered by the occurrence of the events.

For example, let us take Abrial's model of "controlling cars on a bridge" (from [1, Chap. 2]) into account. There is a mainland, an island, a bridge between them, and traffic lights that control cars going to and coming from them in the model. There was only the mainland and island in the initial model. It consisted of a context $Ctx0$ and a machine $Mac0$ shown in Fig. 1. The constant d defined in $Ctx0$ denotes the maximum number of cars allowed to be on the island. The variable n defined in $Mac0$ represents the number of cars on the island. The invariant $inv0\_2$ means that constant d is actually the maximum number of cars on the island. The states of the model consist of two values of variable n and constant d, such as $(n, d) = (0, 1)$. The state space is infinite because d can be an arbitrary natural number that is more than zero. The event $init$ is the initialization event, ML_out is the event corresponding to the transition of a car from the mainland to the island, and ML_in is its inverse event. The guard $grd1$ of the event ML_out is $n < d$ and the action $act1$ is the before-after predicate $n' = n + 1$. The value of the variable just after an event has occurred makes its before-after predicate true. A primed variable, such as n' appearing in $act1$ in a before-after predicate, denotes the value of the variable just after an event has occurred. Thus, the before-after predicate $act1$ means that the value of variable n just after the event is equal to the value of variable n just before it has occurred plus one.

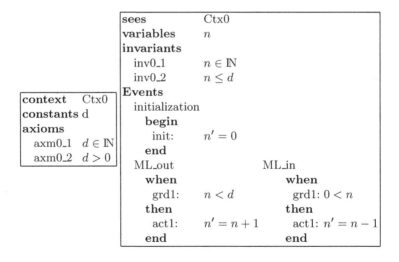

Fig. 1. Event-B specifications of $Mac0$

2.2 Event-B Refinement

The refinement mechanism in Event-B is a way of gradually building a model. An abstract machine is refined by a new machine and new features and details are introduced into the abstract machine. A new machine is called a concrete machine. A sequence of machines linked by a refinement relationship is called a "refinement chain". An increasingly more complicated but accurate model is built through stepwise refinements.

For example, $Mac1$ (Fig. 2) refines $Mac0$. A one-way bridge is introduced into the abstract machine. The variable a is the number of cars on the bridge going to the island, b is the number on the island and c is the number on the bridge coming to the mainland. The variable n defined in $Mac0$ is replaced by these three variables and the invariant $inv1\_2$ denotes the relationship between n and a, b, c. The states consist of four values of the variables a, b, c and the constant d. The invariant $inv1\_3$ denotes that the bridge is one-way. The two events ML_out and ML_in in $Mac0$ are refined as they are events on these

---

refines $Mac0$
sees   Ctx0
variables a, b, c
invariants
   inv1_1 $a \in \mathbb{N} \wedge b \in \mathbb{N} \wedge c \in \mathbb{N}$
   inv1_2 $a + b + c = n$
   inv1_3 $a = 0 \vee c = 0$
Events
  initialization
    begin
     init: $a' = 0 \wedge b' = 0 \wedge c' = 0$
    end

ML_out		ML_in	
refines	ML_out	**refines** ML_in	
when		**when**	
grd1:	$a + b < d$	grd1:	$0 < c$
grd2:	$c = 0$		
then		**then**	
act1:	$a' = a + 1$	act1:	$c' = c - 1$
end		**end**	
IL_in		IL_out	
when		**when**	
grd1:	$0 < a$	grd1:	$0 < b$
		grd2:	$a = 0$
then		**then**	
act1:	$a' = a - 1$	act1:	$b' = b - 1$
act2:	$b' = b + 1$	act2:	$c' = c + 1$
end		**end**	

Fig. 2. Event-B specifications of $Mac1$

three variables. ML_out in $Mac1$ corresponds to the transition of a car from the mainland to the bridge and ML_in is its inverse event. These refined events must simulate the original ones. The before-after predicate of ML_out is $a' = a+1$ but it implicitly contains $b' = b$ and $c' = c$, which means that the values of missing variables in the predicate are equal to the values just before the event. Thus, its precise before-after predicate is $a' = a + 1 \wedge b' = b \wedge c' = c$, and it simulates that of the original because the invariant $a + b + c = n$ holds. Guards of the refined event must not contradict those of the original. The guard of the event ML_out in $Mac1$ is $a + b < d$ and does not contradict the guard $n < d$ of the abstract one because the invariant $a + b + c = n$ holds.

New events can be introduced into abstract models. IL_in and IL_out in $Mac1$ are new events. IL_in corresponds to the transition of a car from the bridge to the island and IL_out is its inverse event. They do not need to modify any abstract variables so that the abstract model is not contradicted, i.e., they need to refine null event $skip$ in which guard is $true$ and action has no meaning.

3   Method

3.1   Construction of Abstract State Graph (CASG)

The goal discussed in this subsection was to construct a state graph from the Event-B specifications, which is useful for understanding the rigorous behavior of an Event-B model. The state space that may be infinite is abstracted by predicate abstraction. We called the graph an "abstract state graph" as in Graf and Saïdi [8] and called our method of constructing the abstract state graph CASG.

Let us assume that we have a machine M. We use symbols Inv_M to denote the conjunction of all invariants and axioms that appear in the refinement chain that precedes M, and BA_{evt} for the conjunction of all before-after predicates of the event evt and the event that refines evt in the refinement chain. For example, $Inv_{Mac1} = (d > 0 \wedge n \le d \wedge a + b + c = n \wedge (a = 0 \vee c = 0))$ and $BA_{ML\_out} = (n' = n + 1 \wedge a' = a + 1 \wedge b' = b \wedge c' = c)$. We also use symbols G_{evt} to denote the conjunction of all the guards of event evt and Evt_M to denote the set of the events of model M.

An abstract state graph of the Event-B machine consists of (S, I, L, δ), where S is a set of abstract states, I is a set of initial abstract states, L is a labeling function of the set S and δ is a transition function with guard conditions. An abstract state that constitutes S is defined by a predicate and a set of states that satisfy the predicate. For example, $0 \le n \le 2$ represents the set of states $\{(n, d) \mid n, d \in \mathbb{N}, 0 \le n \le 2\}$. After this we will use a predicate to denote an abstract state, i.e., we will refer to abstract states and predicates as exchangeable words. We will use "concrete states" to refer to states of the model to distinguish them from abstract states.

First, let us define set S of abstract states. We use $sat(p)$ to denote that the predicate p is satisfiable. Set S is constructed as:

$$S = \left\{ s \mid E \subseteq Evt_M, s = Inv_M \wedge \bigwedge_{evt \in E} G_{evt} \wedge \bigwedge_{evt \in Evt_M \setminus E} \neg G_{evt}, sat(s) \right\}.$$

This definition constructs abstract states by making equivalence classes of concrete states that satisfy invariants in terms of enabled events in each state. The possibly infinite state space is reduced into finite state space. However, note that the state space is approximated by the invariants and it may include some unreachable concrete states. This is discussed in Subsect. 3.5. This approximation is reasonable since invariants are properties that developers require the model to satisfy and they are verified by discharging proof obligations. Although the idea that states that have the same enabled events are regarded as being the same is similar to the method described in Leuschel and Turner [13], our method does not require developers to specify the range of infinite sets. It also provides predicates that explain the conditions of individual states.

The set I of initial abstract states is the set of abstract states that the before-after predicate of the initialization event satisfies.

The labeling function L for each $s \in S$, to specify the events enabled in an abstract state, is defined as:

$$L(s) = \{ evt \in Evt_M \mid sat(G_{evt} \wedge s) \}.$$

The transition function δ is then constructed. We use $after(s)$ for each predicate s to denote a predicate where all variable symbols are replaced with primed variable symbols, which means that they are values just after events have occurred. Note that $after$ only replaces variable symbols, and not constant symbols. The δ for each $s \in S$ and $evt \in L(s)$ is defined as:

$$\delta(s, evt) = \{ (s', g) \mid s' \in S, g = s \wedge BA_{evt} \wedge after(s'), sat(g) \}.$$

The predicate $BA_{evt} \wedge after(s')$ is like the weakest precondition if the state will be s' just after evt has occurred. Thus, g is a guard condition of the transition. Note that even if there is an edge, the corresponding transition cannot always occur in M because of our approximation.

Let us take the model $Mac0$ (Fig. 1) as an example. The graph constructed by using CASG is outlined in Fig. 3. The two lines in each ellipse, such as $\{ML\_out\}$ and $n = 0 \& d > 0$ in the top ellipse, denote the enabled events and abstract states. Type invariants have been omitted. The two lines beside the arrow denote the name of the event and the guard condition.

A graph constructed by using CASG is an abstraction of the actual state graph of an Event-B model, in which state space may be infinite. Solving satisfiability problems enables us to explore the full state space when checking the existence of transitions. An important aspect of this abstraction is that transitions that actually occur in the model are in the graph and transitions that are not in the graph do not occur in the model.

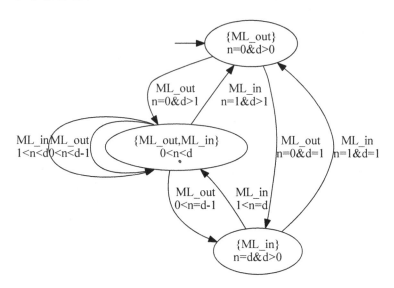

Fig. 3. Abstract state graph of $Mac0$

3.2 Construction of Refinement Abstract State Graph (CRASG)

This subsection explains how we constructed a graph that took refinement into consideration. We assumed that we had machines M_A and M_C, such that M_C refined M_A. Each state in M_A was refined by some states in M_C through the refinement. We wanted to reflect such a relation between the graphs of M_A and M_C. In other words, each abstract state of the graph of M_C constructed with the method described in this subsection corresponds to one abstract state of the graph of M_A constructed with CASG. We called the method of constructing the abstract state graph that took refinement into account CRASG.

Let us define a binary relation R_V to clarify the relation between the abstract state graphs for M_A and M_C. The R_V is a binary relation between the abstract states of the graphs. Let S_A be the abstract states set of the graphs for M_A that is constructed by using CASG, and let S_C be the abstract states set of the graph for M_C that is constructed by using CRASG. The R_V is defined as:

$$R_V = \{(s, s') \in S_C \times S_A \mid sat(s \wedge s')\}.$$

Here, $(s, s') \in R_V$ means that there is a concrete state in s that corresponds to a state in s'. Our main objective in this subsection is to explain how we constructed the graph of M_C, such that R_V is a function, which means each abstract state in S_C corresponds to one abstract state in S_A.

Let us now construct the abstract states set S. Let $Evt = Evt_{M_C} \cup Evt_{M_A}$. The S is constructed as:

$$S = \left\{ s \mid E \subseteq Evt,\, s = Inv_{M_C} \wedge \bigwedge_{evt \in E} G_{evt} \wedge \bigwedge_{evt \in Evt \setminus E} \neg G_{evt},\, sat(s) \right\}.$$

The construction of S splits the state space of M_C by the equivalence relation, where the enabled events of M_C are the same and those of M_A are the same if the space is projected onto the space of M_A. Therefore, the abstract states in the graph of M_A that are constructed by using CASG are divided even more into S, and R_V becomes a function. Note that unlike CASG, there can be states where the same events of M_C are enabled. Then, the remainder of the construction of the graph is similar to that with CASG.

For example, the graph of $Mac1$ that is constructed by using CRASG is given in Fig. 4. The squares denote the abstract states of the $Mac0$ graph. The dashed arrows mean that they correspond to transitions in the $Mac0$ graph. The predicates in the graph have been omitted to the extent that they can be understood. The guard labels have been completely omitted.

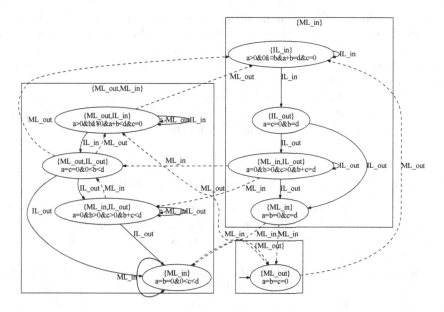

Fig. 4. Abstract state graph of $Mac1$ by taking refinement into consideration

The graph of M_C that is constructed by using CRASG is the refinement of the graph of M_A that is constructed with CASG. The abstract states of the graph of the concrete model can be grouped by which abstract state of the graph for M_A they satisfy because the binary relation R_V is a function. Moreover, the transitions of the graph for M_C can be grouped by which transition of the graph for M_A they simulate. These facts indicate that the graph provides a visualization of how M_C simulates M_A in terms of states and transitions.

3.3   Implementation

These two methods need to solve numerous satisfiability problems. Event-B is based on first-order predicate logic, and thus predicates used in Event-B models are first-order predicates within a practical range. A satisfiability modulo theories (SMT) solver [3] is one of the tools used to solve them automatically. Although the range of problems that it can solve is limited, it can solve many of the predicates in practical Event-B models. We used Z3 [5] as an SMT solver for our implementation. We succeeded in automatically creating a graph for $Mac0$, $Mac1$ and $Mac2$ (as a result of the second refinement of "controlling cars on a bridge") by using CASG and the graph for $Mac1$ and $Mac2$ by using CRASG.

3.4 Checking for Existence of Transitions

A transition corresponding to an edge in the graph of the model constructed with our methods could not always actually occur in the model because of our approximation. However, developers could check whether or not each transition (s, evt, s') could actually occur by using linear temporal logic (LTL [7]) model checking. The properties to check for occurrences could be directly represented by using LTL$^{[e]}$ [15] that introduced the operator $[evt]$, which meant the next executed event was evt. The condition that a transition (s, evt, s') could occur was formulated by LTL$^{[e]}$ in the form $\neg G \neg (s \wedge [evt] \wedge X(s'))$.

The LTL model checker could be used in ProB [14], which also supports LTL$^{[e]}$. Note that it adopts lower approximation and model checking is done under some values of constants. Thus, developers need to appropriately set the range.

3.5 Checking Validity and Strengthening Invariants

This subsection introduces another unique use of the graphs that were constructed with our methods due to our construction. It promotes the enhancement of invariants to strictly represent their expectations on acceptable and unacceptable states.

Sufficient invariants should ideally be given to strictly distinguish expectations on reachable and unreachable state spaces. However, this does not always hold, even at the level of the model appearing in [1], which is one of the most well-known references on Event-B. One reason for this is that developers completely understand the reachable concrete state space and decide not to add invariants to the model because it is redundant and these would not matter. In other words, reachable spaces are indirectly constrained by other means, such as guard conditions in events. This implicit approach may cause problems when other developers try to understand and revise the model in this case. Another reason is that they do not understand the reachable state space and have missed the invariants held in the model. There probably will be unexpected behaviors in the model in this case. Therefore, it is worth suggesting invariants that hold to confirm validity.

Our approach approximates the concrete state space with the invariants of a model. Thus, our graph can provide the difference between the actual reachable state space and the invariants by examining reachability. Other visualization methods [9,11] could not provide this because their construction was based on the simulation of the model and did not take invariants into consideration. Developers then checked each of the suggested invariants; they could be invariants that should have been added or they represented unexpectedly attained unreachable states caused by faults in the model (e.g., overly strong guard conditions). There are three methods that suggest invariants that can be achieved by using our graph.

First, let us assume there is an unreachable abstract state s from the initial abstract state. Then, its negation $\neg s$ is an invariant. This is because if there are no transitions into the abstract state in the model, then all the events preserve the negation.

Second, let us assume that there is a transition $s \xrightarrow{evt,g} s'$ in the graph that actually does not occur in the model (recall Subsect. 3.4). There are some unreachable concrete states in the source abstract state s of the transition in that case. We can find such a state by using an SMT-solver and finding an assignment of the predicate $s \wedge g \wedge BA_{evt} \wedge s'$. Developers can then find some invariants by investigating why this state was unreachable.

Third, let us assume that there is abstract state s that contains a concrete state unreachable from other abstract states. The condition is formulated as:

$$after(s) \Rightarrow \bigvee_{(s',evt):(s,g)\in\delta(s',evt)} s' \wedge BA_{evt}.$$

The predicate $s' \wedge BA_{evt}$ means the possible reachable states from s' just after event evt has occurred. Thus, the formula means that abstract state s is actually included in the possible reachable states from other states. There are unreachable concrete states in s if the condition does not hold, and some invariants can be added.

4   Evaluation

4.1   Setting

We evaluated our methods by using three applications of the graphs. Subsection 4.2 explains how we investigated the graph constructed by using CASG. This is useful for understanding the overall behaviors of the model and the validation of the state space by using predicates. Subsection 4.3 describes how we investigated the graph constructed with CRASG. This would help developers understand the details on changes in behaviors caused by refinement. Subsection 3.5 explains how many we found the stronger invariants than those described in the specifications. We used the $Mac2$ model (from [1, Chap. 2]) that refines $Mac1$ to evaluate our methods in addition to $Mac0$ and $Mac1$.

4.2   Abstract State Space Exploration

Our main objective was to help developers understand the behaviors of an Event-B model. Our methods provide state graphs that represent behaviors. Abstraction of the state space reduces the complexity of the original state graph and makes it easy for developers to comprehend behaviors. In addition, abstract states help them validate the model since these states are represented by predicates on the variables and constants in the model and the predicates can promote developers' understanding of the states.

Here, we will provide an example of exploring the state space by using our graph. Our graph provides the conditions for the variables and constants that represent the set of all concrete states that enable the same events. The example of $Mac0$ (Fig. 1) indicates that, if ML_out and ML_in are concurrently enabled, then developers can see that $0 < n < d$ holds in the state on the left in Fig. 3. The predicate is easy for them to compare with their intentions because they describe various predicates in Event-B modeling and they can perceive the situation with the model from the predicates. If they write the guard of ML_out incorrectly as $n < d - 1$, the predicate of the state with label $\{ML\_out, ML\_in\}$ will be $0 < n < d - 1$. They can then find that the guard is incorrect because their intention is for the number of cars allowed on the island to be d, but this is not achieved. It is also important for d to be symbolic. Due to this, they can validate that this condition holds no matter what the value of constant d is.

The graph of an Event-B model constructed by using CASG helps developers validate the state space by providing predicates on the variables and constants. However, this method does not completely solve the problem of complexity in the graph if the model is very complicated. One possible solution is for developers to focus on the change caused by refinement, which will be discussed in next section. This cannot completely solve the problem, but it is effective for limited ranges of investigation.

It also helps developers to comprehend overall sequences of the executed events by searching paths in the graph, even though not all sequences of events that correspond to the paths can be executed. The guard conditions of the transitions are useful when validating sequences of events because they can express conditions where the sequences can be executed.

4.3   Refinement Abstract State Graph Exploration

A graph constructed by using CRASG helps developers understand changes in behaviors caused by refinement. Changes are not trivial from the specifications because refinement is across the invariants, guards, and actions. The graph is very useful for understanding some aspects of the effect of refinement.

An aspect is how a concrete model simulates its abstract model. In other words, the correspondences between abstract states and transitions in the model before refinement to those in the model after refinement are drawn in the graphs. Let us take model $Mac1$ (Fig. 2) as an example. There is a square in Fig. 4 that is labeled $\{ML\_in\}$, which represents the abstract state of the graph for $Mac0$ that satisfies the predicate $n = d \wedge d > 0$. The predicate $n = d$ means that the number of cars on the island has reached the limit. Thus, the states and transitions in the square describe how the cars are moving between the island and mainland along the bridge to solve traffic jams in the model $Mac1$. There are also two transitions labeled ML_out between the squares labeled $\{ML\_out, ML\_in\}$, and $\{ML\_in\}$

in Fig. 4. Such transitions correspond to the transition in the graph in Fig. 3 that are labeled ML_out between the abstract states labeled $\{ML\_out, ML\_in\}$ and $\{ML\_in\}$, which means that the number of cars on the island has just reached the limit.

Another aspect is how a concrete model does not simulate its abstract model. The abstract model can be refined so that some transitions of the abstract model cannot occur in the concrete model by strengthening guards regardless of whether they have been intended or not. This graph helps developers at such times to discover what transition has occurred and what has not occurred in the concrete model. If developers write the guard of ML_out of $Mac1$ incorrectly as $a + b < d - 1$, the transitions that correspond to the transition labeled ML_out between abstract states labeled $\{ML\_out, ML\_in\}$ and $\{ML\_in\}$ will actually disappear from the graph. Developers can then find the degree of degradation and check whether it is intended.

The graph constructed by using CRASG provides correspondences in the state graph for concrete and abstract models. It is useful for developers to fully understand refinement by comparing it with the models. It also helps developers explore the model by focusing on the change if the model is complicated.

4.4   Checking Validity and Strengthening Invariants

We applied the method described in Subsect. 3.5 to $Mac2$ and found missing invariants that were needed to express precise reachable states. The details are described in Appendix A. As a result, we discovered seven invariants and discharged their proof obligations on the Rodin platform [2], which is equipped with theorem provers. Moreover, we checked that the state space expressed by the invariants was the actual reachable state space of the $Mac2$ model.

However, methods such as these three present several problems. One problem is that the methods are not automatic except for the first one. The methods provide a hint but require some suggestions by developers. Even though they are required to do so, it is difficult for other graph visualization methods to find stronger invariants. Moreover, such suggestions also help them validate the model and the discovered invariants support the building of a more accurate model.

5   Related Work

Our method is classified in terms of abstraction as the predicate abstraction described in Graf and Saidi [8]. Given a set of predicates, it splits the state space of variables appearing in a program or model based on the Boolean value of the predicates. It can solve a state space explosion problem in model checking by providing appropriate predicates. We used it for graph visualization of an Event-B model. We chose the set of guards for the event and the invariants of the model as input for predicate abstraction. This very effectively expressed behaviors of the model because the state space could be approximated by the invariants and the guards determined the sequences of the executed events. The unique feature of our approach is that we first construct apparently reachable states from specified predicates, such as invariants, and we then examine actual (un)reachability.

Graph visualization is one of the primary methods of enabling the behaviors of a finite state machine to be understood. As described in Dulac et al. [6], the readability of formal specifications is a factor that has not widely been used in industry and visualization often helps people understand specifications. Our methods were aimed at visualizing the behaviors of a possibly infinite state machine of Event-B by reducing it into a finite state machine using predicate abstraction.

There are several other methods of visualizing the state space of an Event-B model. ProB [12] can generate a state graph of the model. The number of states and transitions in the graph are sizably large because it tries to generate the original state space. Thus, it is hard to understand behaviors. Moreover, ProB requires the range of infinite sets to be specified. Thus, the generated state space may be restricted and developers may miss unexpected behaviors. However, the state space of our graph is approximated by invariants so that all the behaviors of the model can be expressed in the graph.

Other methods of visualizing the state space are described in Leuschel and Turner [13]. There are two methods called the deterministic finite automaton (DFA)-abstraction algorithm and the signature merge method. These methods have aimed at reducing the complexity of the graph generated by ProB. The method of DFA abstraction is based on the classical minimization algorithm for DFA. This method produces a graph in which the sequences of transitions are equivalent to those in the original state space, but it cannot effectively reduce the state space and its graph still makes it difficult for developers to understand behaviors. The signature merge method is similar to ours in terms of the way abstraction focuses on enabled events. However, there are no predicates to represent the states and developers thus find our graph is more understandable.

Another similar approach described in Ladenberger and Leuschel [11] is creating projection diagrams. A projection diagram is an abstraction of the original graph that is obtained by using some projection function. The method can reduce complexity more effectively than the two approaches explained above and our methods by focusing on certain variables or some expressions in the model. However, it may lose too much information to enable the overall behavior to be understood, unlike that in ours.

In contrast to visualizing the state space of the model, unified modeling language-B (UML-B) [16] is a method of building an Event-B model by drawing a diagram. It is similar to UML and easy to use by developers who are familiar with it. It mitigates the burden in Event-B modeling but is not suitable for building a complex model.

Another method of understanding the behavior of models is model checking [4]. In particular, Hoang et al. [10] stated that proof obligations in Event-B ensure safety properties; on the other hand, LTL model checking ensured temporal liveness properties. However, it is difficult to check overall behavior unlike that in our methods, such as the reachable state space or the sequences of executed events.

6 Conclusion

We proposed two methods of constructing the graph of an Event-B model from the specifications. Our methods are useful for graphically understanding the rigorous behavior of the model. They also allow developers to investigate reachable or unreachable states and transitions that cannot be searched by other graph visualizations [11,13]. The second method is useful for checking the correspondence between the graphs of the abstract and concrete machines and understanding changes in behaviors caused by refinement. Additionally, our methods enable developers to enhance invariants to strictly represent their expectations. We concluded that our methods could help developers to understand the behaviors of the model and validate it from various viewpoints. One possible direction in future work is to develop a more effective way of visualization for large systems.

A Appendix

This appendix explains how we investigated the advanced and unique use described in Subsects. 3.5 and 4.4 to discover invariants that were stronger than the invariants described in the specifications by using a graph constructed with CASG. We used the $Mac2$ model and the specifications are in Abrial [1, Chap. 2].

We applied the first method and discovered three unreachable states from the graph. One of them is represented by

$$a = 0 \wedge 0 < b < d \wedge c = 0 \wedge ml\_tl = il\_tl = red \wedge ml\_pass = il\_pass = true.$$

We then tried to add the predicate

$$\neg(a = 0 \wedge 0 < b < d \wedge c = 0 \wedge ml\_tl = il\_tl = red \wedge ml\_pass = il\_pass = true)$$

as an invariant to the $Mac2$ model on the Rodin platform [2]. As proof obligation is automatically discharged by them, the predicate is actually an invariant of the model. This invariant is equivalent to:

$$(a = c = 0 \wedge ml\_tl = il\_tl = red \wedge ml\_pass = il\_pass = true) \Rightarrow (b = 0 \vee b = d),$$

which means that if all the traffic lights are red, the flags are true and there are no cars on the bridge, then the number of cars on the island is zero or has reached its capacity. Developers can check if the situation is valid in the model.

We investigated the number of transitions in the $Mac2$ graph constructed by using CASG that could occur in the second method. We specified the range of the constant d from one to 10 because it seemed to be sufficient from our investigation of the model. We checked all 58 edges in the graph and discovered

16 edges that did not actually occur. One of them was the transition labeled IL_in from the abstract state represented by:

$$a > 0 \wedge b \geq 0 \wedge a+b < d-1 \wedge c = 0 \wedge ml\_tl = green \wedge il\_tl = red \wedge il\_pass = true$$

to another represented by:

$$a = c = 0 \wedge b < d-1 \wedge ml\_tl = green \wedge il\_tl = red \wedge il\_pass = true$$
$$\wedge (b = 0 \vee (b > 0 \wedge ml\_pass = false)). \quad (1)$$

A concrete state where the transition can occur is:

$$(a, b, c, d, ml\_tl, il\_tl, ml\_pass, il\_pass) = (1, 1, 0, 4, green, red, false, true).$$

However, it is actually unreachable because the condition $ml\_pass = false$ requires the event ML_tl_green to occur and ML_out_1 and ML_out_2 must not subsequently occur. There was some suggestion that the model always satisfies $a > 0 \Rightarrow ml\_pass = true$ because $a > 0$ means ML_out_1 or ML_out_2 has occurred at least once just after ML_tl_green has taken place. Then, we added it as an invariant to the Rodin platform, but its proof obligation was not automatically discharged. Due to an analysis of the failure of the proof, which is often used in Event-B, we added $(ml\_tl = red \wedge a + b \neq d) \Rightarrow a = 0$ as an invariant and all proof obligations were automatically discharged.

Finally, let us take $Mac2$ as an example of the third method. The abstract state represented by the predicate (1) does not satisfy the condition. All the transitions into it are labeled ML_tl_green. Since ML_tl_green makes ml_pass false, all concrete states where ml_pass is true in the abstract state are unreachable. There was some suggestion that the predicate $a = b = 0 \wedge ml\_tl = green \Rightarrow ml\_pass = false$ was an invariant. Therefore, we added it and proof obligation was discharged.

References

1. Abrial, J.R.: Modeling in Event-B: System and Software Engineering. Cambridge University Press, New York (2010)
2. Abrial, J.R., Butler, M., Hallerstede, S., Hoang, T.S., Mehta, F., Voisin, L.: Rodin: an open toolset for modelling and reasoning in Event-B. Int. J. Softw. Tools Technol. Transf. (STTT) **12**(6), 447–466 (2010)
3. Barrett, C., Sebastiani, R., Seshia, S., Tinelli, C.: Satisfiability modulo theories. In: Handbook of Satisfiability, Frontiers in Artificial Intelligence and Applications, vol. 185, Chap. 26, pp. 825–885. IOS Press (2009)
4. Clarke, E.M., Grumberg, O., Peled, D.A.: Model Checking. MIT Press, Cambridge (1999)
5. de Moura, L., Bjørner, N.: Z3: an efficient SMT solver. In: Ramakrishnan, C.R., Rehof, J. (eds.) TACAS 2008. LNCS, vol. 4963, pp. 337–340. Springer, Heidelberg (2008). doi:10.1007/978-3-540-78800-3_24
6. Dulac, N., Viguier, T., Leveson, N.G., Storey, M.D.: On the use of visualization in formal requirements specification. In: Proceedings of the IEEE Joint International Conference on Requirements Engineering, pp. 71–80 (2002)

7. Gabbay, D., Pnueli, A., Shelah, S., Stavi, J.: On the temporal analysis of fairness. In: Proceedings of the 7th ACM SIGPLAN-SIGACT Symposium on Principles of Programming Languages (POPL 1980), pp. 163–173. ACM, New York (1980)

8. Graf, S., Saidi, H.: Construction of abstract state graphs with PVS. In: Grumberg, O. (ed.) CAV 1997. LNCS, vol. 1254, pp. 72–83. Springer, Heidelberg (1997). doi:10. 1007/3-540-63166-6_10

9. Ham, F.V., van de Wetering, H., Van Wijk, J.J.: Visualization of state transition graphs. In: Proceedings of the IEEE Symposium on Information Visualization 2001 (INFOVIS 2001), p. 59. IEEE Computer Society, Washington, DC (2001)

10. Hoang, T.S., Schneider, S., Treharne, H., Williams, D.M.: Foundations for using linear temporal logic in Event-B refinement. Form. Aspects Comput. 28(6), 909–935 (2016)

11. Ladenberger, L., Leuschel, M.: Mastering the visualization of larger state spaces with projection diagrams. In: Butler, M., Conchon, S., Zaïdi, F. (eds.) ICFEM 2015. LNCS, vol. 9407, pp. 153–169. Springer, Cham (2015). doi:10.1007/978-3-319-25423-4_10

12. Leuschel, M., Butler, M.: ProB: an automated analysis toolset for the B method. Int. J. Softw. Tools Technol. Transf. 10(2), 185–203 (2008)

13. Leuschel, M., Turner, E.: Visualising larger state spaces in PROB. In: Treharne, H., King, S., Henson, M., Schneider, S. (eds.) ZB 2005. LNCS, vol. 3455, pp. 6–23. Springer, Heidelberg (2005). doi:10.1007/11415787_2

14. Lichtenstein, O., Pnueli, A.: Checking that finite state concurrent programs satisfy their linear specification. In: Proceedings of the 12th ACM SIGACT-SIGPLAN Symposium on Principles of Programming Languages (POPL 1985), pp. 97–107. ACM, New York (1985)

15. Plagge, D., Leuschel, M.: Seven at one stroke: LTL model checking for high-level specifications in B, Z, CSP, and more. Int. J. Softw. Tools Technol. Transf. 12(1), 9–21 (2010)

16. Snook, C., Butler, M.: UML-B and Event-B: an integration of languages and tools. In: Proceedings of the IASTED International Conference on Software Engineering (SE 2008), pp. 336–341. ACTA Press, Anaheim (2008)

A Framework for Modeling and Verifying IoT Communication Protocols

Maithily Diwan and Meenakshi D'Souza[✉]

International Institute of Information Technology, Bangalore, India
maithily.diwan@iiitb.org, meenakshi@iiitb.ac.in

Abstract. Communication protocols are integral part of the ubiquitous IoT. There are numerous light-weight protocols with small footprint available in the Industry. However, they have no formal semantics and are not formally verified. Since these protocols have many common features, we propose a unified approach to verify these protocols through a framework in Event-B. We begin with an abstract model of an IoT communication protocol which encompasses common features of various protocols. The abstract model is then refined into concrete models for individual IoT protocols using refinement and decomposition techniques of Event-B. Using the above framework, we present models of MQTT, MQTT-SN and CoAP protocols, and verify communication properties like connection-establishment, persistent-sessions, caching, proxying, message ordering, QoS, etc. Our protocol models can be integrated-with or extended-to other formal models of IoT systems using machine-decomposition within Event-B, thus paving way for formal modeling and verification of IoT systems.

Keywords: IoT protocols · MQTT · MQTT-SN · CoAP · Formal modeling and verification · Event-B

1 Introduction

IoT is prevalent in various industries like health care, automotive, manufacturing, power grid and domotics to name a few. IoT not only connects different computing devices but sensors, actuators, people and virtually any object. With the prediction that there will be over 20 billion devices by 2020 [7], IoT will be an integral part of our lives. The end nodes in the IoT are usually sensors or small devices which have limited processing capability and low memory. In such cases, the devices send unprocessed data to cloud which is then shared with other devices/systems subscribing to this large amount of data (either raw data or processed by server), making communication between these devices an important aspect of IoT.

Various protocols are used for communication in an IoTsystem. TCP/IP is a popular protocol used in lower layers. Several protocols are adapted for the application layer in an IoT system - Message Queue Telemetry Transport Protocol

© Springer International Publishing AG 2017
K.G. Larsen et al. (Eds.): SETTA 2017, LNCS 10606, pp. 266–280, 2017.
https://doi.org/10.1007/978-3-319-69483-2_16

(MQTT) [9], Message Queue Telemetry Transport Protocol Sensors (MQTT-SN) [10], The Constrained Application Protocol (CoAP) [11], eXtensible Messaging and Presence Protocol (XMPP) [12], Advanced Message Queuing Protocol (AMQP) [13] to name a few. Most of them are being used for IoT systems as they are bandwidth efficient, light-weight and have small code foot-print [8]. Features like publish-subscribe, messaging layer, QoS(Quality of Service) levels, resource discovery, re-transmission, etc. are prevalent in these protocols.

Our framework for an IoT protocol modeling and verification is realized through an abstract model of the protocol. The abstract model consists of commonalities among various application layer protocols like communication modes, connection establishment procedure, message layer, time tracking and attacker modules. We then decompose these various modules and refine them into more concrete models for individual protocols. Properties that hold true for these protocols are verified in these models. We use Event-B to model the communication channel and the client and server side communication entities, all of which together implement the protocol. By verifying the accuracy of the model through simulations, invariant checking and LTL properties satisfiability, we are able to conclude that our models of various protocols are correct.

Messages/streams are used as basic entities of communication between multiple clients and servers. Structure of a message apart from payload, usually consists of many fields of various types. Event-B provides record datatypes [3] through which complicated message structure with multiple attributes and sub-attributes can be expressed succinctly. All the properties of the protocol to be proved are expressed as invariants which are essentially predicates that are always true. The automatic and interactive proof discharge in Event-B using the Rodin tool [4] verifies if these invariants (properties) are satisfied for all the events in the model.

The paper is organized as follows. IoT protocols and their properties are described in Sect. 2. Section 3 highlights the features of Event-B that we use for modeling. Our Event-B model and their refinements for different protocols are detailed in Sect. 4. Verified properties and their results are presented in Sect. 5. Section 6 discusses related work and Sect. 7 presents the conclusion and on-going work.

2   IoT Communication Protocols: MQTT, MQTT-SN and CoAP

Most of the applications in IoT need a reliable network and use existing Internet to communicate with the cloud/servers and with other nodes. Hence it is common to use the existing TCP/IP stack with underlying physical, DLL, network and transport layer. However TCP/IP is heavy weight as compared to a lighter UDP, in which case reliability has to be built in the application layer protocol. Other IoT communication requirements are: low bandwidth, low memory consumption, small code foot-print, self recovery, resource discovery, light-weight, low message overhead, low power consumption, authentication, security, appropriate QoS.

We briefly describe some of the application protocols highlighting the features and properties which we verify in this paper.

2.1   MQTT

MQTT [9] is a publish-subscribe protocol designed for constrained devices connected over unreliable, low bandwidth networks. It gives flexibility to connect multiple servers to multiple clients. The protocol has low message overhead which makes it bandwidth efficient and can be easily implemented on a low powered device. Significant features offered by MQTT are explained below:

1. 3 levels of QoS: "At most once"- no acknowledgement is expected for a publish message, "at least once" - every message receives an acknowledgement and "exact once" - guaranteed message delivery without duplicates.
2. Subscribe: Clients can subscribe/unsubscribe to a topic with desired QoS.
3. Keep-alive: In absence of application messages within keep-alive time, client sends a ping request to keep the network active.
4. Persistent Session: Persistent session is achieved by storing the session state of channel and can be restored upon re-connection. It includes previous configurations, subscriptions, unacknowledged messages, etc.
5. Retain Message: When a new subscriber or an offline subscriber re-connects, the retained message is immediately published with configured QoS.
6. Will message: Pre-configured "will" message is sent by the server when a publishing client goes offline and wants to inform the subscribing clients.
7. Authentication: A user-password feature is used for authentication. TLS (Transport Layer Security) is optional for data encryption [9].

2.2   MQTT-SN

MQTT-SN is another data centric protocol and is based on MQTT with adaptations to suit the wireless communication environment. Unlike MQTT, MQTT-SN does not require an underlying network like TCP/IP making it a low complexity, light weight protocol. Significant differences between MQTT and MQTT-SN are listed below:

1. Gateway Advertisement and Discovery: A MQTT-SN client conntects to MQTT server via a gateway, implementing translation between the two protocols. A discovery procedure is used by the clients to discover the actual network address of an operating server/gateway.
2. Topic Registration: To reduce bandwidth, a client can use pre-defined short topic names/IDs or register a long topic name with the server and use a corresponding topic ID for further communication.
3. QoS −1: In addition to QoS 0, 1 and 2, MQTT-SN offers QoS −1 where the client communicates with the server without a formal connection establishment and topic registration procedures.
4. Support of Sleeping Clients: Power saving clients can to go to sleep mode and wake up periodically using keep-alive message. A server/gateway buffers messages destined to the client and send them to client when they wake up.

2.3 CoAP

CoAP is a specialized web transfer protocol based on REST architecture, fulfilling Machine to Machine (M2M) requirements in constrained environments. CoAP has low header overhead, parsing complexity, and has uri based addressing. It is stateless HTTP mapping, allowing proxies to be built providing access to CoAP resources via HTTP. Following are significant features of CoAP:

1. Layered Architecture: CoAP implements a request-response model with asynchronous message exchanges at lower layer. The messaging layer deals with UDP and asynchronous nature of interactions, and the request-response interactions use method and response codes. Requests and responses are carried in confirmable and non-confirmable messages. A response can be piggybacked in acknowledgement or separate message.
2. Unicast/multicast requests: For discovering resources and services in the network, CoAP uses multicast request. After a connection is established with a server, unicast mode is used.
3. Reliability: CoAP uses a layer of messages that supports optional reliability of "at least once" with an exponential back-off mechanism.
4. Proxying and Caching: A cache could be located in an endpoint or an intermediary called proxy. Caching is enabled using freshness and validity information carried with CoAP responses. A max-age option in a response indicates its not fresh after its age is greater than the specified time. A proxy can however validate the stored response with server even after max-age expiry.
5. Resource Discovery: Like MQTT-SN, CoAP uses multicast requests to discover services and resources in the network.
6. Observe feature: CoAP can be used in publish-subscribe mode by using observe and notification options.
7. Security: Optional security using Datagram Transport Layer Security (DTLS).

3 Event-B

Event-B [1] is based on B-Method which provides a formal methodology for system-level modeling and analysis. Event-B uses set theory as a modeling notation and first order predicate calculus for writing axioms and invariants. It uses step by step refinement to represent systems at different abstraction levels and provides proofs to verify consistency of refinements. Initially the model is constructed on basis of known requirements. As and when required, one can refine and add the new properties while satisfying the requirements in the underlying model.

An Event-B model has two types of components: contexts and machines. Contexts contain all the data structures required for the system which are expressed as sets, constants and relations over the sets. A machine "sees" a context to use the data structures or types. A machine has several events and can also define

Table 1. Comparison of IoT communication protocols

Sl.no	Protocol feature	MQTT	MQTT-SN	CoAP
1	Architecture	Asynchronous Message exchange	Asynchronous Message exchange	REST architecture Layered Approach
2	Transport Layer	TCP	Any	UDP
3	Communication type	UniCast	UniCast/Multicast	UniCast/Multicast
4	Addressing	ClientID Server address	ClientID Server address	Uri Based
5	Messaging pattern	Publish Subscribe	Publish Subscribe	Request-Response Publish-Subscribe
6	QoS Levels	AtmostOnce, AtleastOnce, ExactOnce	AtmostOnce, AtleastOnce, ExactOnce	AtmostOnce, AtleastOnce
7	Persistent Session	Yes	Yes	Yes
8	Retained Message /Offline/Caching	Yes	Yes	Yes
9	Proxying/Caching	No	Yes	Yes
10	Resource Discovery	No	Yes	Yes
11	Sleep Mode	No	Yes	Yes
12	Security	Optional TLS	Optional TLS	Optional DTLS

variables and its types. A machine can refine another machine to introduce new events, refine events, split events or merge events. An event consists of guards which need to be satisfied before the actions in events are executed. When an event is enabled and executed, the variables are updated as per the actions in the event.

An invariant is a condition on the state variables that must hold permanently. In order to achieve this, it is required to prove that, under the invariant in question and under the guards of each event, the invariant still holds after being modified according to the transition associated with that event [5].

Rodin and ProB

Rodin [4] implements Event-B and is based on Eclipse platform. It provides an environment for modeling refinements and discharges proofs. It has sophisticated automatic provers like PP, ML and SMT, which automatically discharge proofs for refinements, feasibility, invariants and well-definedness of expressions within guards, actions and invariants. Event-B also provides interactive proving mechanism for manual proofs which can be used when the automatic proof discharge fails. Rodin offers various plug-ins for development including different text editors, decomposition/modularization tools, simulator ProB, etc.

ProB [6] provides a simulation environment through animation for Event-B model. A given machine can be simulated with all its events. In the animation environment, one can select and run the given events by selecting parameters or execute with random solution. During simulation, the state of the system before and after every event execution can be observed. The state gives values of all the variables in the machine, evaluates invariants, axioms and guards for all the events. Additionally any expression can be monitored in the animator. The model can also be checked for deadlocks, invariant violations and errors in the model which will help to construct an accurate model.

4   Protocol Modeling and Decomposition Using Event-B

A communication channel is a network connection which is established between a client and a server or between two clients or between two servers. In an IoT system there could be multiple channels connecting several clients and servers. Our Event-B model consists of communication channels of the IoT system which implement a communication protocol. As shown in the Fig. 1, the model has Event-B contexts and machines. The contexts have all the data structures and axioms required to setup a machine. The machine includes communication part of client and server implemented as events, and the properties required to be verified are written as invariants.

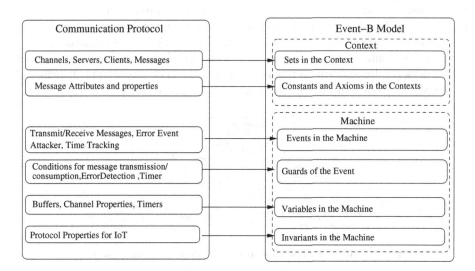

Fig. 1. Mapping between communication protocol and Event-B model

The protocol modeling is done in two major steps:

1. Building a common abstract model encompassing the common features of various protocols.
2. Refining this common abstract model into a concrete model of a particular IoT protocol.

Our modeling is done using the techniques of machine decomposition [14], refinement [2] and atomicity decomposition in Event-B [15].

4.1   Common Abstract Model

The common abstract model implements the commonalities among various protocols as mentioned in Table 1. Figure 2 is a diagrammatic representation of the abstract model.

Context: A basic communication entity is modelled as a message. Set named MSG and all its attributes are defined as relations over the set MSG and the sets defined for the attributes. A projection function is used to extract the value of an attribute for a given message [3].

Machine Refinements: The atomicity of event Communication Channel is broken into two events representing modes of communication: Unicast and Broadcast/Multicast. Similarly a further refinement of the model breaks down the atomicity of these events into Service and Resource Discovery. A UniCast event is broken into ChannelEstablishment and ChannelConversation events. Since these events are not yet atomic, they can be further split as shown in Fig. 3 where ChannelConversation of previous refinement is further broken into many more events. Figures 2 and 3 together show the three refinement steps done in the common abstract model. It is to be noted that our common abstract model does not breakdown to the lowest atomic level of events. This is achieved in the next step of building concrete model for a particular protocol.

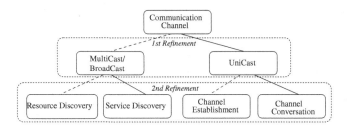

Fig. 2. Atomicity decomposition of common abstract model

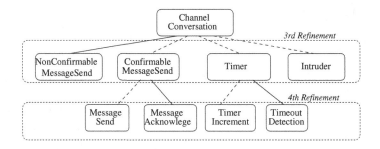

Fig. 3. Atomicity decomposition of ChannelConversation module

Machine Decomposition: The leaves of the atomicity decomposition diagram give us the events of the final refinement of the common model. Further on when we build models of particular protocols, these events further explode into more atomic events blowing up the size of the model. It has been observed that many of these events have very few interfaces among them and they can be independently be refined. This allows us to use the technique of machine decomposition in Event-B. Figure 4 gives such a decomposition of our abstract model. In Sect. 4.2 we give an example of how these modules of decomposed machines are further refined to give more concrete model of MQTT.

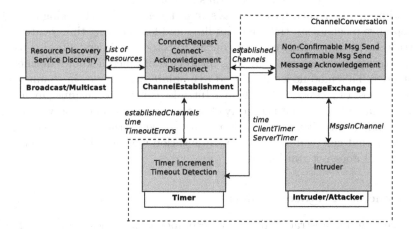

Fig. 4. Machine decomposition of common abstract model

Events in Decomposed Modules

1. Multicast/Broadcast: It is used when a node has to communicate to more than one peer node. The Multicast/Broadcast event is broken down into atomic events Service Discovery and Resource Discovery which are used to find the nodes that can publish the required information on the network. Once the nodes with required resources/services are discovered, the information is shared with ChannelEstablishment module.

2. ChannelEstablishment: The List of Resources/Services is used to establish connection with the desired node. Events ConnectRequest and ConnectAcknowledgement are used for connection establishment. After the communication is over the connection can be disconnected to release the limited resources through Disconnect event. Disconnect event is made convergent to avoid live lock in the model. Error handling events detect errors and appropriately terminate connections as per the session configurations. In our model, error detection events are related to connection time-out and reconnecting an existing channel. Timeout error information is communicated through Timeout interface with Timer module and the channelEstablished interface is shared with ChannelConversation modules.

3. ChannelConversation: This is a pseudo module which contains the Message-Exchange, Timer and Intruder modules.
4. MessageExchange: This module includes all the application message transfer events i.e., all the transmit/receive events for message send and acknowledgement. These events update the message buffers and track time for message transmission and reception.
5. Timers: There is a global time ticking through an event called "Timer" and there are local timers maintained by client and server. These timers are incremented when either there is a send event happening or to just delay time in case of channel inactivity. Every transmission and reception event will store the time at which each message was sent or received. Time tracking is used for keep-alive mechanism, time-out handling and for verifying time related properties. In further refinements of concrete protocols, timers can also be used for strategies like exponential back off in case of failed acknowledgement.
6. Intruder: This module is introduced to emulate disturbance in channel which leads to loss of messages. A malicious Intruder event can consume any message in the channel that is not yet received by the intended client or server. Intruder can simulate attackers, connection drops, or any other disturbances in the network that can lead to loss of the application message. This is a convergent event and does not run forever.

4.2   Concrete Protocol Models

From the common abstract model, the decomposed machines are refined further to add details specific to a protocol. Some of the features which are not used in the protocol need not be used or refined. For example there is no broadcast or multicast support in MQTT protocol. Hence this module does not need any refinement in MQTT model. The contexts from the abstract model are extended to add detailed attributes. Channel variables and internal buffers are introduced to track the dynamic behaviour of the channel that include messages in channel, topics subscribed, payload counters, send and receive buffers, timers, configuration settings, etc. Following is a detailed description of MQTT protocol model created from the abstract model. We then briefly describe the other two protocol models (MQTT-SN and CoAP) which follow similar procedures.

MQTT Protocol Model: MQTT protocol is modeled by abstracting communication network in an IoT system consisting of two channels. For illustrative purpose, we have modeled the channels with two servers and two clients.

ChannelEstablishment Module - From the abstract module containing events ConnectRequest, ConnectAcknowledgement and Disconnect, MQTT specific refinement is done to include configuration details and disconnection due to errors. When a channel is established, the configuration settings of the channel communicated between the client and the server are stored in channel variables.

MessageExchange Module - First refinement of the module introduces publish and subscribe message with their acknowledgement events. These events are further refined to send original message, duplicate message and reception of the message at both client and server sides. Figure 5 gives the refinement steps and atomic decomposition for transmit messages in this module. Similar model is built for acknowledgement messages.

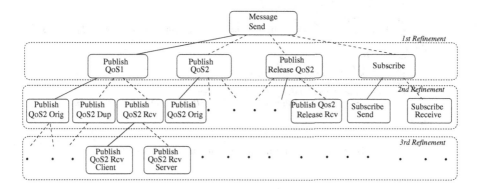

Fig. 5. Atomicity decomposition of confirmable message transmission - QoS1 and QoS2

To track if the correct message is delivered with required QoS and time, the "Payload" is implemented as a counter with a range of 0 to 9 which allows us to uniquely identify every message transmitted. The range of the counter can be extended to any number without affecting our model. By keeping a track of how many times the message with a given payload value is received, we can verify interesting properties related to QoS, message ordering, retained message and persistent sessions. Figure 6 is an example of the QoS0 Publish event transmitted by an MQTT client. The guards ensure that a message of type publish with QoS0 is transmitted on the channel which is already established. In the actions, the channel is populated with a new message carrying unique payload, ClientTimer is initialized, direction of the message is set, PayloadCounter is incremented and Timer-increment event is triggered.

Timer and Intruder modules - Timer Module is refined to include ClientSide and ServerSide Timer, and corresponding Timeout events. Intruder Module does not have any particular refinement for MQTT.

MQTT-SN Protocol Refinement: MQTT-SN model reuses MessageExchange, ChannelEstablishment, Timer and Intruder Modules from MQTT. The Multicast/Broadcast Module is refined from common abstract model to add events related to gateway discovery in the network using search gateway messages. New topic registration procedure is added to the ChannelConversation Module.

CoAP Protocol Refinement: ChannelConversation module from abstract model is refined to include request-response layer by adding events that are

Event Client_Publish_QoS0 ⟨ordinary⟩ ≙
   any
      M10
      ch
      Topic
   where
      grd1: $Msg\_Topic(M10) = Topic$
      grd2: $IncrementTime = FALSE$
      grd3: $M10 \in MSG \wedge Topic \in TOPIC$
      grd4: $PayloadCounter \in 0..9$
      grd5: $(M10 \mapsto ((Publish \mapsto AtmostOnce) \mapsto PayloadCounter)) \in Msg\_Type\_QoS$

   then
      act1: $Client\_timer(ch) := 1$
      act2: $IncrementTime := TRUE$
      act3: $PayloadCounter := PayloadCounter + 1$
      act4: $LimitTimer := 1$
      act5: $Channel\_Direction(ch) := Client\_To\_Server$
      act6: $MsgsInChannel(ch) := MsgsInChannel(ch) \cup \{M10 \mapsto PayloadCounter\}$

   end

Fig. 6. Event for publishing message with QoS0

enabled to send a request and receive a corresponding response either piggy-backed or separate. Each of these events then trigger the message layer events to transmit confirmable or non-confirmable messages and receive corresponding acknowledgements. Token ID matching and message ID matching is carried out to ensure every request receives its response. ChannelEstablishment module is refined to add multi-hop connection consisting of multiple channels. Multi-cast/Broadcast module is refined to discover resources and services in the network. Timer and Intruder modules are directly used from the common abstract model.

4.3 Model Validation

ProB is used for validating our model through simulation of events and checking LTL properties for common abstract model. Accuracy of the model can be obtained by executing different runs and observing the sequence of events and variable values in each of these events. ProB also reports any invariant violation or error in events which is then corrected in the model. Model validation is also done by writing and verifying invariants.

5 Verification of IoT Properties Using Event-B

Following are some of the significant properties that are verified through the model by writing them as invariants that have to be satisfied for all the events in protocol specific models. The property invariant contains two parts well-definedness expressions and the actual property to be proved. We omit the well-definedness conditions and state only the actual property to be proved. Properties 1 to 7 are verified in MQTT and MQTT-SN models and 8 to 11 are verified in CoAP model.

1. Message Ordering: If both client and server make sure that no more than one message is "in-flight" at any one time, then no QoS1 message will be received after any later one. For example a subscriber might receive them in the order $1, 2, 3, 3, 4$ but not $1, 2, 3, 2, 3, 4$.
   Refer to Sect. 4.6 in [9].

$$
\begin{aligned}
\forall ch \cdot \forall pc1 \cdot \forall pc2 \cdot ((pc1 &\in 0 \cdot \cdot 9 \wedge pc2 \in 0 \cdot \cdot 9 \wedge ch \in establishChannel \\
&\wedge (pc1 \in Client\_MsgSentQoS2(ch) \vee pc1 \in Client\_MsgSentQoS1(ch)) \\
&\wedge (pc2 \in Client\_MsgSentQoS2(ch) \vee pc2 \in Client\_MsgSentQoS1(ch)) \\
&\wedge (time > SendTRange(pc2) + Response\_Timeout) \\
&\wedge pc1 \neq pc2 \wedge (SendTRange(pc1) < SendTRange(pc2)) \\
&\Rightarrow (RcvTRange(pc1) \leq RcvTRange(pc2))
\end{aligned} \tag{1}
$$

2. Persistent Session: When a client reconnects with "CleanSession" set to 0, both the client and server must re-send any unacknowledged publish packets (where QoS > 0) and publish release packets using their original packet Identifiers. Refer to Normative Statement number MQTT-4.4.0-1 in [9]. The variable RcvTRange is updated with current time only after the message is received. Hence it should be greater than the SendTRange time.

$$
\begin{aligned}
\forall ch \cdot \forall pc \cdot ((pc &\in 0 \cdot \cdot 9 \wedge ch \in establishChannel \\
&\wedge Channel\_CleanSess(ch) = FALSE \\
&\wedge ((pc \in Client\_MsgSentQoS1(ch)) \vee (pc \in Client\_MsgSentQoS2(ch)) \\
&\wedge (time > (SendTRange(pc) + Response\_Timeout)) \\
&\Rightarrow (RcvTRange(pc) > SendTRange(pc)))
\end{aligned} \tag{2}
$$

3. QoS of a message from Client1 to Client2: The effective QoS of any message received by the subscriber is minimum of QoS with which the publishing client transmits this message and the QoS set by the subscriber while subscribing for the given topic. E.g. Publishing client sends with QoS1 oand subscribing client has subscribed with QoS2, with effective QoS being 1. Refer to Sect. 4.3 in [9].

$$
\begin{aligned}
\forall ch \cdot \forall pc \cdot \forall chnl \cdot \forall msg \cdot ((pc &\in 0 \cdot \cdot 9 \wedge ch \in establishChannel \\
&\wedge msg \in MSG \wedge chnl \in establishChannel \\
&\wedge (pc \in Client\_MsgSentQoS1(ch) \\
&\wedge (msg \mapsto ((PUBLISH \mapsto AtleastOnce) \mapsto pc)) \in Msg\_Type\_QoS \\
&\wedge ((Msg\_Topic(msg) \mapsto ExactOnce) \in Channel\_TopicQoS(chnl)) \\
&\wedge ((time - SendTRange(pc)) \geqslant Response\_Timeout))) \\
&\Rightarrow (\exists QC \cdot ((QC \geq 1) \wedge Client\_MsgReceived\_2(chnl) = QC)))
\end{aligned} \tag{3}
$$

4. Exponential Backoff: The sender retransmits the Confirmable message at exponentially increasing intervals, until it receives an acknowledgement or runs out of attempts. Refer to Sect. 4.2 in [11].

$$\forall ch \cdot \forall pc \cdot ((pc \in 0 \cdot \cdot 11 \wedge ch \in establishChannel \wedge pc \in MsgSent(ch)$$
$$\wedge \, RetransmissionCounter(pc) \geqslant Max\_Retransmit(ch) \tag{4}$$
$$\Rightarrow ((SendTRange(pc) - SendTPrev(pc)) \leq Ack\_Timeout(pc)$$
$$\wedge \, (SendTRange(pc) - SendTPrev(pc)) > 0))$$

5.1   Proof Obligations Results

Our validated models of MQTT, MQTT-SN and CoAP have together discharged 1840 proof obligations, of which 88% proof obligations were automatically discharged through AtlierB, SMT, PP and ML provers. The proof obligations include well-definedness of predicates and expressions in invariants, guards, actions, variant and witnesses of all the events, feasibility checks, variable reuse check, guard strengthening and witness feasibility in refinements, variant checks for natural number and decreasing variants for convergent and anticipated events, theorems in axioms and invariant preservation for refinements and invariants used for verification of required properties. About 30% of proofs discharged in the models are for verification of properties written as invariants. Table 2 gives a summary of the properties verified.

Table 2. Proof obligation statistics for verified properties of IoT protocols

Sl.no	Protocol property	Proof obligations	Result
1	Duplicate Channel	10	Passed
2	Message Ordering	34	Passed
3	Persistent Session	34	Passed
4	QoS1 in single channel	26	Passed
5	QoS2 in single channel	26	Passed
6	Retained QoS1 message	24	Passed
7	Retained QoS2 message	24	Passed
8	Effective QoS0 in Multi channel(3 cases)	66	Passed
9	Effective QoS1 in Multi channel(3 cases)	66	Passed
10	Effective QoS2 in Multi channel(3 cases)	72	Passed
11	Request-Response Matching and Timeout	39	Passed
12	Confirmable Message ID Matching and Timeout	39	Passed
13	Exponential Backoff	39	Passed

6   Related Work

Communication protocols for IoT have been used for over a decade now, but there has been no attempt to provide formal semantics for these protocols. A recent paper shows that there are scenarios where MQTT has failed to adhere to the QoS requirement [16]. However the paper is limited to partial model of MQTT protocol for QoS properties. In another work, a protocol used for IoT - Zigbee is verified for properties related to connection establishment properties [17] using Event-B. In [19] and [20], the authors give methods to evaluate performance of MQTT protocol with regards to different QoS levels used and compare with other IoT protocol CoAP. In [18] the author again tests connection properties using passive testing for XMPP protocol in IoT.

We differ from the above mentioned approaches by proposing a framework comprising of a common model for IoT protocols which can be used to build models of different IoT protocols. These models verify properties required for IoT like connection establishment, persistent sessions, retained-message transmission, will messages, message ordering, proxying, caching and QoS and provide proof obligations for these properties through automatic proof discharge and interactive proof discharge methods.

7   Conclusion and Future Work

In this paper we have proposed a framework using Event-B to model IoT protocols. We then have used this framework and went on to model some of the widely used IoT protocols viz., MQTT, MQTT-SN and CoAP. Through simulation and proof obligation discharge in Rodin, we have formally verified that the properties related to QoS, persistent session, will, retain messages, resource discovery, two layered request-response architecture, caching, proxying and message deduplication. We show that the protocols work as intended in an uninterrupted network as well as with an intruder which consumes messages in the network. The three protocols modeled in this paper implement simple mechanisms to provide reliable message transfer over a lossy network. They are also able to reduce overhead by providing features like persistent connections, retain messages, caching and proxying which are essential for IoT systems. Our work is a stepping stone towards providing formal semantics of IoT protocols and systems.

Future research would focus on modeling the other aspects of protocols like security, user authentication, encryption and different attacker modules. We would also like to move verification of more properties from the concrete protocol models to the common abstract model. We would like to further compare other protocols for IoT like AMQP and XMPP by modeling them using our framework. It would also be interesting to integrate the protocol model into an existing model of IoT system and verify the properties required at the system level.

References

1. Event-B. http://www.Event-B.org/
2. Abrial, J.R.: Modeling in Event-B: System and Software Engineering. Cambridge University Press, Cambridge (2010)
3. Evans, N., Butler, M.: A proposal for records in Event-B. In: Misra, J., Nipkow, T., Sekerinski, E. (eds.) FM 2006. LNCS, vol. 4085, pp. 221–235. Springer, Heidelberg (2006). doi:10.1007/11813040_16
4. Rodin Tool. http://wiki.Event-B.org/index.php/Rodin_Platform
5. Rodin Hand Book. https://www3.hhu.de/stups/handbook/rodin/current/pdf/rodin-doc.pdf
6. ProB tool. https://www3.hhu.de/stups/prob/index.php/Main_Page
7. Gartner newsroom. http://www.gartner.com/newsroom/id/3165317
8. Karagiannis, V., Chatzimisios, P., Vazquez-Gallego, F., Alonso-Zarate, J.: A survey on application layer protocols for the internet of things. Trans. IoT Cloud Comput. **3**(1), 11–7 (2015)
9. MQTT Ver. 3.1.1. http://docs.oasis-open.org/mqtt/mqtt/v3.1.1/os/mqtt-v3.1.1-os.html
10. MQTT-SN Ver. 1.2. http://mqtt.org/new/wp-content/uploads/2009/06/MQTT-SN_spec_v1.2.pdf
11. The Constrained Application Protocol (CoAP) RFC7252. https://tools.ietf.org/html/rfc7252
12. Extensible Messaging and Presence Protocol (XMPP) Core RFC6120. http://xmpp.org/rfcs/rfc6120.html
13. Advanced Message Queuing Protocol ver. 1.0. http://docs.oasis-open.org/amqp/core/v1.0/amqp-core-complete-v1.0.pdf
14. Pascal, C., Renato, S.: Event-B model decomposition, DEPLOY Plenary Technical Workshop (2009)
15. Salehi Fathabadi, A., Butler, M., Rezazadeh, A.: A systematic approach to atomicity decomposition in Event-B. In: Eleftherakis, G., Hinchey, M., Holcombe, M. (eds.) SEFM 2012. LNCS, vol. 7504, pp. 78–93. Springer, Heidelberg (2012). doi:10.1007/978-3-642-33826-7_6
16. Aziz, B.: A formal model and analysis of the MQ telemetry transport protocol. In: Ninth International Conference, Availability, Reliability and Security (ARES), pp. 59–68. Fribourg (2014)
17. Gawanmeh, A.: Embedding and verification of ZigBee protocol stack in Event-B. In: Procedia Computer Science, vol. 5, pp. 736–741. ISSN 1877–0509 (2011)
18. Che, X., Maag, S.: A passive testing approach for protocols in Internet of Things. In: Green Computing and Communications (GreenCom), IEEE and Internet of Things (iThings/CPSCom), IEEE International Conference on and IEEE Cyber, Physical and Social Computing, pp. 678–684. IEEE Press (2013)
19. Lee, S., Kim, H., Hong, D.K., Ju, H.: Correlation analysis of MQTT loss and delay according to QoS level. In: The International Conference on Information Networking(ICOIN), pp. 714–717. IEEE (2013)
20. Thangavel, D., Ma, X., Valera, A., Tan, H.X., Tan, C.K.: Performance evaluation of MQTT and CoAP via a common middleware. In: IEEE Ninth International Conference, Intelligent Sensors, Sensor Networks and Information Processing (ISSNIP), pp. 1–6. IEEE Press (2014)

Formalization

Formal Analysis of Information Flow in HOL

Ghassen Helali[1(✉)], Sofiène Tahar[1], Osman Hasan[1], and Tsvetan Dunchev[2]

[1] Electrical and Computer Engineering, Concordia University, Montreal, Canada
{helali,o_hasan}@encs.concordia.ca,
tahar@ece.concordia.ca
[2] Computer Science and Engineering, University of Bologna, Bologna, Italy
tsvetan.dunchev@unibo.it

Abstract. Protecting information has become very important due to the safety-critical nature of many computer-based applications. Information flow analysis plays a very important role in quantifying information-related properties under external attacks. Traditionally, information flow analysis is performed using paper-and-pencil based proofs or computer simulations but due to their inherent nature, these methods are prone to errors and thus cannot guarantee accurate analysis. As an accurate alternative, we propose to conduct the information flow analysis within the sound core of a higher-order-logic theorem prover. For this purpose, some of the most commonly used information flow measures, including Shanon entropy, mutual information, min-entropy, belief min-entropy, have been formalized. In this paper, we use the Shannon entropy and mutual information formalizations to formally verify the Data Processing and Jensen's inequalities. Moreover, we extend the security model for the case of the partial guess scenario to formalize the gain min-entropy. These formalizations allow us to reason about the information flow of a wide range of systems within a theorem prover. For illustration purposes, we perform a formal comparison between the min-entropy leakage and the gain leakage.

Keywords: Information flow · Entropy · Gain function · g-Leakage · Theorem proving · Higher-order logic

1 Introduction

Information flow analysis mainly consists of using information measures to evaluate the amount of information an attacker could get by observing the low output of a system or a protocol. Examples of this analysis include the evaluation of anonymity protocols [27] and security networks [31]. Protecting the confidentiality of sensitive information and guaranteeing a perfect level of anonymity are increasingly being required in numerous fields such as electronic payments [17], auctioning [32] and voting [7].

Various techniques for analyzing the information flow have been used. The possibilistic approach [3] consists of using non-deterministic behaviors to model

© Springer International Publishing AG 2017
K.G. Larsen et al. (Eds.): SETTA 2017, LNCS 10606, pp. 283–299, 2017.
https://doi.org/10.1007/978-3-319-69483-2_17

the given system. Information flow analysis based on epistemic logic [11] and process calculi [28] fall into the category of possibilistic analysis. This approach is limited in terms of distinguishing between systems of varying degrees of protection [10]. As a solution for this limitation, probabilistic approaches, based on information and statistics, are considered as a more reliable alternative for computing information flow. In a threat model where the secret should be guessed in one try, the main objective of the attacker is to maximize the probability of guessing the right value of the high input (secret), in one try, by betting on the most probable element. To cater for this particular threat model, Renyi's entropy metrics [26], i.e., min-entropy and belief min-entropy are employed [30]. These measures are commonly used to effectively reason about deterministic and probabilistic systems.

Due to the difficulty of preventing the information leakage completely, "small" leaks are usually tolerated [19,29] by the above-mentioned information flow measures. With respect to the partial guess, g-leakage [2] is introduced as a generalization of the min-entropy model. The main idea of this notion is to extend the vulnerability in order to take into consideration the so called *gain function g*. The gain function models the profit that an attacker gets by using a certain guess z over the secret x. The gain value ranges from 0, when the guess has no corresponding secret value, to 1, in the case of an ideal guess. Hence the vulnerability (g-vulnerability) is redefined as the maximum expected gain over all possible guesses [2].

Traditionally, the quantitative analysis of information flow has been conducted using paper-and-pencil and computer simulation. The paper-and-pencil technique cannot cope with complex systems due to the high chances of human error while dealing with large models. On the other hand, the computer simulation approach cannot be considered accurate due to the use of numeric approximations. In order to overcome those shortcomings, formal methods [12] have been proposed as a sound technique to enhance accuracy of safety-critical systems. For instance, in [19], the probabilistic mode checker PRISM has been used to reason about several information systems, e.g., the Dining Cryptographers protocol. However, the state-space explosion problem of model checking limits the scope of its usage in information flow analysis. In contrast, higher-order-logic theorem proving can be used for the analysis of information flow to overcome these limitations.

In [8], Coble has formalized the conditional mutual information in the higher-order-logic theorem prover HOL4 [1] based on the Lebesgue integration. These fundamentals have been later used to formally analyze the privacy and the anonymity guarantees and proposed the Dining Cryptographers. However, Coble's formalization of Lebesgue integrals can only consider finite-valued measures, functions and integrals. Considering this fact, Mhamdi et al. [22] generalized the formalizations of the probability and information theories by introducing the notions of extended real numbers and formalizing Borel sigma algebra that covers larger classes of functions in terms of integrability and convergence. The authors further used these fundamentals to formalize the measures of entropy, relative entropy and mutual information [9]. In the same context, information and conditional information leakage degree have been formalized [23] in HOL4 to assess security and anonymity protocols. Similarly, Hölzl [15,16] formalized a generic version of the measure, probability and information theories in Isabelle/HOL. This definition is very similar to Coble's work. Hölzl used the measure and the probability theories to define the Kullback-Leibler divergence, entropy, conditional entropy, mutual information and conditional mutual information and verify the properties related to the quantification of the information represented by a random variable [24].

Most of our work is based on the probability and information theories, formalized in Mhamdi's work [21], due to their completeness and availability in HOL4. We previously used these fundamentals to develop formal reasoning support for information flow using min-entropy and belief min-entropy [14], which we are extending to the gain min-entropy (g-leakage), which considers the model where the secret is totally guessed based on a partial gain about the secret using a certain guess. We will also use the formalized information measures in [23] to conduct the formal verification of the Data Processing [4] and Jensen's [18] inequalities which are major properties in information flow analysis. In the information flow context, the Data Processing Inequality (DPI) states that any post-processing of data does not increase the information leakage while Jensen's Inequality shows a relation between data averaging and data processing.

To the best of our knowledge, these measures have not been formalized before. We apply them to conduct an information leakage analysis of a threat scenario to compare min-entropy leakage and g-leakage (based on gain min-entropy) and since a small/partial leak can be tolerated, we show that the min-entropy leakage can be arbitrarily greater than the g-leakage.

2 Preliminaries

This section describes the HOL4 environment as well as the formalization of probability and information theories, which we would be building upon to formalize the DPI and Jensen's Inequality as well as the gain Min-Entropy notions.

2.1 HOL Theorem Prover

The HOL system is an environment for interactive theorem proving in higher order logic. Higher-order logic is a system of deduction with a precise semantics and is expressive enough to be used for the specification of almost all classical mathematics theories. In order to ensure secure theorem proving, the logic in the HOL system is represented in the strongly-typed functional programming language ML. An ML abstract data type is used to represent higher-order-logic theorems and the only way to interact with the theorem prover is by executing ML procedures that operate on values of these data types.

Soundness is assured as every new theorem must be verified by applying the basic axioms and primitive inference rules or any other previously verified theorems/inference rules. The HOL system has been used to formalize pure mathematics and verify industrial software and hardware systems.

2.2 Probability and Information Theory

Probability and information theories provide mathematical models to evaluate the uncertainty of random phenomena. These concepts are commonly used in different fields of engineering and computer sciences, such as signal processing, data compression and data communication, to quantify the information. Recently, the probability and information theories have been widely used for cryptographic and information flow analysis [29]. Some foundational notions of these formalizations are described below.

Let X and Y denote discrete random variables, with x and y and $\mathcal{X}$ and $\mathcal{Y}$ denoting their specific values and set of all possible values, respectively. Similarly, the probabilities of X and Y being equal to x and y is denoted by $p(x)$ and $p(y)$, respectively.

- Probability Space: *a measure space such that the measure of the state space is 1.*
- Independent Events: *Two events X and Y are independent iff $p(X \cap Y) = p(X)p(Y)$.*
- Random Variable: *$X : \Omega \to \mathcal{R}$ is a random variable iff X is $(F, \mathcal{B}(\mathcal{R}))$ measurable, where Ω is the state space, F denotes the set of events and $\mathcal{B}$ is the Borel sigma algebra of real valued functions.*
- Joint Probability: *A probabilistic measure where the likelihood of two events occurring together and at the same point in time is calculated. Joint probability is the probability of event Y occurring at the same time event X occurs. It is mathematically expressed as $p(X \cap Y)$ or $p(X, Y)$.*
- Conditional Probability: *A probabilistic measure where an event X will occur, given that one or more other events Y have occurred. Mathematically $p(X|Y)$ or $\frac{p(X \cap Y)}{p(Y)}$.*
- Expected Value: $E[X]$ of a random variable X is its Lebesgue integral with respect to the probability measure. The following properties of the expected value have been verified in HOL4 [22]:
  1. $E[X + Y] = E[X] + E[Y]$
  2. $E[aX] = aE[X]$
  3. $E[a] = a$
  4. $X \leq Y$ then $E[X] \leq E[Y]$
  5. X and Y are independent then $E[XY] = E[X]E[Y]$
- Variance and Covariance: Variance and covariance have been formalized in HOL4 using the formalization of expectation. The following properties have been verified [22]:
  1. $Var(X) = E[X^2] - E[X]^2$
  2. $Cov(X, Y) = E[XY] - E[X]E[Y]$

3. $Var(X) \geq 0$
4. $\forall a \in R, Var(aX) = a^2 Var(X)$
5. $Var(X+Y) = Var(X) + Var(Y) + 2Cov(X,Y)$

The above-mentioned definitions and properties have been utilized to formalize the foundations of information theory in HOL4 [22]. The widely used information theoretic measures can be defined as:

- The Shannon Entropy: *It measures the uncertainty of a random variable*

$$H(X) = -\sum_{x \in \mathcal{X}} p(x) \log p(x)$$

- The Conditional Entropy: *It measures the amount of uncertainty of X when Y is known*

$$H(X|Y) = -\sum_{y \in \mathcal{Y}} p(y) \sum_{x \in \mathcal{X}} p(x|y) \log p(x|y)$$

- The Mutual Information: *It represents the amount of information that has been leaked*

$$I(X;Y) = I(Y;X) = H(X) - H(X|Y)$$

- The Relative Entropy or Kullback Leiber Distance: *It measures the inaccuracy or information divergence of assuming that the distribution is q when the true distribution is p*

$$D(p\|q) = \sum_{x \in \mathcal{X}} p(x) \log \frac{p(x)}{q(x)}$$

- The Guessing Entropy: *It measures the expected number of tries required to guess the value of X optimally*

$$G(X) = \sum_{1 \leq i \leq n} i p(x_i)$$

- The Rényi Entropy: *It is related to the difficulty of guessing the value of X*

$$H_\alpha(X) = \frac{1}{1-\alpha} \log \left(\sum_{x \in \mathcal{X}} P[X = x]^\alpha \right)$$

Among the measures listed above, Mhamdi [21] and Coble [8] formalized the Entropy, Conditional Entropy, Relative Entropy and Mutual Information in HOL4 and Hölzl [15] formalized similar concepts in Isabelle/HOL.

3   Shannon Based Information Flow

In this section, we will use the most common measures to quantify information flow, such as Shannon entropy, related entropy and mutual information formalized in [23] to formally verify the Data Processing Inequality as well as Jensen's Inequality properties.

3.1   Data Processing Inequality

According to the Data Processing Inequality (DPI), post-processing cannot increase information. Quantitatively, considering three random variables X, Y and Z satisfying the Markov property [5], the DPI states that Z cannot have more information about X than Y has about X; which is

$$I(X, Z) \leq I(X, Y)$$

Our formalization is based on the Discrete Time Markov Chain formalization (DTMC) [20], formalized information measures and probability theory [21].

The motivation behind this definition relies on the fact that the three random variables X, Y and Z satisfy the Markov property and thus

$$p(x, y, z) = p(x).p(y, z|x) = p(x).p(y|x).p(z|x, y)$$

Similarly, we can also deduce that

$$p(z|x, y) = p(z|y)$$

In order to formally verify the DPI, we first formalized the conditional mutual information

Definition 1. *(Conditional Mutual Information)*
For discrete random variables X, Y, and Z, conditional mutual information is defined as

$$
\begin{aligned}
I(X; Y|Z) &= \sum_{z \in Z} \sum_{y \in Y} \sum_{x \in X} P_{X,Y,Z}(x, y, z) \log \frac{P_z(Z).P_{X,Y,Z}(x, y, z)}{P_{X,Z}(x, z).P_{Y,Z}(y, z)} \\
&= H(X, Z) + H(Y, Z) - H(X, Y, Z) - H(Z) \\
&= H(X|Z) - H(X|Y, Z)
\end{aligned}
$$

Then, using the commutativity of the distribution function which says that $P_{Y,Z}((y, z)) = P_{Z,Y}((z, y))$, we get the following equality: $I(X; Y, Z) = I(X; Z, Y)$ Therefore, the following result can be deduced:

$$I(X; Y|Z) + I(X; Z) = I(X; Y, Z) = I(X; Z, Y) = I(X; Z|Y) + I(X; Y)$$

Theorem 1. *(Symmetry of Mutual Information Property)*

```
⊢ ∀ b p X Y Z.
  (POW (p_space p) = events p) ∧ prob_space p ∧
  random_variable X p s1 ∧ random_variable Y p s2 ∧
  random_variable Z p s3 ∧ random_variable (λ x.(Z x, Y x)) p s32 ∧
  FINITE (p_space p) ∧
  (mutual_information b p s1 s2 X Y ≠ −∞ ∧
  mutual_information b p s1 s2 X Y ≠ +∞ ∧
  mutual_information b p s1 s3 X Z ≠ −∞ ∧
```

```
mutual_information b p s1 s3 X Z ≠ +∞) ⇒
  conditional_mutual_information b p s1 s3 s2 X Z Y +
  mutual_information b p s1 s2 X Y =
  conditional_mutual_information b p s1 s2 s3 X Y Z +
  mutual_information b p s1 s3 X Z
```

where POW and FINITE refer to the *power set* operator and *finiteness* tester in HOL4 respectively.

The proof of the property above relies on the associativity of the joint distribution, namely $P(X, (Y, Z)) = P(X, (Z, Y))$ as well as the symmetry of the additivity. Now we formally verify our main goal, DPI, as follows

Theorem 2. *(Data Processing Inequality: DPI)*
For all random variables X, Y and Z satisfying the Markov property, the DPI states that $I(X; Z) \leq I(X; Y)$

which is formalized in HOL as follows:

```
⊢ ∀b p X Y Z.(POW (p_space p) = events p) ∧ prob_space p ∧
  random_variable X p s1 ∧ random_variable Y p s2 ∧
  random_variable Z p s3 ∧
  random_variable (λ x.(Y x, Z x)) p s23 ∧
  random_variable (λ x.(Z x, Y x)) p s32 ∧
  FINITE (p_space p) ∧    mc p X Y Z∧
  (mutual_information b p s1 s2 X Y ≠ −∞ ∧
  mutual_information b p s1 s2 X Y ≠ +∞ ∧
  mutual_information b p s1 s3 X Z ≠ −∞ ∧
  mutual_information b p s1 s3 X Z ≠ +∞) ⇒
    mutual_information b p s1 s2 X Y ≥
    mutual_information b p s1 s3 X Z
```

where mc p X Y Z denotes the Markov property and the assertions related to the mutual information are constraints to avoid the infinite bounds of the information leakage.

For proving this theorem, we first need to prove the following two properties:

- $\forall X, Y$ random variables X and Y, the mutual information between X and Y is non-negative, $I(X; Y) \geq 0$
- if X, Y and Z form a Markov chain, then $I(X; Z|Y) = 0$

Applying the above properties to the equality:

$$I(X; Y|Z) + I(X; Z) = I(X; Z|Y) + I(X; Y)$$

as well as the previously verified property which states

$$I(X; Y, Z) = I(X; Z, Y)$$

our result can be proved.

The above result states that any transformation of the output channel Y will not give more information about the input X than itself. This concept also states that the information content of a signal cannot be increased via a local physical operation: post-processing cannot increase information. The main challenges of proving this result in HOL is to use the formalized notions of probability and information theories and reason about one of the major applications of the information theory. By proving the DPI, we show the usefulness of the theoretic information framework formalized in HOL.

3.2   Jensen's Inequality

Jensen's inequality has applications in many fields of applied mathematics and specifically information theory. For example, it plays a key role in the proof of the information inequality, $0 \le D(p\|q)$. In the following, we prove Jensen's inequality in its measure theoretic form as an application for information theory formalized in HOL. We first formalize in HOL4 the notion of convex functions:

Definition 2. *(Convex function)*
```
⊢ conv_func = ∀ x y z. (x<y ∧ y<z) ⇒
    ((f(y)-f(x))/(y-x) ≤ (f(z)-f(y))/(z-y))
```

Now, let Ω be a probability space, μ is a measure funcion on Ω, and g and f be arbitrary convex functions on the real numbers, respectively. Then according to Jensen's inequality: $\int_\Omega f(g(x)) \, d\mu \ge f(\int_\Omega g(x) \, d\mu)$.

The most challenging part of the proof of Jensen's inequality is to prove the existence of subderivatives a and b of f, such that for all x, $a.x + b \le f(x)$, where for $x_0 = \int_\Omega g(x) \, d\mu$ we reach the equality $a.x_0 + b = f(x_0)$. This follows from the following two facts:

- According to the Mean value theorem, there exists ν such that if $x < \nu < \xi$, then: $\frac{f(x)-f(\xi)}{x-\xi} = f'(\nu)$
- Since f is convex, then its derivative increases, i.e. $f'(\nu) \le f'(\xi)$

Having a and b, the proof of Jensen's inequality is straightforward:
$$\int_\Omega f(g(x)) \, d\mu \ge \int_\Omega (a.g(x) + b) \, d\mu$$
$$\ge a. \int_\Omega g(x) \, d\mu + b. \int_\Omega 1 \, d\mu$$
$$\ge a.x_0 + b$$
$$\ge f(x_0)$$
$$= f(\int_\Omega g(x) \, d\mu)$$

Since μ is a measure, it holds that $\mu(\Omega) = 1$. Therefore $\int_\Omega 1 \, d\mu = 1$.

Using the monotonicity of sub-derivatives and the existence of a convex function properties, we formalize Jensen's inequality for the continuous case:

Theorem 3. *(Jensen's Inequality)*
⊢ ∀f g m. measure_space m ∧ integrable m g ∧ (b = integral m λy.b)∧
  (a *(integral m λx.g(x)) + b = f((integral m λx. g(x)))) ∧
  (∀x.a * x + b ≤ f x) ⇒
    f (integral m λx.g(x)) ≤ integral m λx.(f(g(x)))

The result verified above is a relation between the integral of a convex function and the value of a convex function of an integral. In the information theoretic context, Jensen's inequality relates the averaging of data to the transformation of data. This result is then formally verified in HOL4.

4   Partial Guess, Gain Function and g-Leakage

In this section, we analyze the threat scenarios where the secret is totally guessed in one try by using min-entropy measures [14]. This model is extended with the presence of the attacker's belief leading to the concept of the belief min-entropy [14]. Since the guess of the sensitive information can be partial, we formalize the gain function and the gain min-entropy. We first start by the formalization of the gain function and the related leakage properties. Compared to min-entropy and belief min-entropy measures, where the secret is assumed to be guessed in one try, the new model assumes that the secret can be partially guessed. We then introduce the notion of gain functions, which range from 0 to 1 and operate over a guess z and a secret x. Then $g(z, x)$ models the gain that an attacker gets about the secret x using the guess z.

Definition 3. *(Gain Function)*
Given a set $\mathcal{X}$ of possible secrets and a finite and non-empty set of guesses $\mathcal{Z}$, a gain function is defined as: $g : \mathcal{Z} \times \mathcal{X} \rightarrow [0, 1]$

For the rest of the paper, H_g, V_g and IL_g will respectively denote *gain min entropy* (also called as g-min-entropy), *(prior/posterior)gain vulnerabilty* (also called g-vulnerability) and *gain information leakage* (also called g-leakage). Based on the gain function, we define the prior vulnerability:

Definition 4. *(Prior g-Vulnerability)*
Given a gain function g, a random variable X modeling the a-priori behavior, the prior g-vulnerability is

$$V_g(X) = \max_{z \in \mathcal{Z}} \sum_{x \in \mathcal{X}} p(X = x).g(z, x)$$

which is formalized in HOL4 as follows

⊢ ∀ p X g Z. prior_g_vulnerability = extreal_max_set
        (IMAGE (λz. $\sum_{x \in \mathcal{X}}$ distribution p X {x}.g(z,x))
          (IMAGE X (p_space p)) Z)

where `IMAGE f s` in HOL denotes the image of the set s by the function f which in our case is $X(\Omega)$ and `extreal_max_set (IMAGE f s)` refers to the *max* of the set `IMAGE f s` which in our case is the maximum probability over the distributions set.

Compared to the previous definition of vulnerability, the above definition shows that the gain is weighted by the probability of the secret itself, which means that the adversary $\mathcal{A}$ tries to make a guess maximizing the gain about every x from $\mathcal{X}$.

Definition 5. *(Posterior g-Vulnerability)*
Given a gain function g, a high input behavior modeled by the random variable X and a low output modeled by Y, the posterior g-vulnerability is

$$V_g(X|Y) = \sum_{y \in \mathcal{Y}} \max_{z \in \mathcal{Z}} \sum_{x \in \mathcal{X}} p(X = x).p(Y = y|X = x).g(z, x)$$

$$= \sum_{y \in \mathcal{Y}} \max_{z \in \mathcal{Z}} \sum_{x \in \mathcal{X}} p(X = x, Y = y).g(z, x)$$

This definition can be formalized in HOL4 as

```
⊢ ∀p X Y g Z. posterior_g_vulnerability =
  ∑ extreal_max_set(IMAGE (λz. ∑ distribution p Y {y}.
  y∈𝒴                              x∈𝒳
    conditional_distribution p X Y ({x},{y}).g(z,x))
      (IMAGE X (p_space p)))
```

Now we define the uncertainty measures; g-min-entropy (initial uncertainty), and conditional g-min-entropy, (remaining uncertainty) which will be used to define the g-leakage.

Definition 6. *(g-Min-Entropy, g-Conditional-Min-Entropy and g-Leakage)*
```
⊢ g_min_entropy p X g Z = -log(prior_g_vulnerability p X g Z)
⊢ g_conditional_min_entropy p X Y g Z =
   -log(posterior_g_vulnerability p X Y g Z)
⊢ g_information_leakage p X Y g Z =
   g_min_entropy p X g Z - g_conditional_min_entropy p X Y g Z
```

We next consider a model where the attacker can get partial knowledge about the secret using a certain guess. The gain function models the benefit that the attacker gets about the secret. We then verify that the prior g-vulnerability cannot exceed the posterior g-vulnerability. Thus the g-leakage is positive:

Theorem 4. *(Positive g-Leakage)*

```
⊢ ∀ p X Y g Z. prob_space p ∧ FINITE (p_spac p) ∧ FINITE Z ∧
   Z ≠ ∅ ∧ p_space p ≠ ∅ ∧   ∀x. x ∈ p_space p ⇒
   {x} ∈ events p ∧ events p = POW(p_space p) ∧
   ∀x z. 0 ≤ g(z, x) ∧ g(z, x) ≤ 1⇒
   0≤ g_information_leakage p X Y g Z
```

Proof. First, note that the gain information leakage (g-leakage) is $IL_g = H_g(X) - H_g(X|Y) = log(V_g(X|Y)) - log(V_g(X))$. After simplification, our goal will be reduced to $V_g(X) \le V_g(X|Y)$. Then

$$V_g(X) = \max_z \sum_x \sum_y P(X = x, Y = y).g(z, x)$$

$$\le \sum_y \max_z \sum_x P(X = x, Y = y).g(z, x)$$

$$\le V_g(X|Y)$$

Next, we will study the case when the g-leakage is equal to zero. We will evaluate the condition under which this result occurs. Before stating this property formally we need first to define the notion of the expected gain of a guess z. With respect to the same configuration, the prior and posterior expected gains are defined as:

Definition 7. *(Prior Expected Gain)*

$$E_g(z) = \sum_x P(X = x).g(z, x)$$

which is formalized in HOL4 as follows

⊢ ∀p X g z. prior_expected_gain p X g Z =
$$\sum_{x \in X(\Omega)}$$ distribution p X {x}.g(z,x)

Definition 8. *(Posterior Expected Gain)*
Given an output y the expected gain of a guess z is
$$E_g(z, y) = \sum_x P(X = x)P(Y = y \mid X = x).g(z, x)$$

The HOL4 formalization of this definition is

⊢ ∀p X Y y g z. posterior_expected_gain p X Y y g z =
$$\sum_{x \in X(\Omega)}$$ distribution p X {x}.
conditional_distribution p Y X ({y},{x}).g(z,x)

In the context of vulnerabilities and information flow, these definitions satisfy the following properties:

Theorem 5. *(Expected Gain and Vulnerabilities)*

- $V_g(X) = \max_z E_g(z)$
- $V_g(X \mid Y) = \sum_y \max_z E_g(z, y)$
- $E_g(z) = \sum_y E_g(z, y)$

We prove these results in the HOL4 theorem prover as follows

⊢ ∀p X g Z. prior_g_vulnerability p X g Z =
   extreal_max_set (IMAGE (λz. prior_expected_gain p X g z) Z)

⊢ ∀p X Y g Z. FINITE Ω ∧ prob_space p ∧
   (∀x. x ∈ Ω ⇒ {x} ∈ events p) ⇒
      posterior_g_vulnerability p X Y g Z = $\sum_{y \in Y(\Omega)}$ extreal_max_set
      (IMAGE (λz. posterior_expected_gain p X Y y g z) Z)

⊢ ∀p X Y y g z. prob_space p ∧ FINITE Ω ∧ Ω ≠ ∅ ∧
   (∀x. x∈ Ω ⇒{x} ∈ events p) ∧
   (∀x. x∈(X(Ω)) ⇒ 0≤g(z,x) ∧ g(z,x)≤1) ⇒
      prior_expected_gain p X g z =
      $\sum_{y \in Y(\Omega)}$ posterior_expected_gain p X Y y g z

We will later use these properties in order to verify the zero valued g-leakage result. We prove the fact that the g-leakage of 0 is related to the expected gain of all outputs, i.e., this statement occurs if there exists a guess z' maximizing the expected gain for all outputs y.

Theorem 6. *(Zero Gain Information Leakage)*
Given a random variable X modeling the initial uncertainty, a random variable Y modeling the remaining uncertainty and a gain function g, the g-leakage is 0 if there exists a guess $z' \in Z$ such that: $\forall z \; y. \; E_g(z', y) \geq E_g(z, y)$

In the HOL4 environment, this property is formalized as follows:

⊢ ∀p X Y y g Z. (prob_space p ∧ FINITE (p_space p) ∧
   ((p_space p) ≠ ∅) ∧
   (∀x. x ∈ p_space p ⇒ {x} ∈ events p) ∧ (FINITE Z) ∧
   (events p = POW (p_space p)) ∧ (Z ≠ ∅) ∧
   (∀x z. (0 ≤ (g (z,x))) ∧ ((g (z, x)) ≤ 1)) ∧
   (0 < prior_g_vulnerability p X g Z)) ⇒
   ((∃z'. (z'∈Z) ∧ (∀z y. (posterior_expected_gain p X Y y g z) ≤
   (posterior_expected_gain p X Y y g z'))) ⇒
      (g_information_leakage p X Y g Z = 0))

Proof. If such a guess exists, then we first prove that it corresponds to the maximum prior expected gain $\forall z. \; E_g(z') \geq E_g(z)$. Then, using the previous results, it follows that the posterior g-vulnerability is equal to the prior expected gain of the best guess

$$V_g(X, Y) = \sum_y E_g(z, y) = \sum_y \max_z E_g(z, y) = \sum_y E_g(z', y) = E_g(z')$$

However, since $E_g(z')$ is the prior g-vulnerability $V_g(X)$, from Theorem 5, so it follows from the definition of the g-leakage that this measure is 0.

Based on the soundness of theorem proving, the above-mentioned formally verified theorems are guaranteed to be accurate and contain all the required assumptions. Moreover, these results can be built upon to reason about information flow analysis of various applications within the sound core of a theorem prover.

5   Min-Entropy Leakage and g-Leakage

In this section, we illustrate the practical usefulness of the theoretical foundations developed in this paper so far. We will present a threat scenario in which we conduct a comparison between the min-entropy leakage and the g-leakage and show that the g-leakage can be smaller than min-entropy leakage. Consider the channel (Matrix of transitional probabilities), described in Table 2, where x_i are the high inputs and y_i are the outputs modelled, respectively, with the random variables X and Y. We assume for this example that inputs and outputs are *uniformly* distributed.

Table 1. Transition channel

	y_1	y_2
x_1	$\frac{1}{2}$	$\frac{1}{2}$
x_2	1	0
x_3	0	1

Table 2. Gain function

g_d	x_1	x_2	x_3
z_1	1	0	0
z_2	0	1	0.98
z_3	0	0.98	1

For our particular example, we consider the gain function called the distance gain function between the secrets, assuming that $\mathcal{X} = \mathcal{Z}$, $g_d(z,x) = 1 - \mathrm{d}(z,x)$ where $\mathrm{d}(z,x)$ is the normalized distance between z and x. Using this configuration, we prove that the min-entropy leakage is equal to $log\ 2 = 1$ and g-leakage is equal to $log\ \frac{2}{1.98}$. The formalization of this theorem in HOL4 is

Theorem 7. *(Comparing Min-Entropy Leakage and g-Leakage)*

```
⊢ ∀p X Y Z g. (prob_space p) ∧ (FINITE (p_space p)) ∧
    (∀x. (x ∈ p_space p) ⇒ {x} ∈ events p) ∧
    ((IMAGE X (p_space p)) = {0;1;2}) ∧
    ((IMAGE Y (p_space p)) = {0;1}) ∧ (Z = {0;1;2}) ∧
    (∀x. x ∈ (IMAGE X (p_space p)) ⇒ distribution p X {x} =
      (1/|IMAGE X (p_space p)|) ∧
    (∀y. y ∈ (IMAGE Y (p_space p)) ⇒
    distribution p Y {y} = (1/|IMAGE Y (p_space p)|) ∧
    (conditional_distribution p Y X ({0},{0}) = (1/2)) ∧
    (conditional_distribution p Y X ({0},{1}) = 1) ∧
    (conditional_distribution p Y X ({0},{2}) = 0) ∧
    (conditional_distribution p Y X ({1},{0}) = (1/2)) ∧
```

```
(conditional_distribution p Y X ({1},{1}) = 0) ∧
(conditional_distribution p Y X ({1},{2}) = 1) ∧
(g(0,0) = 1) ∧ (g(0,1) = 0) ∧ (g(0,2) = 0) ∧ (g(1,0) = 0) ∧
(g(1,1) = 1) ∧ (g(1,2) = Normal (0.98)) ∧ (g(2,0) = 0) ∧
(g(2,1) = Normal (0.98)) ∧ (g(2,2) = 1)) ⇒
  ((information_leakage p X Y = 1) ∧
  (g_information_leakage p X Y g Z = log (2/(Normal(1.98)))))
```

where g refers to the gain function, `conditional_distribution` denotes the transition distribution with respect to Tables 1 and 2 and the term `Normal` is used for the extended real numbers theory.

The proof of this result is conducted using the vulnerability properties, probability reasoning and real analysis. The first part of the theorem is proved using Theorem 4.2. The second part of the goal is verified by computing the values of the initial and remaining vulnerabilities. We find that

$$V_{gd}(X) = \frac{1}{3}\max\{1+0+0, 0+1+0.98, 0+0.98+1\} = 0.66$$

Now for the posterior vulnerability, we calculate the posterior distribution $P(X = x|Y = y_1) = (\frac{1}{3}, \frac{2}{3}, 0)$.

$$V_{gd}(X = x|Y = y_1) = max \left\{ \begin{array}{l} \frac{1}{3}.1 + \frac{2}{3}.0 + 0.0, \\ \frac{1}{3}.0 + \frac{2}{3}.1 + 0.0.98, \\ \frac{1}{3}.0 + \frac{2}{3}.0.98 + 0.1 \end{array} \right\} = \frac{2}{3}$$

Similarly, we prove that $V_{gd}(X = x|Y = y_2) = \frac{2}{3}$ and then by rewriting these values on their corresponding quantities, we get $IL_{gd}(X, Y) = log\frac{\frac{2}{3}}{0.66} \approx log\frac{2}{1.98}$. Here the min-entropy leakage is greater than the g-leakage. Perceptively, this example differentiates between x_2 and x_3. Due to the relations (z_2, x_3) and (z_3, x_2), under the distance gain function, it follows that x_2 and x_3 are so close (a gain of 0.98). Thus the g-vulnerability hardly increases. The proof of this result required 550 lines of HOL4 code [13] and around 60 man-hours in terms of reasoning effort. These results are considered to be accurate and the analysis covers any type of systems (in terms of state space size).

6   Conclusion

This paper presents a formalization of some of the most commonly used properties of information flow in higher-order logic. These properties, depending on the threat model, are based on Shannon entropy and gain min-entropy. This formalization provides a more reliable and richer information flow analysis framework compared to the traditional definitions of quantitative information flow analysis as formalized measures cover a wide variety of threat scenarios. We used the formalized notions of Shannon entropy to verify the Data Processing Inequality, which states that leakage cannot be increased by post processing of the information, and Jensen's Inequality, which is a relation between the averaging and the processing of information flow. The g-leakage is introduced as a generalization of the min-entropy leakage and belief min-entropy leakage to assess the case where the secret is guessed partially using a gain function, which models the benefit that an attacker gets about the secret. Gain functions engender the possibility to cover a variety of operational scenarios. The proposed formalization can be built upon to conduct the information flow analysis within the sound core of a theorem prover and thus the analysis is guaranteed to be free of approximation and precision errors. For illustration purposes, we performed a comparison analysis between the min-entropy leakage and the g-leakage using the HOL4 theorem prover and the analysis results were found to be generic and accurate.

This work is conducted as a formal framework that can be used to formally verify many information flow aspects depending on the threat model. It provides a reasonable foundation for information flow in HOL. Many applications can be analyzed using our formalization, such as the Crowds protocol [25] and Freenets [6]. We are aiming to extend this work for the formal analysis of channel capacity (min-capacity and gain-capacity) and compare them with Shanon capacity. Starting from a specific leakage bound, our work can be used to evaluate the input set based on the output set. This formalization can in turn be used to formally ensure a specific level of security of critical information.

References

1. HOL4, hol.sourceforge.net (2017)
2. Alvim, M.S., Chatzikokolakis, K., Palamidessi, C., Smith, G.: Measuring information leakage using generalized gain functions. In: IEEE Symposium on Computer Security Foundations, pp. 265–279 (2012)
3. Andrea, S.: Possibilistic information theory: a coding theoretic approach. Fuzzy Sets Syst. **132**(1), 11–32 (2002)
4. Beaudry, N.J., Renner, R.: An intuitive proof of the data processing inequality. Quantum Inform. Comput. **12**(5–6), 432–441 (2012)
5. Chung, K.L.: Markov Chains with Stationary Transition Probabilities (1967)
6. Clarke, I., Sandberg, O., Wiley, B., Hong, T.W.: Freenet: a distributed anonymous information storage and retrieval system. In: Federrath, H. (ed.) Designing Privacy Enhancing Technologies. LNCS, vol. 2009, pp. 46–66. Springer, Heidelberg (2001). doi:10.1007/3-540-44702-4_4
7. Clarkson, M.R., Chong, S., Myers, A.C.: Civitas: toward a secure voting system. In: IEEE Symposium on Security and Privacy, pp. 354–368. IEEE Computer Society (2008)
8. Coble, A.R.: Anonymity, Information, and Machine-Assisted Proof. Ph.D. thesis, King's College, University of Cambridge, UK (2010)
9. Cover, T.M., Thomas, J.: Entropy, relative entropy and mutual information. In: Elements of Information Theory. Wiley-Interscience (1991)
10. Dubois, D., Nguyen, H.T., Prade, H.: Possibility theory, probability and fuzzy sets: misunderstandings, bridges and gaps. In: Dubois, D., Prade, H. (eds.) Fundamentals of Fuzzy Sets. The Handbooks of Fuzzy Sets Series, pp. 343–438. Kluwer, Boston (2000)
11. Halpern, J., O'Neill, K.: Anonymity and information hiding in multiagent systems. J. Comput. Secur. **13**(3), 483–514 (2005)

12. Hasan, O., Tahar, S.: Formal verification methods. In: Encyclopedia of Information Science and Technology, pp. 7162–7170. IGI Global Pub. (2015)
13. Helali, G., Dunchev, C., Hasan, O., Tahar, S.: Towards The Quantitative Analysis of Information Flow in HOL, HOL4 code (2017). http://hvg.ece.concordia.ca/projects/prob-it/gainMinEntropy.php
14. Helali, G., Hasan, O., Tahar, S.: Formal analysis of information flow using min-entropy and belief min-entropy. In: Iyoda, J., de Moura, J. (eds.) SBMF 2013. LNCS, vol. 8195, pp. 131–146. Springer, Heidelberg (2013). doi:10.1007/978-3-642-41071-0_10
15. Hölzl, J.: Construction and Stochastic Applications of Measure Spaces in Higher-Order Logic. Ph.D. thesis, Institut für Informatik, Technische Universität München, Germany (2012)
16. Hölzl, J., Heller, A.: Three chapters of measure theory in Isabelle/HOL. In: van Eekelen, M., Geuvers, H., Schmaltz, J., Wiedijk, F. (eds.) ITP 2011. LNCS, vol. 6898, pp. 135–151. Springer, Heidelberg (2011). doi:10.1007/978-3-642-22863-6_12
17. Hua, J., Jing, Y.: On-line payment and security of e-commerce. In: International Conference on Computer Engineering and Applications, pp. 545–550. CEA, WSEAS (2007)
18. Jebara, T., Pentland, A.: On Reversing Jensen's Inequality. In: Advances in Neural Information Processing Systems 13. MIT Press (2000)
19. Chatzikokolakis, K., Palamidessi, C., Panangaden, P.: Anonymity protocols as noisy channels. Inf. Comput. $206(2–4)$, 378–401 (2008)
20. Liu, L.: Formalization of Discrete-time Markov Chains in HOL. Ph.D. thesis, Dept. of Electrical and Computer Engineering, Concordia University, Canada (2013)
21. Mhamdi, T.: Information-Theoretic Analysis using Theorem Proving. Ph.D. thesis, Dept. of Electrical and Computer Engineering, Concordia University, Canada (2012)
22. Mhamdi, T., Hasan, O., Tahar, S.: Formalization of entropy measures in HOL. In: Eekelen, M., Geuvers, H., Schmaltz, J., Wiedijk, F. (eds.) ITP 2011. LNCS, vol. 6898, pp. 233–248. Springer, Heidelberg (2011). doi:10.1007/978-3-642-22863-6_18
23. Mhamdi, T., Hasan, O., Tahar, S.: Quantitative analysis of information flow using theorem proving. In: Aoki, T., Taguchi, K. (eds.) ICFEM 2012. LNCS, vol. 7635, pp. 119–134. Springer, Heidelberg (2012). doi:10.1007/978-3-642-34281-3_11
24. Mhamdi, T., Hasan, O., Tahar, S.: Formalization of measure theory and lebesgue integration for probabilistic analysis in HOL. ACM Trans. Embedded Comput. Syst. $12(1)$ (2013)
25. Reiter, M.K., Rubin, A.D.: Crowds: anonymity for web transactions. ACM Trans. Inform. Syst. Secur. $1(1)$, 66–92 (1998)
26. Rényi, A.: On measures of entropy and information. In: Berkeley Symposium on Mathematics, Statistics and Probability, pp. 547–561 (1961)
27. Sassone, V., ElSalamouny, E., Hamadou, S.: Trust in crowds: probabilistic behaviour in anonymity protocols. In: Wirsing, M., Hofmann, M., Rauschmayer, A. (eds.) TGC 2010. LNCS, vol. 6084, pp. 88–102. Springer, Heidelberg (2010). doi:10.1007/978-3-642-15640-3_7
28. Schneider, S., Sidiropoulos, A.: CSP and anonymity. In: Bertino, E., Kurth, H., Martella, G., Montolivo, E. (eds.) ESORICS 1996. LNCS, vol. 1146, pp. 198–218. Springer, Heidelberg (1996). doi:10.1007/3-540-61770-1_38
29. Smith, G.: On the foundations of quantitative information flow. In: Alfaro, L. (ed.) FoSSaCS 2009. LNCS, vol. 5504, pp. 288–302. Springer, Heidelberg (2009). doi:10.1007/978-3-642-00596-1_21

30. Smith, G.: Quantifying information flow using min-entropy. In: IEEE International Conference on Quantitative Evaluation of Systems, pp. 159–167 (2011)
31. Syverson, P., Goldschlag, D., Reed, M.: Anonymous connections and onion routing. In: IEEE Symposium on Security and Privacy, Oackland, California, pp. 44–54 (1997)
32. Trevathan, J.: Privacy and Security in Online Auctions. Ph.D. thesis, School of Mathematics, Physics and Information Technology, James Cook University, Australia (2007)

Formalizing SPARCv8 Instruction Set Architecture in Coq

Jiawei Wang[1], Ming Fu[1], Lei Qiao[2], and Xinyu Feng[1(✉)]

[1] University of Science and Technology of China, Hefei, China
xyfeng@ustc.edu.cn
[2] Beijing Institute of Control Engineering, Beijing, China

Abstract. The SPARCv8 instruction set architecture (ISA) has been widely used in various processors for workstations, embedded systems, and space missions. In order to formally verify the correctness of embedded operating systems running on SPARCv8 processors, one has to formalize the semantics of SPARCv8 ISA. In this paper, we present our formalization of SPARCv8 ISA, which is faithful to the realistic design of SPARCv8. We also prove the determinacy and isolation properties with respect to the operational semantics of our formal model. In addition, we have verified that a trap handler function handling window overflows satisfies the user's expectations based on our formal model. All of the formalization and proofs have been mechanized in Coq.

Keywords: SPARCv8 · Coq · Verification · Operational semantics

1 Introduction

Computer systems have been widely used in national defense, finance and other fields. Building high-confidence systems plays a significant role in the development of computer systems. Operating system kernel is the most foundational software of computer systems, and its reliability is the key in building high-confidence computer system.

In aerospace and other security areas, the underlying operating system is usually implemented in C and assembly languages. In existing OS verification projects, *e.g.*, Certi*μ*C/OS-II [20] and seL4 [17], the assembly code is usually not modeled in order to simplify the formalization of the target machine. They use abstract specifications to describe the behavior of the assembly code to avoid exposing the details of underlying machines, *e.g.*, register and stack. Therefore, the assembly code in OS kernels is not actually verified. To verify whether the assembly code satisfies its abstract specifications, it is inevitable to formalize the semantics of the assembly instructions.

This work is supported in part by grants from National Natural Science Foundation of China (NSFC) under Grant Nos. 61632005, 61379039 and 61502031.

© Springer International Publishing AG 2017
K.G. Larsen et al. (Eds.): SETTA 2017, LNCS 10606, pp. 300–316, 2017.
https://doi.org/10.1007/978-3-319-69483-2_18

As a highly efficient and reliable microprocessor, the SPARCv8 [6] instruction set architecture has been widely used in various processors for workstations, embedded systems, and space missions. For instance, SpaceOS [19] running on SPARCv8 processors is an embedded operating system developed by Beijing Institute of Control Engineering (BICE) and deployed in the central computer of Chang'e-3 lunar exploration mission. On the one hand, to formally verify SpaceOS, we need to formalize the SPARCv8 instruction set and build the mathematical semantic model of the assembly instructions. On the other hand, to ensure the consistency between the behavior of the target assembly code and the C source code, we hope to use the certified compiler CompCert [18] to compile SpaceOS. However, CompCert only supports translating Clight, which is an important subset of C, into ARM [1], x86 [10], PowerPC [5] instruction set currently. It does not support SPARCv8 at the backend. Extending CompCert to support SPARCv8 requires us to formalize the SPARCv8 instruction set. In this paper, we make the following contributions:

- We formalize the SPARCv8 ISA. Our formal model is faithful to the behaviors of the instructions described in the SPARCv8 manual [7], including most of the features in SPARCv8, *e.g.*, windowed registers, delayed control transfer, interrupts and traps.
- We prove that the operational semantics of our formal model satisfy the determinacy property, and the execution in the user mode or the supervisor mode satisfies the isolation property.
- We take the trap handler for window overflows as an example, and give its pre-condition and post-condition to specify the expected behaviors. Like proving programs with Hoare triples, we prove that the trap handler satisfies the given pre-/post-conditions and does not throw any exceptions.
- All of the formalization and proofs have been mechanized in Coq [2]. They contain around 11000 lines of coq scripts in total. The source code can be accessed via the link [3].

Related Work. Fox and Myreen gave the ARMv7 ISA model [14], they used monadic specification and formalized the instruction decoding and operational semantics. Narges Khakpour et al. proved some security properties of ARMv7 in the proof assistant tool HOL4, including the kernel security property, user mode isolation property, and so on. Andrew Kennedy et al. formalized the subset of x86 in Coq [16], and they used type classes, notations and the mathematics library Ssreflect [8]. The CompCert compiler also has the formal modeling of ARM and x86. There are lots of modeling work related to the x86 and ARM, but due to the specific features of SPARCv8, these x86 and ARM ISA models can not be used directly for the SPARCv8 ISA.

Zhe Hou et al. modeled the SPARCv8 ISA in the proof assistant tool Isabelle [15], which is close to our work. But their work is focused on the SPARCv8 processor itself, instead of the assembly code running on it. To verify the assembly code, we need a better definition on the syntax and operation semantics. And the definition of machine state needs to be hierarchical and easy to use when we verify the code running on it. Additionally, they did not model the interrupt feature in SPARCv8, hence their model could not describe the uncertainty of the operational semantics caused by interrupt. Besides, our formalization of the SPARCv8 ISA is implemented in Coq, while CompCert is implemented in Coq too. We can use our Coq implementation to extend the CompCert at the backend to support SPARCv8 in the future.

There are some other verification work at assembly level [11–13], which give the formal models of different subset of x86 the instruction set and the behavior of the x86 interrupt management. They mainly study the verification technology of x86 assembly code, the instruction set is relatively small. In the meanwhile, the model is simple. We formalized the SPARCV8 ISA by considering all the features of SPARCv8. In the next section, we will give a brief overview of these features.

2   Overview of SPARCv8 ISA

The Scalable Processor Architecture (SPARC) is a reduced instruction set computing (RISC) instruction set architecture (ISA) originally developed by Sun Microsystems [9]. It is widely used in the electronic systems of space devices for its high performance, high reliability and low power consumption. Compared to other architectures, SPARCv8 has the following unique mechanisms:

- A variety of control-transfer instructions (CTIs) and annulled delay instructions for more flexible function jumps.
- The register window and window rotation mechanism for swapping context more efficiently.
- Two modes, user mode and supervisor mode, for separating the application code and operating system code at the physical level.
- A variety of traps for swapping modes through a special trap table that contains the first 4 instructions of each trap handler.
- Delayed-write mechanism for delaying the execution of register write operation for several cycles.

These characteristics pose quite a few challenges for formal modeling. We use the example below to demonstrate the subtle control flow in SPARCv8.

Example. The following function CALLER calls the function SUM3 to add three variables together.

```
CALLER:                              SUM3:
      . . .
1     mov 1, %o0                 6        save %sp, -64, %sp
2     mov 2, %o1                 7        add %i0, %i1, %l7
3     call SUM3                   8        add %l7, %i2, %l7
4     mov 3, %o2                 9        ret
5     mov %o0, %l7              10        restore %l7, 0, %o0
      . . .
```

The function SUM3 requires three input parameters. When the CALLER calls SUM3, it places the first two arguments, then calls SUM3 (Line 3) before placing the third argument (Line 4). In other words, the call instruction will be executed before the mov instruction which places the last argument. The reason is that when we call an another function by using instructions such as call, it will record the address that is going to jump to in the current execution cycle. The real transfer procedure is executed in the next instruction cycle. This feature is called "delayed transfer", which also happens at lines 9 and 10.

In SUM3, we use save and restore instructions (Lines 6 and 10) to save and restore the caller's context. When this program is running, both CALLER and SUM3 have register windows as their contexts, and their windows are overlapping. When the CALLER needs to save the context and pass the parameters to SUM3, it will put the parameters in the overlapping section and rotates the window so that the SUM3's register window is exposed. At this point, the non-overlapping portion of the CALLER's window is hidden. These steps are implemented by the save instruction. When SUM3 needs to pass the return value to the CALLER, it will put the return value in the overlap section and rotate the window to destroy its own space. These steps are implemented by the restore instruction.

The semantics of the delayed transfer and the window rotation mechanism are quite tricky in SPARCv8. In addition, other special mechanisms of SPARCv8 mentioned above are complicated and their behaviors are non-trivial. Therefore, it is necessary to give a formal model of the SPARCv8 ISA, which is the basis of verifying the SPARCv8 code.

3 Modeling SPARCv8 ISA

The SPARCv8 instruction set provides programmers with a hardware-oriented assembly programming language. To formalize it, first we need to provide the abstract syntax of the given language. Then we define the machine state. Finally, we give the operational semantics for the instructions.

3.1 Syntax

Figure 1 shows the syntax of the SPARCv8 assembly language. Here we only give some typical instructions i that show the key features introduced in Sect. 2. **bicca** makes a delayed control transfer if the condition η holds, otherwise it annuls the next instruction and executes the following code (unless η is al, as explained in Sect. 3.3). The conditional expression η can be always (al), equal (eq), not equal (ne), etc.. **save** (or **restore**) saves (or restores) the caller's context by rotating the register window. **ticc** triggers a software trap, and **rett** returns from traps. **wr** writes some specific registers, which are defined as $Symbol$. $Symbol$ contains the processor state register (psr), window invalid mask register (wim), trap base register (tbr), multiply/divide register (y) and ancillary state registers (asr). asr are used to store the processor's ancillary state. The write by **wr** may be delayed for several cycles, as explained in Sects. 3.2 and 3.3.

(*Word*)	w	$\in$	$Int32$	(*OpExp*)	α	$::=$	$r \mid w$
(*GenReg*)	r	$::=$	$r_0 \mid \ldots \mid r_{31}$	(*AddrExp*)	β	$::=$	$\alpha \mid r + \alpha$
(*AsReg*)	asr	$::=$	$asr_0 \mid \ldots \mid asr_{31}$	(*TrapExp*)	γ	$::=$	$r \mid r + r \mid r + w \mid w$
(*Symbol*)	ς	$::=$	$psr \mid wim \mid tbr \mid y \mid asr$	(*TestCond*)	η	$::=$	$al \mid eq \mid ne \mid \ldots$
(*SparcIns*)	i	$::=$	\multicolumn{5}{l	}{$\mathbf{bicca}\ \eta\ \beta \mid \mathbf{save}\ r_s\ \alpha\ r_d \mid \mathbf{restore}\ r_s\ \alpha\ r_d \mid \mathbf{rett}\ \beta \mid$}
			\multicolumn{5}{l	}{$\mathbf{ticc}\ \eta\ \gamma \mid \mathbf{wr}\ r_d\ \alpha\ \varsigma \mid \ldots$}

Fig. 1. The syntax of the SPARCv8 assembly language

The address expressions, operand expressions and trap expressions in these instructions are defined as *OpExp*, *AddrExp* and *TrapExp*. w stands for 32-bit integer constants (*Word*). r stands for general registers (*GenReg*).

Note that the `call`, `mov`, and `ret` instructions in the example in Sect. 2 are not given in the syntax, since they are all synthetic instructions, which can be defined from the basic instructions [7].

3.2   Machine States

Register File. Here we give the definition of the register files (*RegFile*).

$$(\textit{RegName})\ q\ ::=\ r \mid \varsigma \mid pc \mid npc \mid \kappa \mid \tau \qquad (\textit{RegFile})\ R \in \textit{RegName} \rightarrow \textit{Word}$$

We use q to represent the register name (*RegName*), including *GenReg* and *Symbol*, which were explained in Sect. 3.1. It also includes the program counter pc, the next program counter npc, the trap flag τ and annulling flag κ. A register file R is modeled as a total function mapping register names to 32-bit integers.

Program Counters. SPARCv8 uses two program counters, *viz.*, pc and npc to control the execution. pc contains the address of the instruction currently being executed, while npc holds the address of the next instruction (assuming a trap does not occur). The function next below defines the change of program counters when no transfer occurs. It updates pc with npc and increases npc by 4.

$$\mathsf{next}(R)\ \stackrel{def}{=\!=\!=}\ R\{pc \rightsquigarrow R(npc)\}\{npc \rightsquigarrow R(npc) + 4\}$$

If transfer occurs during the instruction execution, for example, if the evaluation of conditional expression returns **true** when we execute the instruction bicca, the function djmp will be executed:

$$\mathsf{djmp}(w, R)\ \stackrel{def}{=\!=\!=}\ R\{pc \rightsquigarrow R(npc)\}\{npc \rightsquigarrow w\}$$

djmp updates pc with npc and sets npc to the target address. As mentioned in the example in Sect. 2, when we call a function, the target address w is stored in npc in the current execution cycle. Because the next instruction is fetched from pc, the transfer is not made immediately and is delayed to the next cycle instead. The *delayed transfer* is applied for all transfer instructions in SPARCv8.

Window Registers. We use the frame and the frame list to describe the window registers and window rotating. The definitions are given as follows:

$$(Frame) \ f ::= [w_0, \ldots, w_7] \qquad (FrameList) \ F ::= \mathbf{nil} \mid f::F$$
$$(RState) \ Q ::= (R, F)$$

A frame is an array that contains 8 words, and a frame list is a list of frames and its length is 2N-3 (N is the number of windows).

We divide the general registers ($r_0 \ldots r_{31}$) in the register file R into four groups, global out, local and in, as shown in Fig. 2(1). They represent the current view of the accessible general registers. There are also registers unaccessible, which are grouped into frames and stored on the frame list. We pair the register file and the frame list together as the register state Q.

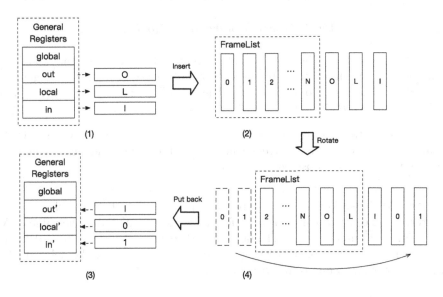

Fig. 2. Left rotation of the window

The view of currently accessible registers can be changed by rotating the window, which exchanges the data between the register file and the frame list. This is done to save and restore execution contexts, as what the `caller` and `sum3` do in the example in Sect. 2. Below we demonstrate the left rotation of the window in Fig. 2. The formal definition is given as left_win(Q) in Fig. 3. The rotation takes the following steps:

- We convert three groups of the general registers (out, local and in) into a frame list consisting of 3 frames, as shown in Fig. 2(1). The conversion is formalized as fetch(R) in Fig. 3.
- As shown in Fig. 2(2) and (4), we can insert these 3 frames at the end of the frame list, then left rotate the frame list.

$$R[r_i, \ldots, r_{i+7}] \overset{def}{=\!=} [R(r_i), \ldots, R(r_{i+7})]$$

$$R\{[r_i, \ldots, r_{i+7}] \rightsquigarrow f\} \overset{def}{=\!=} R\{r_i \rightsquigarrow w_0\} \ldots \{r_{i+7} \rightsquigarrow w_7\}$$
$$\textbf{where } f = [w_0, \ldots, w_7]$$

$$\mathsf{fetch}(R) \overset{def}{=\!=} R[r_8, \ldots, r_{15}] :: R[r_{16}, \ldots, r_{23}] :: R[r_{24}, \ldots, r_{31}] :: \textbf{nil}$$

$$\mathsf{left}(F_L, F_l) \overset{def}{=\!=} (\ F'_L \ ++ \ (p :: q :: \textbf{nil})\ ,\ F'_l \ ++ \ (m :: n :: \textbf{nil})\ )$$
$$\textbf{where } F_L = m :: n :: F'_L, F_l = p :: q :: F'_l$$

$$\mathsf{replace}(l, R) \overset{def}{=\!=} R\{[r_8, \ldots, r_{15}] \rightsquigarrow f_o\}\{[r_{16}, \ldots, r_{24}] \rightsquigarrow f_l\}\{[r_{24}, \ldots, r_{31}] \rightsquigarrow f_i\}$$
$$\textbf{where } l = f_o :: f_l :: f_i :: \textbf{nil}$$

$$\mathsf{left\_win}(Q) \overset{def}{=\!=} \textbf{let } (F', l) := \mathsf{left}(F, \mathsf{fetch}(R)) \textbf{ in}$$
$$\textbf{let } R' := \mathsf{replace}(l, R) \textbf{ in } (R'\{cwp \rightsquigarrow \mathsf{post\_cwp}(R)\}, F')$$
$$\textbf{where } Q = (R, F),\ \mathsf{post\_cwp}(R) \overset{def}{=\!=} (R(cwp) + 1) \textbf{ mod } N$$

Fig. 3. Definition of the window rotation

– Finally, as shown in Fig. 2(4) and (3), we remove 3 frames from the tail of the frame list, and insert them to the corresponding positions in the 32 general registers. The last two steps are modeled as $(F', l) := \mathsf{left}(F, \mathsf{fetch}(R))$ and $R' := \mathsf{replace}(l, R)$ in Fig. 3.

Since the left rotation of the window increases the label of current window by 1, we also need to update the current window pointer (cwp, a segment of psr) to the new value, namely $\mathsf{post\_cwp}$.

Delayed Writes. When we execute the **wr** instruction to write the symbol register, the execution will be delayed for X cycles ($0 \leq X \leq 3$). The value of X is implementation-dependent. The delay list D consists of a sequence of delayed writes d. Each d is a triple consisting of the remaining cycles to be delayed, the target register and value to be written.

$$\begin{array}{ll} (InitDC) \quad X \in [0..3] & (DelayItem) \quad d ::= (c, \varsigma, w) \\ (DelayCycle) \quad c \in [0..X] & (DelayList) \quad D ::= \textbf{nil} \mid d :: D \end{array}$$

There are 2 operations defined on the delay list, as shown bellow.

– When we execute the **wr** instruction, we will insert a delayed write into the delay list using function $\mathsf{set\_delay}$:

$$\mathsf{set\_delay}(\varsigma, w, D) \overset{def}{=\!=} (X, \varsigma, w) :: D$$

– At the beginning of each instruction cycle, we scan the delay list, remove the delayed writes whose delay cycles are 0 and execute them, and then decrement the delay cycles of the remaining delayed writes, as shown below:

$$\text{exe\_delay}(Q, D) \quad \stackrel{def}{=\!=\!=} \quad \begin{cases} (\text{write\_symbol}(\varsigma, w, Q), D') & \text{if } D = (0, \varsigma, w)::D', \\ \text{let } (Q', D'') := \text{exe\_delay}(Q, D') \\ \text{in } (Q', (n-1, \varsigma, w)::D'') & \text{if } D = (n, \varsigma, w)::D', \\ & n \neq 0 \\ (Q, D) & \textbf{otherwise} \end{cases}$$

$$\textbf{where } Q = (R, F)$$

write_symbol writes the value w into the register ς. The details can be found in the technical report [4].

Machine States and Code Heap. We use M to represent the memory (*Memory*), which maps the addresses (*Address*) to words. The full memory is split into two parts for the user mode and the supervisor mode respectively. We formalize the memory as a pair that consists of the user memory M_u and the supervisor memory M_s. The machine state S contains the memory pair Φ, the register state Q and the delay list D.

(*Address*)	a	$\in$	$Word$	(*MemPair*) Φ	$::= (M_u, M_s)$
(*Memory*)	M	$\in$	$Address \rightharpoonup Word$	(*State*) S	$::= (\Phi, Q, D)$

Besides the machine state, we also define the code heap C, the pair of code heap Δ and the event e, shown as below.

(*Label*)	l	$\in$	$Word$	(*World*) W	$::= (\Delta, S)$
(*CodeHeap*)	C	$\in$	$Label \rightharpoonup SparcIns$	(*Event*) e	$::= w \mid \bot$
(*CodePair*)	Δ	$::=$	(C_u, C_s)	(*EventList*) E	$::= \textbf{nil} \mid e::E$

C represents the code heap, which maps the labels to the instructions. The code heap of user mode and supervisor mode together form the pair of code heap Δ. The whole world W consists of the code heap Δ of two modes and the machine state S. e stands for events. If a trap occurs, the corresponding trap label w is recorded as an event, otherwise it is $\bot$. An event list E is introduced for producing events of the multi-step execution.

3.3   Operational Semantics

We define the operational semantics with multiple layers as shown in Fig. 4, where the main features of SPARCv8 are introduced at different layers. This layered operational semantics is good for our verification work, for example, when we verify some instructions such as **bicca**, **ticc**, *etc.*, we will only consider the register file and memory. If we put the exposed window register and the hidden window register on the same layer as [7] or [15] does, all the registers will always show up in the verification process.

In Fig. 4, from the top to the bottom, we first define the operational semantics of some simple instructions which only access the register file and memory using the transition $(M, R) \stackrel{i}{\longrightarrow} (M', R')$.

$$(M, R) \xrightarrow{\ i\ } (M', R')$$ Simple Instructions

$$\Downarrow$$

$$(M, Q, D) \circ\!\!\xrightarrow{\ i\ } (M', Q', D')$$ Window Registers and Delayed Write

$$\Downarrow$$

$$C \vdash (M, Q, D) \bullet\!\!\longrightarrow (M', Q', D')$$ Executing Delay and Handling Annulling Flag

$$\Downarrow$$

$$\Delta \vdash S \overset{e}{=\!\!=\!\!=\!\!\Longrightarrow} S'$$ Interrupts, Traps and Mode Switch

Fig. 4. The structure of operational semantics

Secondly, we lift the first layer and give the operational semantic of specific instructions about the window register and delayed write features using the transition $(M, Q, D) \circ\!\!\xrightarrow{\ i\ } (M', Q', D')$.

Thirdly, we use the transition $C \vdash (M, Q, D) \bullet\!\!\longrightarrow (M', Q', D')$ to define the operational semantics of delay execution and annulling flag handling.

Finally, we give the operational semantic rules of interrupt, trap execution and mode switch as the transition $\Delta \vdash S \overset{e}{=\!\!\Longrightarrow} S'$, which defines the whole behavior of the entire program. Next, we will introduce some rules of the operational semantics of each layer. The omitted rules can be found in the technical report [4] and our Coq implementations [3].

Simple Instructions. The **bicca** η β instruction evaluates the address expression β to get the value w, and requires the address w to be *word-aligned*. It decides whether to transfer and whether to annul the next instruction by the conditional expression.

- If the value of the conditional expression is *false*, it makes no transfer and *sets the annulling flag* only.

$$\frac{[\![\beta]\!]_R = w \quad \text{word\_aligned}(w) \quad [\![\eta]\!]_R = false}{(M, R) \xrightarrow{\ \textbf{bicca}\ \eta\ \beta\ } (M, \text{set\_annul}(\text{next}(R)))} \ (\text{BICCA-FALSE})$$

- If the type of the conditional expression is not *al* and the value is *true*, it executes the delayed transfer but does not annul the next intruction.

$$\frac{[\![\beta]\!]_R = w \quad \text{word\_aligned}(w) \quad \eta \neq al \quad [\![\eta]\!]_R = true}{(M, R) \xrightarrow{\ \textbf{bicca}\ \eta\ \beta\ } (M, \text{djmp}(w, R))} \ (\text{BICCA-TRUE})$$

- If the type of the conditional expression is *al*, we will execute the delayed transfer and set the annulling flag. The rule is omitted here (see TR [4]).

Recall the definition of next and djmp in Sect. 3.2. Other functions not defined here can be found in the TR [4].

ticc η γ evaluates the trap expression γ. If the condition η is true, it *sets the trap flag* (set_user_trap). We use $w_{<6:0>}$ to represent the lowest 7 bits of w. **ticc** η γ does nothing if η is false (the corresponding rule omitted).

$$\frac{[\![\,\gamma\,]\!]_R = w \qquad [\![\,\eta\,]\!]_R = true}{(M, R) \xrightarrow{\text{ticc } \eta \ \gamma} (M, \text{set\_user\_trap}(w_{<6:0>}, R))} \quad \text{(TICC-TRUE)}$$

Rules for other simple instructions are given in the TR [4].

Window Registers and Delayed Writes. Here we give semantics for instructions that manipulate the frame list and delay list. First, we use the LIFT1 rule to lift the transition $(M, R) \xrightarrow{i} (M', R')$ to $(M, Q, D) \overset{i}{\circ\!\!\longrightarrow} (M', Q', D')$.

$$\frac{(M, R) \xrightarrow{i} (M', R')}{(M, (R, F), D) \overset{i}{\circ\!\!\longrightarrow} (M', (R', F), D)} \quad \text{(LIFT1)}$$

When the **wr** r_d α ς instruction is executed in user mode (usr_mode), since it does not have permissions for access the register wim, tbr and psr, so ς must be y or asr_i. Then it executes the XOR operation of α and r_d to get the value w. Next we insert the triple (X, ς, w) into the delay list D using the function set_delay (see Sect. 3.2). Finally, it resets pc and npc with the function next.

$$\text{inc\_win}(Q) \overset{def}{=\!=\!=} \begin{cases} \text{left\_win}(Q) & \text{if } \neg\text{win\_masked}(\text{post\_cwp}(R), R) \\ \bot & \textbf{otherwise} \end{cases}$$
$$\textbf{where } Q = (R, F), \text{win\_masked}(w, R) \overset{def}{=\!=\!=} 2^w \&\& R(wim) \neq 0$$

$$\text{rett\_f}(Q) \overset{def}{=\!=\!=} \begin{cases} (\text{restore\_mode}(\text{enable\_trap}(R')), F') & \text{if } \text{inc\_win}(Q) = (R', F') \\ \bot & \textbf{otherwise} \end{cases}$$

$$\text{exe\_trap}(Q) \overset{def}{=\!=\!=} \begin{cases} \textbf{let } (R', F') := \text{right\_win}(Q) \textbf{ in} \\ \textbf{let } R'' := \text{to\_sup}(\text{save\_mode}(\text{disable\_trap}(R'))) \textbf{ in} \\ (\text{tbr\_jmp}(\text{clear\_trap}(\text{save\_pc\_npc}(r_{17}, r_{18}, R''))), F') \\ \qquad\qquad\qquad\qquad\qquad\quad \text{if trap\_enabled}(R) \\ \bot \qquad\qquad\qquad\qquad\qquad\quad \textbf{otherwise} \end{cases}$$
$$\textbf{where } Q = (R, F)$$

$$\text{tbr\_jmp}(R) \overset{def}{=\!=\!=} R\{pc \rightsquigarrow R(tbr)\}\{npc \rightsquigarrow R(tbr) + 4\}$$

$$\text{save\_pc\_npc}(r_m, r_n, R) \overset{def}{=\!=\!=} \begin{cases} R\{r_m \rightsquigarrow R(pc)\}\{r_n \rightsquigarrow R(npc)\} & \text{if } \neg\text{annuled}(R) \\ \text{clear\_annul}(R\{r_m \rightsquigarrow R(npc)\} \\ \qquad\qquad \{r_n \rightsquigarrow R(npc + 4)\}) & \textbf{otherwise} \end{cases}$$

Fig. 5. Auxiliary definitions

$$\frac{\mathsf{usr\_mode}(R) \quad \varsigma = y \text{ or } asr_i \quad [\![\, r_d \,]\!]_R \text{ xor } [\![\, \alpha \,]\!]_R = w \quad D' = \mathsf{set\_delay}(\varsigma, w, D)}{(M, (R, F), D) \circ\!\xrightarrow{\ \mathbf{wr}\ r_d\ \alpha\ \varsigma\ } (M, (\mathsf{next}(R), F), D')} \text{(WRUSR)}$$

When the **wr** $r_d\ \alpha\ \varsigma$ instruction is executed in supervisor mode, it has fully access to all symbol registers. The rule of it is given in the TR [4].

For the rules SAVE and RESTORE, we first decrease or increase the label of the window using the function dec_win or inc_win. Then we evaluate the operand expression α to get the value a. Next we assign the value of $[\![\, r_s \,]\!]_R + a$ to r_d. The definition of the inc_win can be found in Fig. 5. The increasing operation is allowed if the post-window (post_cwp) is not masked ($\neg$win_masked). The function dec_win is similar to the inc_win and therefore it is not given here.

$$\frac{\mathsf{dec\_win}(R, F) = (R', F') \quad [\![\, \alpha \,]\!]_R = a \quad R'' = R'\{r_d \rightsquigarrow [\![\, r_s \,]\!]_R + a\}}{(M, (R, F), D) \circ\!\xrightarrow{\ \mathbf{save}\ r_s\ \alpha\ r_d\ } (M, (\mathsf{next}(R''), F), D)} \text{(SAVE)}$$

$$\frac{\mathsf{inc\_win}(R, F) = (R', F') \quad [\![\, \alpha \,]\!]_R = a \quad R'' = R'\{r_d \rightsquigarrow [\![\, r_s \,]\!]_R + a\}}{(M, (R, F), D) \circ\!\xrightarrow{\ \mathbf{restore}\ r_s\ \alpha\ r_d\ } (M, (\mathsf{next}(R''), F'), D)} \text{(RESTORE)}$$

For the rule RETT, we first require that the trap is not enabled ($\neg$trap_enabled) and the system is in supervisor mode (sup_mode). Then we evaluate the address expression β to get the value w and require the address w to be *word-aligned*. Then we increase the label of the window (inc_win), enable the trap (enable_trap) and restore the previous mode (restore_mode) by using function rett_f, which is defined in Fig. 5. Finally, the system transfers to the address w by using djmp.

$$\frac{\begin{array}{ccc} \neg\mathsf{trap\_enabled}(R) & \mathsf{sup\_mode}(R) & [\![\, \beta \,]\!]_R = w \\ \mathsf{word\_aligned}(w) & \mathsf{rett\_f}(R, F) = (R', F') \end{array}}{(M, (R, F), D) \circ\!\xrightarrow{\ \mathbf{rett}\ \beta\ } (M, (\mathsf{djmp}(w, R'), F'), D)} \text{(RETT)}$$

Exceptions. If some of the conditions (e.g., *word-aligned*) in the above rules are not satisfied, the system will throw exceptions. Exceptions include traps and abortions. Traps such as divided by zero, memory not aligned, window overflow, and so on, will put the trap type label into the trap type register (a segment of tbr), then the system will execute this trap in the next cycle (see the explanation below). The abortions make the system to get stuck.

Executing Delay and Handling Annulling Flag. Here we check the delay list and handle the annulling flag.

- The exe_delay function executes the delayed writes (as described in Sect. 3.2). If the annulled flag has not been setted ($\neg$annulled), it will pick up an instruction from the code heap and execute it.

$$\frac{\mathsf{exe\_delay}(Q, D) = (Q', D') \quad \neg\mathsf{annulled}(Q')}{C(Q'.pc) = i \quad (M, Q', D') \circ\!\xrightarrow{\ i\ } (M', Q'', D'')}{C \vdash (M, Q, D) \bullet\!\longrightarrow (M', Q'', D'')}$$

- Otherwise, If the annulled flag has been setted (annulled), it will skip one instruction and unset the annulling flag (clear_annul).

$$\frac{\text{exe\_delay}(Q, D) = (Q', D') \qquad \text{annulled}(Q') \qquad \text{next}(\text{clear\_annul}(Q')) = Q''}{C \vdash (M, Q, D) \bullet\!\!\longrightarrow (M, Q'', D')}$$

Interrupts, Traps and Mode Switch. In each instruction cycle, we deal with interrupts and traps first.

- If there is an interrupt request with level w and it is allowed (interrupt), the system triggers a trap after this external interrupt happens. It will record the trap type (get_tt) and execute this trap (exe_trap), then it will dispatch an instruction. The definition of interrupt and get_tt can be found in the technical report [4].

$$\frac{\text{interrupt}(w, Q) = Q' \qquad \text{get\_tt}(Q') = w' \qquad \text{exe\_trap}(Q') = Q''}{C_s \vdash (M_s, Q'', D) \bullet\!\!\longrightarrow (M'_s, Q''', D')}{(C_u, C_s) \vdash ((M_u, M_s), Q, D) \overset{w'}{=\!=\!\Longrightarrow} ((M_u, M'_s), Q''', D')}$$

When we execute the trap (exe_trap), first we need to make sure that the system allows traps to occur (trap_enabled). Then we rotate the window to the right (right_win), forbid traps to occur (disable_trap), save the current mode (save_mode) and enter the supervisor mode (to_sup). Finally we save pc and npc to register r_{17} and r_{18} by using function save_pc_npc, unset the trap flag (clear_trap) and jump to the address of the trap handler (tbr_jmp). Function exe_trap, tbr_jmp and save_pc_npc are defined in Fig. 5. The function right_win is similar to the left_win and therefore it is not given here.

- If the system has a trap, it will record and execute this trap, and then dispatch an instruction.

$$\frac{\text{has\_trap}(Q) \qquad \text{get\_tt}(Q) = w \qquad \text{exe\_trap}(Q) = Q'}{C_s \vdash (M_s, Q', D) \bullet\!\!\longrightarrow (M'_s, Q'', D')}{(C_u, C_s) \vdash ((M_u, M_s), Q, D) \overset{w}{=\!=\!\Longrightarrow} ((M_u, M'_s), Q'', D')}$$

- If the system does not have traps, it will select the code heap and the memory according to the mode (usr_mode or sup_mode) and dispatch an instruction.

$$\frac{\neg\text{has\_trap}(Q) \qquad \text{usr\_mode}(Q) \qquad C_u \vdash (M_u, Q, D) \bullet\!\!\longrightarrow (M'_u, Q', D')}{(C_u, C_s) \vdash ((M_u, M_s), Q, D) =\!=\!\Longrightarrow ((M'_u, M_s), Q', D')}$$

$$\frac{\neg\text{has\_trap}(Q) \qquad \text{sup\_mode}(Q) \qquad C_s \vdash (M_s, Q, D) \bullet\!\!\longrightarrow (M'_s, Q', D')}{(C_u, C_s) \vdash ((M_u, M_s), Q, D) =\!=\!\Longrightarrow ((M_u, M'_s), Q', D')}$$

Multi-step Execution. In a single step, the system changes from the state S to the state S' and produces an event e. The event e is used to record whether the system has a trap in an instruction cycle. If the trap occurs, it will record the trap type into the event list. Otherwise e is $\bot$ for steps where no trap occurs. The transition of zero-or-multiple steps is defined as below.

$$\frac{}{\Delta \vdash S \overset{nil}{\Longrightarrow}^0 S} \qquad \frac{\Delta \vdash S \overset{e}{\Longrightarrow} S'' \quad \Delta \vdash S'' \overset{E}{\Longrightarrow}^n S'}{\Delta \vdash S \overset{e::E}{\Longrightarrow}^{n+1} S'}$$

4 Determinacy and Isolation Properties

In this section, we will prove that our formal model satisfies the determinacy and isolation properties. The determinacy property explains that the execution of the machine is deterministic with the given sequence of external interrupts. The isolation property characterizes separation of the memory space of the user mode and the supervisor mode, which guarantees the space security of the entire system.

We use $\Delta \vdash S \overset{E}{\Longrightarrow}^* S_1$ to represent zero-or-multiple steps of the execution under the given sequence of external interrupts E. Theorem 4.1 says that, if two executions start from the same initial states and both of them produce the same sequence of external interrupts, then they should arrive at the same final states.

Theorem 4.1 (Determinacy). *If* $\Delta \vdash S \overset{E}{\Longrightarrow}^* S_1$, $\Delta \vdash S \overset{E}{\Longrightarrow}^* S_2$, *then* $S_1 = S_2$. *where* $\Delta \vdash S \overset{E}{\Longrightarrow}^* S'$ *is defined as* $\exists n, \Delta \vdash S \overset{E}{\Longrightarrow}^n S'$.

In SPARCv8 ISA, triggering a trap is the only way of switching to the supervisor mode. We will prove this property first. That is, if a system is running in the user mode at the beginning, it will run in the user mode forever if there is no trap. First, we give the conditions of running n steps in user mode as below:

$$\Delta \vdash S \overset{}{\bullet\!\!\Longrightarrow}^n S' \overset{def}{=\!\!=\!\!=} \mathsf{usr\_mode}(S) \wedge \mathsf{empty\_DL}(S) \wedge \Delta \vdash S \overset{E}{\Longrightarrow}^n S' \\ \wedge \mathsf{no\_trap\_event}(E)$$

where

$$\mathsf{empty\_DL}(S) \overset{def}{=\!\!=\!\!=} D = \mathbf{nil} \quad \text{where } S = ((M_u, M_s), Q, D) \\ \mathsf{no\_trap\_event}(E) \overset{def}{=\!\!=\!\!=} \forall e \in E, e = \bot$$

We first require the system to be in the user mode initially (usr_mode). Second, because of the delayed write feature, we need to require the delay list to be empty (empty_DL), otherwise the system may enter the supervisor mode if there is a delayed write item in the delay list that will modify the S segment of PSR. Finally, we require there is no trap in the system after several steps (no_trap_event).

After giving these conditions, we need to prove that the system is always running in the user mode under these conditions, as shown in Theorem 4.2:

Theorem 4.2 (In User Mode). *If* $\Delta \vdash S \bullet\!\!\Longrightarrow^n S'$, *then* usr_mode($S'$).

It says that, if the system satisfies all the conditions defined in $\Delta \vdash S \bullet\!\!\Longrightarrow^n S'$, it will be in the user mode after n steps. Since this theorem is true for all n, the system should be in the user mode after arbitrary steps. So we can call $\Delta \vdash S \bullet\!\!\Longrightarrow^n S'$ as "the system is running in the user mode for n-steps". This property will be used in proving the isolation property later.

Based on Theorem 4.2, we apply it to prove if a system is running in user mode, it does not have the permission to read and write the resource that belongs to the supervisor mode. The isolation property is shown bellow:

Theorem 4.3 (Write Isolation). *If* $\Delta \vdash S \bullet\!\!\Longrightarrow^n S'$, *then* sup_part_eq($S, S'$)

where

$$\text{sup\_part\_eq}(S, S') \stackrel{def}{=\!=} M_s = M_s'$$
$$\textbf{where } S = ((M_u, M_s), Q, D) \ , \ S' = ((M_u', M_s'), Q', D')$$

Theorem 4.4 (Read Isolation). *If* usr_code_eq(Δ_1, Δ_2), usr_state_eq(S_1, S_2), *and* $\Delta_1 \vdash S_1 \bullet\!\!\Longrightarrow^n S_1'$, $\Delta_2 \vdash S_2 \bullet\!\!\Longrightarrow^n S_2'$, *then* usr_state_eq($S_1', S_2'$)

where

$$\text{usr\_state\_eq}(S, S') \stackrel{def}{=\!=} Q = Q' \wedge M_u = M_u'$$
$$\textbf{where } S = ((M_u, M_s), Q, D) \ , \ S' = ((M_u', M_s'), Q', D')$$
$$\text{usr\_code\_eq}(\Delta, \Delta') \stackrel{def}{=\!=} C_u = C_u'$$
$$\textbf{where } \Delta = (C_u, C_s) \ , \ \Delta' = (C_u', C_s')$$

Theorem 4.3 shows that if the system is running in the user mode, it does not modify the resource that belongs to the supervisor mode. Theorem 4.4 shows that if a particular part of two systems are the same at the beginning, they will always be the same when the system is running in the user mode for several steps. The above two theorems show the isolation property of SPARCv8.

5   Verifying a Window Overflow Trap Handler

In this section, we verify a trap handler, which is used to handle exception of the window overflow. The number of windows provided by SPARCv8 is finite. If we execute the **save** instruction to save the context when all the windows have already been used, it will cause a window overflow trap. The window overflow trap handler will be executed to handle the trap. We give the code of the trap handler as below.

First, the handler takes the next window as the masked window, which is implemented by loop shift operation (Lines 1–5 and 7–10). Then the pointer (named *cwp*, a segment of *psr*) that always points to the current window points to the next window, and we store the value of the current window into the memory (Lines 6 and 11–26). Finally, the handler restores *cwp* and returns (Lines 27–30). The window overflow trap handler saves the oldest element of the window into the memory and makes the window available for the upcoming **save** operation.

```
WINDOW OVERFLOW:
    1    mov %wim,%13              16    st %15,[%sp+20]
    2    mov %g1,%17               17    st %16,[%sp+24]
    3    srl %13,1,%g1             18    st %17,[%sp+28]
    4    sll %13,NWINDOWS-1,%14    19    st %i0,[%sp+32]
    5    or %14,%g1,%g1            20    st %i1,[%sp+36]
    6    save                     21    st %i2,[%sp+40]
    7    mov %g1,%wim              22    st %i3,[%sp+44]
    8    nop                      23    st %i4,[%sp+48]
    9    nop                      24    st %i5,[%sp+52]
   10    nop                      25    st %i6,[%sp+56]
   11    st %10,[%sp+0]           26    st %i7,[%sp+60]
   12    st %11,[%sp+4]           27    restore
   13    st %12,[%sp+8]           28    mov %17,%g1
   14    st %13,[%sp+12]          29    jmp %11
   15    st %14,[%sp+16]          30    rett %12
```

To verify this window overflow trap handler, we first need to give its specifications, namely, the precondition and the postcondition shown as below:

$$\mathsf{overflow\_pre\_cond}(W) \stackrel{def}{=\!=} \mathsf{single\_mask}(R(cwp), R(wim)) \wedge \mathsf{handler\_context}(R)$$
$$\wedge \mathsf{normal\_cursor}(R) \wedge \mathsf{align\_context}(Q) \wedge$$
$$\mathsf{set\_function}(R(pc), \mathsf{windowoverflow}, C_s) \wedge$$
$$D = \mathbf{nil} \wedge \mathsf{length}(F) = 2N - 3$$
$$\mathbf{where}\ W = (\Delta, (\Phi, Q, D)),\ \Delta = (C_u, C_s),\ Q = (R, F)$$
$$\mathsf{overflow\_post\_cond}(W) \stackrel{def}{=\!=} \mathsf{single\_mask}(\mathsf{pre\_cwp}(2, R), R(wim))$$
$$\mathbf{where}\ W = (\Delta, (\Phi, Q, D)), Q = (R, F)$$

In the pre-condition, $\mathsf{single\_mask}(w, R(wim))$ indicates that the system simply masks the window w, and the rest of the window is all available. $\mathsf{pre\_cwp}(n, R)$ gives the window in front of the current window and the distance of them is n. $\mathsf{handler\_context}$ contains the unique state of the system after the $\mathsf{exe\_trap}$ function is executed. For example, the system must be in the supervisor mode, the trap must be disabled, and so on. $\mathsf{normal\_cursor}$ and $\mathsf{handler\_context}$ illustrate the requirements for pc and npc before entering the overflow trap handler. $\mathsf{align\_context}$ requires the address to be *word-aligned*. The rest gives the requirements for the delay list and the frame list.

In the post-condition, when we finish running the trap handler and return to the original function where the trap occurs, cwp will point to the window used by the original function. At this point, the next window is no longer masked, which means the next window is available. In SPARCv8, when we execute the save instruction and enter the next window, the label of the window is decreased. So we use $\mathsf{pre\_cwp}(2, R)$ to represent the label of the window that has been masked, which also means that we have an available window now.

Then we verify the correctness of the handler by showing that the handler can be safely executed under the given pre-condition. As shown in Theorem 5.1, it says, if the initial state satisfies the precondition, then we can safely execute to a resulting state satisfying the postcondition within 30 steps, and no trap occurs during the execution. More details about the specification and proofs can be found in the technical report [4] and our Coq implementation [3].

Theorem 5.1 (Correctness of the Window Overflow Trap Handler).
If overflow_pre_cond(Δ, S), *then forall S' and E, if $\Delta \vdash S \overset{E}{\Longrightarrow}{}^{30} S'$, then* overflow_post_cond(Δ, S') *and* no_trap_event(E).

6 Conclusion and Future Work

In this paper, we have formalized the SPARCv8 instruction set in Coq, which provides the formal model for verifying SpaceOS at the assembly level. Also the formalization can help us to add SPARCv8 into the backend of CompCert in the future. Since the correctness and availability are also critical in formal modeling of SPARCv8, we prove the determinacy and isolation properties to validate the model, and we also verify the window overflow handler to show the availability of our formalization.

For the future work, we will give the syntax and operational semantics of the remaining instructions, including integer arithmetic instructions, floating point instructions, and coprocessor instructions. To facilitate the code verification process, we will develop a program logic for reasoning about the assembly code, instead of doing verification in terms of the operational semantics directly. We hope to extend CompCert backend to support the SPARCv8 assembly language.

References

1. Arm architecture. https://en.wikipedia.org/wiki/ARM_architecture
2. The coq proof assistant. https://coq.inria.fr
3. Formalizing sparcv8 instruction set architecture in coq (project code). https://github.com/wangjwchn/sparcv8-coq
4. Formalizing sparcv8 instruction set architecture in coq (technical report). https://wangjwchn.github.io/pdf/sparc-coq-tr.pdf
5. Powerpc. https://en.wikipedia.org/wiki/PowerPC
6. Sparc. https://en.wikipedia.org/wiki/SPARC
7. The sparc architecture manual v8. http://gaisler.com/doc/sparcv8.pdf
8. Ssreflect. http://ssr.msr-inria.inria.fr
9. Sun microsystems. https://en.wikipedia.org/wiki/Sun_Microsystems
10. x86. https://en.wikipedia.org/wiki/X86
11. Feng, X., Shao, Z.: Modular verification of concurrent assembly code with dynamic thread creation and termination. In: International Conference on Functional Programming (ICFP), pp. 254–267. ACM (2005)

12. Feng, X., Shao, Z., Dong, Y., Guo, Y.: Certifying low-level programs with hardware interrupts and preemptive threads. In: Conference on Programming Language Design and Implementation (PLDI), pp. 170–182. ACM (2008)

13. Feng, X., Shao, Z., Vaynberg, A., Xiang, S., Ni, Z.: Modular verification of assembly code with stack-based control abstractions. In: Conference on Programming Language Design and Implementation (PLDI), pp. 401–414. ACM (2006)

14. Fox, A., Myreen, M.O.: A trustworthy monadic formalization of the ARMv7 instruction set architecture. In: Kaufmann, M., Paulson, L.C. (eds.) ITP 2010. LNCS, vol. 6172, pp. 243–258. Springer, Heidelberg (2010). doi:10.1007/978-3-642-14052-5_18

15. Hou, Z., Sanan, D., Tiu, A., Liu, Y., Hoa, K.C.: An executable formalisation of the SPARCv8 instruction set architecture: a case study for the LEON3 processor. In: Fitzgerald, J., Heitmeyer, C., Gnesi, S., Philippou, A. (eds.) FM 2016. LNCS, vol. 9995, pp. 388–405. Springer, Cham (2016). doi:10.1007/978-3-319-48989-6_24

16. Kennedy, A., Benton, N., Jensen, J.B., Dagand, P.E.: Coq: the world's best macro assembler?. In: Proceedings of the 15th Symposium on Principles and Practice of Declarative Programming (PPDP), pp. 13–24. ACM (2013)

17. Klein, G., Elphinstone, K., Heiser, G., Andronick, J., Cock, D., Derrin, P., Elkaduwe, D., Engelhardt, K., Kolanski, R., Norrish, M., et al.: sel4: formal verification of an os kernel. In: Proceedings of the ACM SIGOPS 22nd Symposium on Operating Systems Principles (SOSP), pp. 207–220. ACM (2009)

18. Leroy, X.: Formal certification of a compiler back-end, or: programming a compiler with a proof assistant. In: 33rd Symposium Principles of Programming Languages (POPL), pp. 42–54. ACM (2006)

19. Qiao, L., Yang, M., Gu, B., Yang, H., Liu, B.: An embedded operating system design for the lunar exploration rover. In: Proceedings of the 2011 Fifth International Conference on Secure Software Integration and Reliability Improvement-Companion (SSIRI-C), pp. 160–165. IEEE Computer Society (2011)

20. Xu, F., Fu, M., Feng, X., Zhang, X., Zhang, H., Li, Z.: A practical verification framework for preemptive OS kernels. In: Chaudhuri, S., Farzan, A. (eds.) CAV 2016. LNCS, vol. 9780, pp. 59–79. Springer, Cham (2016). doi:10.1007/978-3-319-41540-6_4

Tools

How to Efficiently Build a Front-End Tool for UPPAAL: A Model-Driven Approach

Stefano Schivo[1], Buğra M. Yildiz[1], Enno Ruijters[1(✉)], Christopher Gerking[2],
Rajesh Kumar[1], Stefan Dziwok[3], Arend Rensink[1], and Mariëlle Stoelinga[1]

[1] Formal Methods and Tools, University of Twente, Enschede, The Netherlands
{s.schivo,b.m.yildiz,e.j.j.ruijters,r.kumar,
a.rensink,m.i.a.stoelinga}@utwente.nl
[2] Software Engineering, Heinz Nixdorf Institute,
Paderborn University, Paderborn, Germany
christopher.gerking@upb.de
[3] Software Engineering, Fraunhofer IEM, Paderborn, Germany
stefan.dziwok@iem.fraunhofer.de

Abstract. We propose a model-driven engineering approach that facilitates the production of tool chains that use the popular model checker UPPAAL as a back-end analysis tool. In this approach, we introduce a metamodel for UPPAAL's input model, containing both timed-automata concepts and syntax-related elements for C-like expressions. We also introduce a metamodel for UPPAAL's query language to specify temporal properties; as well as a metamodel for traces to interpret UPPAAL's counterexamples and witnesses. The approach provides a systematic way to build *software bridging tools* (i.e., tools that translate from a domain-specific language to UPPAAL's input language) such that these tools become easier to debug, extend, reuse and maintain. We demonstrate our approach on five different domains: cyber-physical systems, hardware-software co-design, cyber-security, reliability engineering and software timing analysis.

1 Introduction

UPPAAL [3] is a leading model checker for real-time systems, allowing one to verify automatically whether a system meets its timing requirements. UPPAAL and its extensions have been applied to a large number of domains, ranging from communication protocols [28], over planning [4] to systems biology [31]. As such, UPPAAL is a popular back-end for various other real-time analysis tools, such as ANIMO [31], sdf2ta [13] and STATE [19]. Typically such tools take their inputs in a domain-specific language (DSL) and translate these inputs into timed automata, which are then fed into UPPAAL to perform the analysis. In this way, domain experts can write their models in a DSL that they are familiar with, while still using UPPAAL's powerful analysis algorithms behind the scenes.

A disadvantage of this approach is, however, that the tools that translate from a DSL to UPPAAL's input language, i.e., *software bridging tools*, are often implemented ad hoc, and hence difficult to debug, reuse, extend and maintain.

© Springer International Publishing AG 2017
K.G. Larsen et al. (Eds.): SETTA 2017, LNCS 10606, pp. 319–336, 2017.
https://doi.org/10.1007/978-3-319-69483-2_19

To overcome this problem, we advocate to develop these tools with *model-driven engineering* (MDE) techniques, which studies [26] have demonstrated can lead to faster software development, with higher levels of interoperability and lower cost. MDE is an approach that uses models as first-class citizens, rather than as by-product of intermediate steps. In MDE, a *metamodel* captures core concepts and behavior of a certain domain. Then, domain-specific models are instances of this metamodel and can be transformed to other models, formats or formalisms via *model transformations.*

In this paper, we propose an MDE approach for tools that use UPPAAL as a back-end. In the context of our approach, we introduce metamodels for UPPAAL timed automata, UPPAAL's query language and its diagnostic traces, in order to transform the domain-specific models to UPPAAL, analyze them and transform the results back to a domain-specific representation, respectively. Our metamodels also support UPPAAL's extensions with cost [4] and probability [7].

We show our approach on five diverse application domains: cyber-physical systems, namely, coordination protocols of MECHATRONICUML; hardware-software co-design, namely, scheduling of synchronous dataflow graphs; cyber-security, namely, analysis of attack trees; reliability engineering, namely, analysis of fault trees; and software timing analysis, namely, timing analysis of Java applications.

Our contributions. To summarize, our main contribution is an MDE approach for building software bridging tools around the UPPAAL model checker. Concretely, we introduce (1) metamodels[1] for UPPAAL's timed automata, queries and traces, providing all the ingredients needed to construct UPPAAL models, verify relevant properties and interpret the results; (2) model transformations from several domain-specific models to the UPPAAL models and back; and (3) five case studies demonstrating how the approach is applied in practice and supports a wide range of application domains.

Overview of our MDE approach. The proposed approach can be seen in Fig. 1. Taking into consideration the analysis of a (generic) domain-specific model, the most important steps involving a bridging software tool that implements our approach are the following:

- In *Step 1*, a domain-specific model is generated/created by the domain expert. This model is an instance of the metamodel of a particular domain of interest. Such a metamodel defines the concepts and their relationships in that domain. For some domains, it may be more convenient to define multiple related metamodels targeting distinct concerns.
- In *Step 2*, the domain-specific model is transformed to a timed-automata model, conforming to the UPPAAL Timed Automata metamodel (UTA) we propose as part of the contribution of this paper. A snippet of such a transformation can be found in Fig. 7.

---

[1] The metamodels are available at https://github.com/uppaal-emf/uppaal.

– In *Step 3*, the property against which the domain-specific model is to be
checked is specified in a query language specific to the domain.

– In *Step 4*, the query specified in the domain-specific query language is transformed to a corresponding UPPAAL query, in turn conforming to the UPPAAL
Query metamodel (UQU) we propose as part of the contribution of this paper.

– In *Step 5*, UPPAAL checks if the timed-automata model (a UTA model) satisfies
the property specified by the generated query (a UQU model). The result of
this operation is usually a diagnostic trace. As part of this step, the UTA and
UQU models are transformed into the native UPPAAL input formats; moreover,
the diagnostic trace natively produced by UPPAAL is transformed into yet
another model, conforming to the UPPAAL Trace metamodel (UTR) that we
also propose as part of the contribution of this paper.

– In *Step 6*, the UTR model is transformed back to a domain-specific representation. This representation can conform to a metamodel that is designed to
express the analysis results in an understandable way by the domain experts.

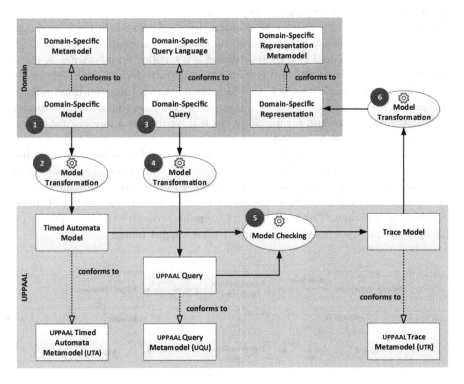

Fig. 1. The generic model-driven engineering approach for building front-end tools that
use UPPAAL as a back-end analysis engine.

Organization of the paper. Section 2 provides some background information about MDE and the timed-automata formalism. Section 3 introduces the three metamodels and their transformations. Section 4 discusses the case studies. Section 5 discusses the related work and Sect. 6 concludes the paper.

2   Background

In this section, we provide some background information about model-driven engineering (cf. Sect. 2.1) and the timed-automata formalism (cf. Sect. 2.2).

2.1   Model-Driven Engineering

Models are powerful tools to express structure, behavior and other properties in domains such as engineering, physics, architecture and other fields. *Model-Driven Engineering (MDE)* is a software engineering approach that considers models not only as documentation, but also adopts them as basic abstractions to be used directly in development processes [33].

To define models of a particular domain, we need to specify their language. In MDE, such a language (often referred to as a *domain-specific language*, DSL) is also specified as a model at a more abstract level, called a *metamodel*. A metamodel captures core concepts and behavior of a certain domain, and defines the permitted structure and behavior, to which its instances (models) must adhere. Another way of saying this is that metamodels describe the syntax of models [34]. Following the common terminology, we will write that a model *conforms to* or *is an instance of* its metamodel.

MDE provides interoperability between domains (and tools in these domains) via *model transformations*. The concept of model transformation is shown in Fig. 2. Model transformations are usually defined in a language designed specifically to this aim and map the elements of a source metamodel to the elements of a target metamodel. The transformation engine executes the transformation definition on the input model and generates an output model.

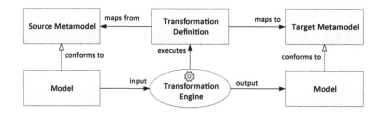

Fig. 2. The concept of *model transformation*.

Benefits of MDE. MDE provides a range of important benefits [36], some of which we briefly discuss below:

- *Interoperability:* As we have mentioned before, there can be multiple domains in a project where various tools are used, each with its own I/O formats. MDE provides interoperability between these domains (and tools in these domains) via model transformations.
- *Higher level of reusability:* The metamodels, models and tools from a domain can be reused by many projects targeting the same domains. Such reuse also increases the quality of the final product since the reused units are revised and improved continuously.
- *Faster tool development:* Domain experts only focus on the concepts of the domain while creating models. Transformations on these models are implemented using languages designed specifically for model transformations rather than using general-purpose languages. Because of these advantages of MDE, the development time of tools decreases.

Tool Choice. There are a number of tools for realizing MDE. The case studies presented in this paper are implemented using the Eclipse Modeling Framework (EMF) [35], a state-of-art tool for implementing MDE techniques. EMF provides the *Ecore* format for defining metamodels and many plug-ins to support various functionalities, such as querying, validation and transformation of models.

2.2   Timed Automata and UPPAAL

Timed automata are finite-state automata with the addition of real-valued *clocks* and synchronization *channels*. In Fig. 3, we show an example timed-automata model (from [5]), with clocks x and y. *Locations* are indicated by circles (double circle for the initial location), and *transitions* are represented by edges. Conditions on clocks can enable transitions (e.g., x > 10 in Fig. 3b, from dim to off) or allow residence in locations (y < 5 in Fig. 3a). Synchronizations can occur when two automata perform complementary actions on the same channel: in the example, outputs press! synchronize with inputs press?. When taking a transition, clocks can be reset (x:=0, y:=0).

Timed-automata models are verified with UPPAAL [3] through queries expressed in a subset of CTL [12]. In Fig. 3c, we show the trace resulting from the verification of the reachability query E<>lamp.bright, which asks whether a state where the lamp automaton is in the bright location is reachable. The verification returns a positive outcome, together with a witness trace, listing the sequence of states and transitions leading to the desired target.

In addition to the standard version of UPPAAL, some of the models presented in this paper are intended for analysis by UPPAAL CORA [4], which allows to compute cost-optimal traces (see Sect. 4.2), and UPPAAL-SMC [7], which allows to perform statistical model checking (see Sect. 4.4).

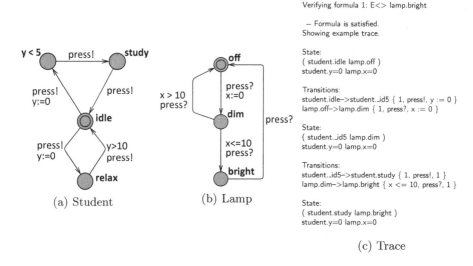

(a) Student   (b) Lamp

(c) Trace

Fig. 3. An example of a timed-automata model (a, b) and the textual output (c) of verifying the reachability query E<>lamp.bright as provided by UPPAAL's command-line tool.

3   Metamodels for the Approach

We use metamodeling to represent the domain of timed automata and enable the back-end analysis of domain-specific models. Our approach extends the work by Greenyer and Rieke [17] towards full-fledged metamodels, covering all language features accepted by the UPPAAL model checker. Thereby, we make sure that model transformations may freely use any of UPPAAL's concepts when translating domain-specific models into timed-automata models.

In Sect. 3.1, we present the metamodel for UPPAAL timed automata (UTA). Section 3.2 describes a metamodel extension for UPPAAL's query language (UQU). A metamodel for traces obtained from UPPAAL (UTR) is given in Sect. 3.3.

3.1   The UPPAAL Timed Automata Metamodel

Figure 4a shows an excerpt from our UPPAAL Timed Automata metamodel (UTA), extending the metamodeling approach proposed in [17]. This metamodel reflects the basic structure of timed automata accepted by UPPAAL.

At the core of UTA is a network of timed automata (NTA). An NTA includes a set of global Declarations, containing instances of the abstract base class Declaration. A declaration is used to introduce elements such as clocks or synchronization channels. Primarily, an NTA includes a non-empty set of templates where each Template represents a type of timed automaton. Moreover, an NTA contains a separate set of system declarations. These are specific TemplateInstances (omitted from the figure), which constitute the set of concrete timed automata that make up the system to be model-checked.

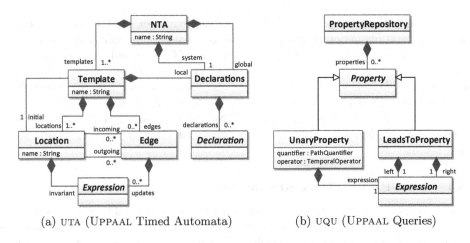

(a) UTA (UPPAAL Timed Automata)   (b) UQU (UPPAAL Queries)

Fig. 4. Partial views from the UTA and UQU.

Templates include locations and edges, and every Template refers to one particular initial location. Templates may also include local declarations (e.g., for clocks that should not be reset from outside the automaton). Every Location refers to its incoming and outgoing edges. In addition, a Location specifies an invariant which is a boolean expression as an instance of the abstract base class Expression. An Edge may contain expressions as well to specify updates of variables (e.g., clock resets). The metamodel also contains syntax-related elements for the C-like expressions supported by UPPAAL.

UTA models are not the native input format of UPPAAL and, therefore, are not directly processable. We have implemented a model-to-text transformation, which takes a UTA model as input and transforms it into UPPAAL native XML.

3.2 The UPPAAL Query Metamodel

Figure 4b depicts an excerpt from our UPPAAL Query metamodel (UQU). Queries are temporal logic properties to be verified using model checking. Multiple queries are bundled by a PropertyRepository, which is the root class of the metamodel. A repository contains a set of properties, where every Property represents one query. Every property is either a UnaryProperty or a LeadsToProperty.

A UnaryProperty is a temporal formula that conforms to the *computation tree logic* (CTL, [12]). First, such a property includes a quantifier (one of *universal* or *existential* quantification) to describe whether the property must hold on all execution paths, or at least one path. Second, it consists of a modal operator (one of *globally* or *finally*) to describe if the property needs to hold in all states of a certain execution path, or needs to hold eventually in some state. Third, unary properties include an expression to be evaluated in the context of the quantifier and the operator. For example, this expression could represent an active location inside an automaton, or a clock value. To this end, UQU extends UTA and reuses the Expression class introduced in Sect. 3.1.

A LeadsToProperty represents a binary property connecting two expressions by means of the *leads-to* operator supported by UPPAAL. Please note that, according to the restrictions imposed by UPPAAL on the set of CTL formulas supported, our metamodel does not allow nested properties. However, we introduce dedicated classes for logical connections of expressions (omitted from Fig. 4b), precisely reflecting the range of functions actually supported by UPPAAL.

Like UTA models, also UQU queries have to be transformed to UPPAAL's native format before they can be actually processed. For this purpose, we provide another model-to-text transformation.

3.3   The UPPAAL Trace Metamodel

The outcome of evaluating a query in UPPAAL can be twofold: either a simple "yes" (for a universally quantified query claiming that a given property holds for all paths) or "no" (for an existentially quantified query asking whether a path with a given property exists), and possibly a trace through the state-space of the timed-automata model along which the query fails to hold (for a universal query) or that is a witness (for an existential query). Queries are very often formulated in such a way that it is known *a priori* whether they hold or not, the interesting part of the outcome is then that diagnostic trace.

UPPAAL outputs its traces in a native textual format that is not too well documented. From [6], we have taken a metamodel (UTR) to capture the information in a tractable way and a parser that produces UTR models from UPPAAL's output. Like UQU, also UTR depends on UTA itself, so that the traces can refer back to their constituent components. Figure 5 gives a high-level overview of UTR.

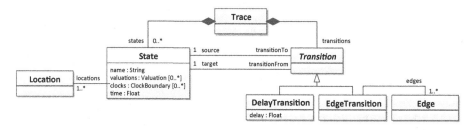

Fig. 5. A partial view from UTR (UPPAAL Trace metamodel).

A Trace consists of States and Transitions; every State except the final one has a single outgoing Transition. A State refers to a set of Locations (one for every TemplateInstance in the system, though that cannot be seen from the provided metamodel fragment), together with Valuations, i.e., bindings for all the variables to concrete values, as well as boundaries for all the clocks in the system (the Valuation and ClockBoundary classifiers are omitted from the figure). Finally, a State stores the absolute time at which the system arrived in that state. A Transition can either be a DelayTransition, in which only time passes, or an EdgeTransition, in which a number of Edges (one for every TemplateInstance involved) fire in synchrony. Location and Edge are imported from UTA.

4   Case Studies

The general MDE approach we propose for bridging software tools has been introduced in Sect. 1. In this section, we present five case studies that have put this approach into practice.

In Table 1, an overview of these case studies is given. After the section number, the second column shows to which domain the approach is applied. The third column contains the list of the metamodels that are used to describe that domain. The fourth column gives the motivation why model checking is used for the particular case study. The fifth column shows which steps from the approach (given in Fig. 1) are implemented in the particular case study. The following subsections describe these case studies in more detail.

The transformations for the cyber-physical systems case study are specified in the QVTo [27] language, for the other cases in the Epsilon Transformation Language [21]. Translation of the timed-automata models to the XML input files for UPPAAL is performed via the Xtend [11] language, using its template expressions for model-to-text transformations.

4.1   Coordination Protocols of CPSs

Future cyber-physical systems (CPS; e.g., cars, railway systems, smart factories) will heavily interact with each other to contribute to aspects like safety, efficiency, comfort and human health. They may achieve this by coordinating their actions via asynchronous message exchange. However, such a coordination must be safe and has to obey hard real-time constraints because any (timing) error may lead to severe damage and even loss of human life. Consequently, the development of so-called coordination protocols that specify the allowed message exchange sequences requires formal verification like model checking to guarantee the functional correctness of the coordination.

Model checkers like UPPAAL are appropriate for verifying such coordination protocols but their language has no built-in support for domain-specific aspects like asynchronous communication including message buffers and quality-of-service (QoS) assumptions (e.g., message delay and reliability). Consequently, the domain expert has to encode these aspects manually, which is a complex and error-prone task. Therefore, the model-driven method MECHATRONICUML [10] defines a DSL for specifying coordination protocols of CPS at a more abstract level. Among others, this DSL enables to specify hierarchical state machines, real-time constraints, message buffers and the QoS assumptions of the protocol. Furthermore, MECHATRONICUML defines a domain query language to ease the specification of formal verification properties that a coordination protocol of MECHATRONICUML shall fulfill. For example, the requirement "At least one instance per message type of the coordination protocol can be in transit" may be specified as follows: `forall(m : MessageTypes) EF messageInTransit(m)`.

In [9,15], we have achieved to fully hide the model checker UPPAAL from the domain expert by specifying domain-specific model checking for coordination protocols of MECHATRONICUML using UPPAAL. Our approach requires all six

Table 1. An overview of the case studies applying the proposed approach.

Sect.	Domain	Domain Metamodels	Motivation for using a model checker	Steps of the approach
4.1	Cyber-Physical Systems	Protocol, Query	To verify whether a coordination protocol fulfills all stated properties	1, 2, 3, 4, 5, 6
4.2	Hardware-Software Co-Design	Synchronous Data Flow Graph, Hardware Platform, Allocation	To obtain a schedule for the execution of the tasks considering optimization objectives of resource and energy	1, 2, 5, 6
4.3	Cyber-Security	Attack-Fault tree	To obtain a schedule of attack steps optimizing objectives like time and cost, or stochastic values, e.g., probability of attack within mission time	1, 2, 5, 6
4.4	Reliability Engineering	Attack-Fault tree	To obtain the probability of failure within mission time	1, 2, 5
4.5	Software Timing Analysis	Java Bytecode, Timing Analysis Extension	To validate Java applications to ensure that they fulfill their timing specifications	1, 2, 5

steps that we introduce in Sect. 1. In particular, we assume that the coordination protocol and its domain queries are specified in Steps 1 and 3. Then, in Step 2, we transform a coordination protocol of MECHATRONICUML into a set of timed automata that conform to UTA. Moreover, in Step 4, we transform our domain query language into properties that conform to UQU. We automate UPPAAL in Step 5 and parse the textual trace into a model that conforms to the UTR metamodel. Finally, in Step 6, we apply a model transformation to translate the trace back to the level of MECHATRONICUML in order to show the trace to the domain expert. We have implemented our concepts successfully into the MECHATRONICUML Tool Suite.

4.2   Synchronous Dataflow Graphs

Hardware-software (HW-SW) co-design is an engineering approach to simultaneously design the hardware and software components of a system to meet optimization objectives. *Synchronous dataflow (SDF) graphs* [25] are a frequently used formalism in the HW-SW co-design domain to represent streaming and dataflow applications in terms of their computation tasks and the data relationships among them. Tasks are represented as nodes, and data input-output relationships between these tasks are represented as edges. SDF graphs can be used to calculate an (energy- or time-) optimal schedule of an application allocated on a particular hardware platform.

In [1], we have applied the generic approach presented in this paper for scheduling analysis of SDF graphs with an energy-optimization objective. Three metamodels are introduced as domain metamodels: The SDF metamodel representing SDF graphs, the hardware platform metamodel representing multiprocessor hardware platforms on which SDF graphs can be mapped, and the allocation metamodel representing such mappings. The domain-specific model, which consists of one instance of each metamodel, is transformed to a timed-automata model and is analyzed with UPPAAL CORA [4]. The trace resulting from this analysis, which is an instance of the trace metamodel given in Sect. 3.3, represents an energy-optimal schedule. In order to make the result available to the domain experts, we have implemented a model transformation from trace models to *schedule* models. Schedule models conform to the *Schedule* metamodel (see Fig. 6) that we have developed and described below.

Schedule is the root of the metamodel. It consists of Executors, Executables and Tasks. An Executor represents a processing unit (which is usually a processor or a core) that executes a task. An Executable is a computation unit that can be executed while a Task is one execution instance of an Executable. A Task has a start time and an optional end time, which are both Time references. The end time is optional since a Schedule may contain Tasks that have not finished.

4.3   Attack Tree Analysis

Modern day infrastructures are frequently faced with cyber attacks. A key challenge is to identify the most dangerous security vulnerabilities, estimate their likelihood and prioritize investments to protect the system from the most riskful scenarios. Security experts often model threat scenarios and perform quantitative risk assessment using attack trees (ATs). These describe how atomic *attack steps* (the tree leaves) combine into complex attacks (intermediate nodes, also called *gates*), leading to the *security breach* represented by the root of the tree. Over the years, numerous formalisms inspired by ATs have been proposed [22]. As they

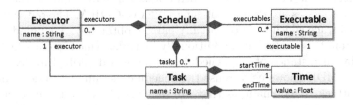

Fig. 6. The Schedule metamodel.

all share the same basic structure, we have developed a metamodel [20] to support interoperability between the different tools made to analyze attack trees. Furthermore, as attack trees resemble fault trees, we enriched the AT metamodel with fault tree constructs, resulting in the *attack-fault tree* (AFT) metamodel [29].

A piece of the transformation from attack trees to UPPAAL can be seen in Fig. 7. This section produces the overall structure (i.e., *system declaration* in UPPAAL) from the class called `AttackTree` in the metamodel `AFT`. The `.equivalent()` function transforms each node into an UPPAAL template and declaration, automatically selecting the transformation rule for that node.

```
rule Base transform at : AFT!AttackTree to out : Uppaal!NTA {
    out.systemDeclarations = new Uppaal!SystemDeclarations();
    out.systemDeclarations.system = new Uppaal!System();
    var iList = new Uppaal!InstantiationList();
    out.systemDeclarations.system.instantiationList.add(iList);
    for (node : AFT!Node in at.Nodes) {
        var converted = node.equivalent();
        if (converted <> null) {
            out.template.add(converted.get(0));
            out.systemDeclarations.declaration.add(converted.get(1));
            iList.template.add(converted.get(1).declaredTemplate);
        }
    }
    out.addTopLevel(at.Root);
}
rule andGate transform node : AFT!Node to ret : List {
    guard : node.nodeType.isKindOf(AFT!AND)
...
```

Fig. 7. Snippet of the translation from the Attack Tree metamodel to UTA.

Traditional ATs are static, and their leaves are decorated with single attributes like cost or time. In order to account for multiple attributes and temporal dependencies we defined transformations from AFT models to UTA models. The security properties that can be checked require either *optimization*, like "What is the cost-optimal path taken by an attacker? [24]", or the use of *stochastic values*, like "What is the probability of an attack within m months? [23]". Similar to what we did for Synchronous Dataflow models, the results of optimization queries are computed using UPPAAL CORA. The outcome of such analysis is a trace which is automatically parsed, obtaining a UTR model. A trace obtained from this analysis can additionally be transformed into a schedule, represented by an instance of the Schedule metamodel described in Sect. 4.2. The adoption of MDE allows us to reuse the Schedule metamodel to describe results from the attack tree domain, as they are semantically close to the SDF results. The stochastic values are computed using UPPAAL-SMC. Plotting these results over time yields graphs similar to the one in Fig. 8.

Currently, the optimization and stochastic security properties are expressed as queries specific for UPPAAL CORA and UPPAAL-SMC, making them incompatible with the current query metamodel.

4.4 Fault Tree Analysis

As society becomes ever more dependent on complex technological systems, the failure of these systems can have disastrous consequences. The field of reliability engineering uses various methods to analyze such systems, to ensure that they meet the required high standards of dependability.

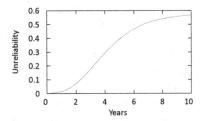

A popular formalism to perform such an analysis is *fault tree* analysis. Faults trees (FTs) are similar to attack trees (described in Sect. 4.3), however rather than modeling deliberate steps in executing an attack, they model component failures (called *basic*

Fig. 8. Example plot of reliability over time as produced by automatic analysis of a fault tree using the UPPAAL-SMC metamodel.

events) that may combine to cause system failures or other undesired events.

Standard FTs were developed in the 1960 s and describe only boolean combinations of faults. Since then, a large number of variations and extensions of fault trees have been developed [30], covering aspects such as timing dependencies, uncertainty, and maintenance. Most of these extensions were developed independently and traditional tools do not support combinations. MDE simplifies the combination of models of different kinds and the analysis of those aspects that are shared between the different formalisms.

Fault trees are described in a unified attack-fault tree (AFT) metamodel also used for attack trees. The main difference from ATs is in the attributes of the basic events. Where attack steps are controlled by an external attacker who makes deliberate decisions based on factors such as cost, faults are inherently stochastic in nature: The failure time is not externally decided, but rather governed by a probability distribution attached to the fault.

The AFT metamodel supports basic events governed by hypoexponential distributions, and gates from standard fault trees, dynamic fault trees [8] and fault maintenance trees [29], as well as gates from attack trees.

As one of the analysis back-ends of the AFT metamodel, we provide a model transformation to a UTA model. Unlike most applications described in this paper, the analysis of this model does not result in a trace or a schedule, nor can its queries be expressed in the current query metamodel. Queries are usually probabilistic in nature, asking questions such as "What is the probability of the system failing within 5 y". Results are then numeric values answering such queries. While it is possible to extract a trace from an FT, its value is limited due to the stochastic nature of the fault tree.

Instead, the typical use of the fault tree metamodel is to produce one UPPAAL-SMC model and automatically query the failure probability at different times. The results of these queries can than be used to produce a plot of the system reliability over time, such as the one shown in Fig. 8.

4.5 Analysis of Java Programs

Model-based verification techniques for software applications require the existence of expressive models. Typically, these models are derived manually, which is a labor-intensive and error-prone task. Also, models need to be maintained and kept consistent with the software application, lest they become outdated.

The framework we have introduced in [38] adopts the generic approach presented in this paper for automatically deriving timed-automata models to validate Java applications, timing requirements in particular, using model checking. In this framework, the bytecode metamodel [37] and its timing analysis extension are introduced as the domain metamodels. The instance of the bytecode metamodel (*bytecode model*) is generated from the target Java application automatically using the JBCPP plug-in. Following this, the bytecode metamodel is enriched through a number of model transformations with additional information necessary for analysis; this includes recursion handling, loop detection, loop iteration bounding, timing information, etc. The additional information is represented as an instance of the extension metamodel. The enriched bytecode model is then transformed to a UTA model to be analyzed with UPPAAL.

Queries are currently manually written and results of the model checking process are not translated back to a domain-specific representation such as to a source-code view. However, the implementation of these points using MDE is suggested in the generic approach is a future direction of the study in [38].

5   Related Work

There are many studies that use UPPAAL to verify systems. We limit this section to the studies that automatically transform domain-specific models to timed automata, or map the results of model checking back to the domain of interest.

The tool ANIMO (Analysis of Networks with Interactive MOdeling) [32] has been introduced to analyze complex biological processes in living cells. ANIMO transforms the domain-specific models defined by biologists to UPPAAL models; then the results of the model checking process are presented back in a domain-specific fashion. The transformations in ANIMO are implemented in a general-purpose language, i.e., Java, whereas the case studies reported in this paper use languages specifically designed for model management tasks.

Frost et al. [14] have introduced a tool for static analysis of timing properties of Java programs. The tool transforms the domain-specific model, which consists of the program, the virtual machine, and the hardware models, to an UPPAAL timed-automata model. The paper does not report any use of MDE techniques.

A toolset to support design-space exploration of embedded systems was introduced by Basten et al. [2]. It aims for the reuse of models between various domains, by providing Java libraries to read design models written in its own specification language and then transform them for use with other tools including UPPAAL for design-space exploration. If the toolset needs to support a new tool, one has to implement new transformations using these libraries. Using a language not specifically designed for such transformations leads to challenges in maintaining the toolset, which are in fact stated as a future direction of research.

In the study by Fakih et al. [13], a tool named sdf2ta has been introduced for analyzing timing bounds of SDF graphs. The tool takes an SDF graph defined using the tool SDF3 and a hardware model defined separately, and automatically generates an UPPAAL timed-automata model. Similar to our tooling choice, they have used EMF for the implementation of sdf2ta, however, it is not reported how the generation of the timed-automata model is achieved.

Herber and Glesner [19] proposed a framework to verify hardware-software co-designs using timed automata. It translates the co-design implemented in SystemC to UPPAAL's timed automata format. This translation is automatically achieved by the SystemC Timed Automata Transformation Engine (STATE) that is specifically designed for SystemC-to-UPPAAL transformations. STATE is implemented directly in Java, which limits interoperability with other tools.

In the work by Hartmanns and Hermanns [18], a toolset has been introduced to facilitate the reuse of various model checkers targeting the stochastic hybrid automata formalism. The toolset uses a high-level compositional modeling language that serves as an interoperability point among existing languages and tools. Conceptually, this language is similar to a metamodel and the transformations from/to this language are implemented using traditional compiler techniques.

The study by Glatz et al. [16] uses model checking to test distributed control systems. The authors mathematically define a mapping from concepts in the control systems domain to the timed-automata domain. In their approach, they suggest implementing this mapping as a translation between the XML formats of these domains, which can be seen as a textual model-based transformation.

6   Conclusions

We have demonstrated the use of MDE in the development of software bridging tools that use UPPAAL as a back-end analysis tool. Our approach uses metamodels as the foundation to translate domain-specific concepts into timed-automata models and queries; the results delivered by UPPAAL are similarly translated back to the original domain, providing experts with access to formal analysis techniques without requiring additional training. We have presented five case studies in different domains to demonstrate how our approach has been applied in practice with the aim of a higher level of interoperability, faster software development and easier maintainability.

The principles we have presented here can be applied to formalisms and analysis tools different from timed automata and UPPAAL by replacing the central metamodels UTA, UQU and UTR with suitable counterparts. Thus we expect our approach to be generally applicable in the development of more software bridging tools which act between DSLs and formal methods.

Acknowledgements. This research was partially funded by STW and ProRail under the project ArRangeer (grant 12238), STW project SEQUOIA (15474), NWO projects BEAT (612001303) and SamSam (628.005.015), and EU project SUCCESS.

References

1. Ahmad, W., Yildiz, B.M., Rensink, A., Stoelinga, M.: A model-driven framework for hardware-software co-design of dataflow applications. In: Berger, C., Mousavi, M.R., Wisniewski, R. (eds.) CyPhy 2016. LNCS, vol. 10107, pp. 1–16. Springer, Cham (2017). doi:10.1007/978-3-319-51738-4_1
2. Basten, T., Hamberg, R., Reckers, F., Verriet, J.: Model-Based Design of Adaptive Embedded Systems. Springer Publishing Company, New York (2013). doi:10.1007/978-1-4614-4821-1
3. Behrmann, G., David, A., Larsen, K.G., Håkansson, J., Petterson, P., Yi, W., Hendrink, M.: UPPAAL 4.0. In: Proceedings of 3rd International Conference on Quantitative Evaluation of Systems (QEST), pp. 125–126 (2006). https://doi.org/10.1109/QEST.2006.59
4. Behrmann, G., Larsen, K.G., Rasmussen, J.I.: Optimal scheduling using priced timed automata. SIGMETRICS Perform. Eval. Rev. **32**(4), 34–40 (2005)
5. Bengtsson, J., Yi, W.: Timed automata: semantics, algorithms and tools. In: Desel, J., Reisig, W., Rozenberg, G. (eds.) ACPN 2003. LNCS, vol. 3098, pp. 87–124. Springer, Heidelberg (2004). doi:10.1007/978-3-540-27755-2_3
6. Brandt, J.: Understanding attacks: modeling the outcome of attack tree analysis. In: 25th Twente Student Conference on IT, vol. 25. University of Twente (2016), BSc. Thesis; see. http://referaat.cs.utwente.nl/conference/25/paper
7. Bulychev, P., David, A., Larsen, K.G., Mikučionis, M., Poulsen, D.B., Legay, A., Wang, Z.: UPPAAL-SMC: statistical model checking for priced timed automata. In: Proceedings of 10th Wks. Quantitative Aspects of Programming Languages (2012). https://doi.org/10.4204/EPTCS.85.1
8. Dugan, J.B., Bavuso, S.J., Boyd, M.A.: Fault trees and sequence dependencies. In: Proceedings of Annual Reliability and Maintainability Symposium, pp. 286–293, January 1990
9. Dziwok, S., Gerking, C., Heinzemann, C.: Domain-specific Model Checking of MechatronicUML Models Using UPPAAL. Technical report tr-ri-15-346, Paderborn University, Jul 2015. https://www.hni.uni-paderborn.de/pub/9121
10. Dziwok, S., Pohlmann, U., Piskachev, G., Schubert, D., Thiele, S., Gerking, C.: The mechatronicUML design method: process and language for platform-independent modeling. Technical report tr-ri-16-352, Software Engineering Department, Fraunhofer IEM / Software Engineering Group, Heinz Nixdorf Institute , version 1.0, December 2016
11. Eclipse foundation Inc: XTend - modernized Java. https://www.eclipse.org/xtend/index.html

12. Emerson, E.A., Clarke, E.M.: Using branching time temporal logic to synthesize synchronization skeletons. Sci. Comput. Program. **2**(3), 241–266 (1982)
13. Fakih, M., Grüttner, K., Fränzle, M., Rettberg, A.: State-based Real-time analysis of SDF applications on MPSoCs with shared communication resources. J. Syst. Archit. **61**(9), 486–509 (2015)
14. Frost, C., Jensen, C., Luckow, K.S., Thomsen, B.: WCET analysis of java bytecode featuring common execution environments. In: Proceedings of 9th International Wks. Java Technologies for Real-Time and Embedded Systems, pp. 30–39. ACM (2011)
15. Gerking, C., Schäfer, W., Dziwok, S., Heinzemann, C.: Domain-specific model checking for cyber-physical systems. In: Proceedings of 12th Wks. Model-Driven Engineering, Verification and Validation (MoDeVVa 2015). Ottawa, September 2015
16. Glatz, B., Cleary, F., Horauer, M., Schuster, H., Balog, P.: Complementing testing of IEC61499 function blocks with model-checking. In: Proceedings of 12th IEEE/ASME International Conference on Mechatronic, Embedded Systems and Applications (MESA) (2016)
17. Greenyer, J., Rieke, J.: Applying advanced TGG concepts for a complex transformation of sequence diagram specifications to timed game automata. In: Schürr, A., Varró, D., Varró, G. (eds.) AGTIVE 2011. LNCS, vol. 7233, pp. 222–237. Springer, Heidelberg (2012). doi:10.1007/978-3-642-34176-2_19
18. Hartmanns, A., Hermanns, H.: The modest toolset: an integrated environment for quantitative modelling and verification. In: Ábrahám, E., Havelund, K. (eds.) TACAS 2014. LNCS, vol. 8413, pp. 593–598. Springer, Heidelberg (2014). doi:10.1007/978-3-642-54862-8_51
19. Herber, P., Glesner, S.: A HW/SW co-verification framework for systemC. ACM TECS **12**(1s), 61:1–61:23 (2013)
20. Huistra, D.: A unifying model for attack trees. Research Project. University of Twente (2015). http://essay.utwente.nl/69399/
21. Kolovos, D.S., Paige, R.F., Polack, F.A.C.: The epsilon transformation language. In: Vallecillo, A., Gray, J., Pierantonio, A. (eds.) ICMT 2008. LNCS, vol. 5063, pp. 46–60. Springer, Heidelberg (2008). doi:10.1007/978-3-540-69927-9_4
22. Kordy, B., Piètre-Cambacédès, L., Schweitzer, P.: DAG-based attack and defense modeling: don't miss the forest for the attack trees. Comput. Sci. Rev. **13–14**, 1–38 (2014)
23. Kumar, R., Stoelinga, M.: Quantitative security and safety analysis with attack-fault trees. In: Proceedings of IEEE 18th International Symposium High Assurance Systems Engineering (HASE), pp. 25–32, January 2017
24. Kumar, R., Ruijters, E., Stoelinga, M.: Quantitative attack tree analysis via priced timed automata. In: Sankaranarayanan, S., Vicario, E. (eds.) FOR-MATS 2015. LNCS, vol. 9268, pp. 156–171. Springer, Cham (2015). doi:10.1007/978-3-319-22975-1_11
25. Lee, E.A., Messerschmitt, D.G.: Synchronous data flow. Proc. IEEE **75**(9), 1235–1245 (1987)
26. Mohagheghi, P., Dehlen, V.: Where Is the proof? - a review of experiences from applying MDE in industry. In: Schieferdecker, I., Hartman, A. (eds.) ECMDA-FA 2008. LNCS, vol. 5095, pp. 432–443. Springer, Heidelberg (2008). doi:10.1007/978-3-540-69100-6_31
27. Object Management Group (OMG): Meta Object Facility (MOF) 2.0 Query/View/Transformation Specification, Version 1.2. OMG Document Number formal/01 Feb 2015. http://www.omg.org/spec/QVT/1.2

28. Ravn, A.P., Srba, J., Vighio, S.: A formal analysis of the web services atomic transaction protocol with UPPAAL. In: Margaria, T., Steffen, B. (eds.) ISoLA 2010. LNCS, vol. 6415, pp. 579–593. Springer, Heidelberg (2010). doi:10.1007/978-3-642-16558-0_47

29. Ruijters, E., Guck, D., Drolenga, P., Stoelinga, M.: Fault maintenance trees: reliability contered maintenance via statistical model checking. In: Proceedings IEEE 62nd Annual Reliability and Maintainability Symposium (RAMS). IEEE, January 2016

30. Ruijters, E., Stoelinga, M.: Fault tree analysis: a survey of the state-of-the-art in modeling, analysis and tools. Comput. Sci. Rev. **15–16**, 29–62 (2015)

31. Schivo, S., Scholma, J., Wanders, B., Camacho, R.A.U., van der Vet, P.E., Karperien, M., Langerak, R., van de Pol, J., Post, J.N.: Modeling Biological Pathway Dynamics With Timed Automata. IEEE J. Biomed. Health Inform. **18**(3), 832–839 (2014)

32. Schivo, S., Scholma, J., van der Vet, P.E., Karperien, M., Post, J.N., van de Pol, J., Langerak, R.: Modelling with ANIMO: between fuzzy logic and differential equations. BMC Syst. Biol. **10**(1), 56 (2016)

33. da Silva, A.R.: Model-driven engineering: A survey supported by the unified conceptual model. Comput. Languages, Systems & Structures 43, 139–155 (2015)

34. Sprinkle, J., Rumpe, B., Vangheluwe, H., Karsai, G.: Metamodelling. In: Model-Based Engineering of Embedded Real-Time Systems, pp. 57–76. Springer (2010)

35. Steinberg, D., Budinsky, F., Paternostro, M., Merks, E.: EMF: Eclipse modeling framework 2.0. Addison-Wesley Professional, 2nd edn. (2009)

36. Völter, M., Stahl, T., Bettin, J., Haase, A., Helsen, S.: Model-driven software development: technology, engineering, management. John Wiley & Sons (2006)

37. Yildiz, B.M., Bochisch, C.M., Rensink, A., Aksit, A.: An MDE approach for modular program analyses. In: Proc. Modularity in Modelling Workshop (2017)

38. Yildiz, B.M., Rensink, A., Bockisch, C., Aksit, M.: A Model-Derivation Framework for Software Analysis. In: Proc. 2nd Wks. Models for Formal Analysis of Real Systems (MARS) (2017)

PranCS: A Protocol and Discrete Controller Synthesis Tool

Idress Husien, Sven Schewe, and Nicolas Berthier$^{(\boxtimes)}$

Department of Computer Science, University of Liverpool, Liverpool, UK
nicolas.berthier@liverpool.ac.uk

Abstract. PranCS is a tool for synthesizing protocol adapters and discrete controllers. It exploits general search techniques such as simulated annealing and genetic programming for homing in on correct solutions, and evaluates the fitness of candidates by using model-checking results. Our **Proctocol and Controller Synthesis** (PranCS) tool uses NuSMV as a back-end for the individual model-checking tasks and a simple candidate mutator to drive the search.

PranCS is also designed to explore the parameter space of the search techniques it implements. In this paper, we use PranCS to study the influence of turning various parameters in the synthesis process.

1 Introduction

Discrete Controller Synthesis (DCS) and *Program Synthesis* have similar goals: they are automated techniques to infer a control strategy and an implementation, respectively, that is correct by construction.

There are mild differences between these two classes of problems. DCS typically operates on the model of a plant. It seeks the automated construction of a strategy to control the plant, such that its runs satisfy a set of given objectives [2,22]. Similarly, program synthesis seeks to infer an implementation, often of a reactive system, such that the runs of this system satisfy a given specification [21]. Program synthesis is particularly attractive for the construction of protocols that govern the intricate interplay between different threads; we use mutual exclusion and leader election as examples.

Apart from their numerous applications to manufacturing systems [19,22, 24], DCS algorithms have been used to enforce fault-tolerance [11], deadlock avoidance in multi-threaded programs [23], and correct resource management in embedded systems [1,3].

Foundations of DCS and program synthesis are similar to principles of model-checking [5,8]. Model-checking refers to automated techniques that determines whether or not a system satisfies a number of specifications. Traditional DCS algorithms are inspired by this approach. Given a model of the plant, they first *exhaustively* compute an unsafe portion of the state-space to avoid for the desired

This work was supported by the Ministry of Higher Education in Iraq through the University of Kirkuk and by the EPSRC through grant EP/M027287/1.

© Springer International Publishing AG 2017

K.G. Larsen et al. (Eds.): SETTA 2017, LNCS 10606, pp. 337–349, 2017.
https://doi.org/10.1007/978-3-319-69483-2_20

objectives to be satisfied, and then derive a strategy that avoids entering the unsafe region. Finally, a controller is built that restricts the behaviour of the plant according to this strategy, so that it is guaranteed to always comply with its specification. Just as for model-checking, *symbolic* approaches for solving DCS problems have been successfully investigated [2,4,10,20].

Techniques based on genetic programming [7,12–17], as well as on simulated annealing [13,14], have been tried for program synthesis. Instead of performing an exhaustive search, these techniques proceed by using a measure of the fitness—reflecting the question *"How close am I to satisfying the specification?"*—to find a short path towards a solution. Among the generic search techniques that look promising for this approach, we focus on *genetic programming* [18] and *simulated annealing* [7,12]. When applied to program synthesis, both search techniques work by successively mutating candidate programs that are deemed "good" by using some measure of their fitness. We obtain their fitness for meeting the desired objectives by using a model-checker to measure the share of objectives that are satisfied by the candidate program, *cf.* [13,14,16,17].

Simulated annealing keeps one candidate solution, and a "cooling schedule" describes the evolution of a "temperature". In a sequence of iterations, the algorithm mutates the current candidate and compares the fitness of the old and new candidate. If the fitness increases, the new candidate is always maintained. If it decreases, a random process decides if the new candidate replaces the old one in the next iteration. The chances of the new candidate to replace the old one then decrease with the gap in the fitness and increase with the temperature; thus, a lower temperature makes the system "stiffer".

Genetic programming maintains a population of candidate programs over a number of iterations. In each iteration, new candidate programs are generated by mutation or by mixing randomly selected candidates ("crossover"). At the end of each iteration, the number of candidates under consideration is shrunken back to the original number. A higher fitness makes it more likely for a candidate to survive this step.

In Sect. 2, we describe the tool PranCS, which implements the simulated annealing based approach proposed in [13,14] as well as approaches based on similar genetic programming from [16,17]. PranCS uses quantitative measures for partial compliance with a specification, which serve as a measure for the fitness (or: quality) of a candidate solution. Furthering on the comparison of simulated annealing with genetic programming [13,14], we extend the quest for the best general search technique in Sect. 3 by:

1. looking for good cooling schedules for simulated annealing; and
2. investigating the impact of the population size and crossover ratio for genetic programming.

2   Overview of PranCS

PranCS implements several generic search algorithms that can be used for solving DCS problems as well as for synthesising programs.

2.1 Representing Candidates

The representation of candidates depends on the kind of problems to solve. Candidate programs are represented as abstract syntax trees according to the grammar of the sought implementation. They feature conditional and iteration statements, assignments to one variable taken among a given set, and expressions involving such variables. Candidates for DCS only involve a series of assignments to a given subset of Boolean variables involved in the system (called "*controllables*").

2.2 Structure of PranCS

The structure of PranCS is shown in Fig. 1. Via the user interface, the user can select a search technique, and enter the problem to solve along with values for relevant parameters of the selected algorithm. For program synthesis, the user enters the number, size, and type of variables that candidate implementations may use, and whether thay may involve complex conditional statements ("if" and "while" statements). DCS problems are manually entered as a series of assignments to state variables involving expressions expressed on state and input variables; the user also lists the subset of input variables that are "controllable". In both cases, the user also provides the specification as a list of objectives.

Generator. The Generator uses the parameters provided to either generate new candidates or to update them when required during the search.

Translator & NuSMV. We use NuSMV [6] as a model-checker. Every candidate is translated into the modelling language of NuSMV using a method suggested by Clark and Jacob [7]. (We detail this translation for programs and plants in [14]

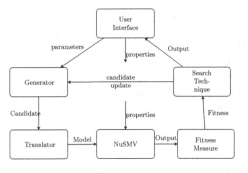

Fig. 1. Overview of PranCS.

and [13] respectively, and give an example program translation in Appendix A.) The resulting model is then model-checked against the desired properties. The result forms the basis of a fitness function for the selected search technique.

Fitness Measure. To design a fitness measure for candidates, we make the hypothesis that the share of objectives that are satisfied so far by a candidate is a good indication of its suitability *w.r.t.* the desired specification. We additionally observe that weaker properties that can be mechanically derived are useful to identify good candidates worth selecting for the generation of further potential solutions. For example, if a property shall hold on all paths, it is better if it holds on some path, and even better if it holds almost surely.

Search Technique. The fitness measure obtained for a candidate is used as a fitness function for the selected search technique. If a candidate is evaluated as correct, we return (and display) it to the user. Otherwise, depending on the search technique selected and the old and new fitness measure/s, the current candidate or population is updated, and one or more candidates are sent for change to the Generator. The process is re-started if no solution has been found in a predefined number of steps (genetic programming) or when the cooling schedule expires (simulated annealing).

2.3   Selecting and Tuning Search Techniques

In terms of search techniques, PranCS implements the following methods: *genetic programming*, and *simulated annealing*. Katz and Peled [17] extend genetic programming by considering the fitness as a pair of "safety-fitness" and "liveness-fitness", where the latter is only used for equal values of "safety-fitness". Building upon this idea, we define two flavours for both simulated annealing and genetic programming: *rigid* (where the classic fitness function is used) and *safety-first*, which uses the two-step fitness approach as above. Further, genetic programming can be used with or without crossovers between candidates [13,14].

Depending on the selected search technique, the tool allows the user to input parameters that control the dynamics of the synthesis process. These parameters determine the likelihood of finding a correct program in each iteration and the expected running time for each iteration, and thus heavily influence the overall search speed. For the genetic

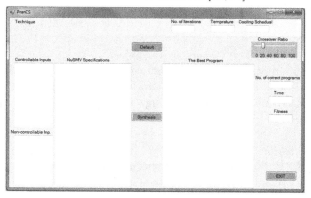

Fig. 2. Graphical User Interface. PranCS allows the user to fine-tune each search technique by means of dedicated parameters.

programming approach, the parameters include the population size, the number of selected candidates, the number of iterations, and the crossover ratio. For simulated annealing, the user chooses the initial temperature and the cooling schedule. Figure 2 shows the graphical user interface of PranCS.

Parameters for Simulated Annealing. In simulated annealing (SA), the intuition is that, at the beginning of the search phase, the temperature is high, and it cools down as time goes by. The higher the temperature, the higher is the likelihood that a new candidate solution with inferior fitness replaces the previous solution. While this allows for escaping local minima, it can also happen that the candidates develop into an undesirable direction. For this reason, simulated annealing does not continue for ever, but is re-started at the end of the cooling schedule. Consequently, there is a sweet-spot in just how long a cooling schedule should be and when it becomes preferable to re-start, but this sweet-spot is difficult to find. We report our experiments with PranCS for tuning the cooling schedule in Sect. 3.1.

Parameters for Genetic Programming. For Genetic Programming (GP), the parameters are the initial population size, the crossover vs mutation ratio, and the fitness measure used to select the individuals. The population size affects the algorithm in two ways: a larger population size could provide better diversity and reduce the number of iterations required or, for a fixed number of iterations, increase the likelihood of finding a solution. However, it also increases the time spent for each individual iteration. The crossover ratio describes the amount of new candidates that are generated by mating. Crossovers allow for the appearance of solutions that synthesise the best traits of good candidates, and a high crossover ratio promises to make this more likely. This requires, however, a high degree of diversity in the population, where these traits need to draw from different parts of the program tree, and it comes to the cost of creating diversity through a reduction of the number of mutations applied in each iteration.

We investigate how the population size and crossover ratio affect the performance of these algorithms in Sects. 3.2 and 3.3.

3 Exploration of the Parameter Space

Besides serving as a synthesis tool, PranCS provides the user with the ability to compare various search techniques. In [13,14], we have carried out experiments by applying our algorithms to generate correct solutions on benchmarks comprising mutual exclusion, leader election, and DCS problems of growing size and complexity. With parameter values borrowed from [16,17], we could already accelerate synthesis significantly using simulated annealing compared to genetic programming (by 1.5 to 2 orders of magnitude).

In this paper, our aim is to further explore the performance impact of the parameters for each search technique. We thus reuse the same scalable benchmarks as in [13,14]: program synthesis problems consist of mutual exclusion ("2 or 3 shared bits") and leader election ("3 or 4 nodes"); DCS problems compute controllers enforcing mutual exclusions and progress between 1 to 6 tasks modelled as automata ("1 through 6-Tasks").

In all Tables, execution times are in seconds; $\bar{t}$ is the mean execution time of single executions (succeeding or failing), and columns T extrapolate $\bar{t}$ based on the success rate obtained in 100 single executions (columns "%").

3.1 Exploring Cooling Schedules for Simulated Annealing

In order to test if the hypothesis from [9] that simulated annealing does most of its work during the middle stages—while being in a good temperature range—

holds for our application, we have developed the tool to allow for "cooling schedules" that do not cool at all, but use a constant temperature. In order to be comparable to the default strategy, we use up to 25,001 iterations in each attempt.

We have run 100 attempts to create a correct candidate using various constant temperatures, and inferred expected overall running times T based on the success rates and average execution time of single executions $\bar{t}$. We first report the results for program synthesis and DCS problems in Tables 1 and 2 respectively.

Table 1. Impact of search temperature (θ) for program synthesis with safety-first simulated annealing

θ	3 nodes			4 nodes			2 shared bits			3 shared bits		
	$\bar{t}$	%	T	$\bar{t}$	%	T	$\bar{t}$	%	T	$\bar{t}$	%	T
0.7	316	0	∞	521	0	∞	147	0	∞	155	0	∞
400	285	0	∞	493	0	∞	143	0	∞	148	0	∞
4,000	196	11	1,781	368	10	3,680	129	3	4,300	121	4	3,025
7,000	97	14	692	314	13	2,415	77	12	641	81	11	252
10,000	73	21	347	138	18	766	15	22	68	17	24	70
13,000	78	22	354	146	19	768	16	23	69	18	24	75
16,000	83	20	415	150	17	882	17	21	80	19	22	86
20,000	87	19	457	153	15	1,020	21	20	105	23	22	104
25,000	94	17	494	167	13	1,284	23	19	121	25	21	191
30,000	108	15	720	184	11	1,672	28	18	155	30	19	157
40,000	117	15	780	193	11	1,754	31	16	193	34	17	200
50,000	129	13	992	201	10	2,010	37	15	246	41	16	256
100,000	193	12	1,608	287	9	3,188	52	11	472	58	13	446

The findings support the hypothesis that some temperatures are much better suited than others: low temperatures provide a very small chance of succeeding, and the chances also go down at the high temperature end.

While the values for low temperatures are broadly what we had expected, the high end performed better than we had thought. This might be because some small guidance is maintained even for infinite temperature, as a change that is decreasing the fitness is taken with an (almost) 50% chance in this case, while increases are always selected. However, the figures for high temperatures are much worse than the figures for the good temperature range of 10,000 to 16,000.

In the majority of cases, the best results have been obtained at a temperature of 10,000. Notably, these results are better than the running time for the cooling schedule that uses a linear decline in the temperature as used and reported in [13, 14]. They indicate that it seems likely that the last third of the improvement cycles in this cooling schedule had little avail, especially for smaller problems.

Table 2. Impact of search temperature (θ) for DCS with Safety-first simulated annealing

θ	1-Task			2-Tasks			3-Tasks			4-Tasks			5-Tasks			6-Tasks		
	$\bar{t}$	%	T	$\bar{t}$	%	T	$\bar{t}$	%	T	$\bar{t}$	%	T	$\bar{t}$	%	T	$\bar{t}$	%	T
0.7	163	0	∞	177	0	∞	192	0	∞	332	0	∞	298	0	∞	613	0	∞
400	93	0	∞	99	0	∞	163	0	∞	167	0	∞	153	0	∞	598	0	∞
4,000	54	7	771	58	6	966	88	6	1,466	98	3	3,266	98	4	2,450	278	3	9,266
7,000	39	12	325	47	9	522	45	9	500	65	6	1,083	79	6	1,316	125	5	2,500
10,000	18	19	94	29	14	207	26	11	236	39	9	433	61	9	677	99	8	1,237
13,000	22	20	110	33	15	220	31	11	281	43	11	390	67	10	670	115	9	1,277
16,000	29	19	152	39	13	300	37	10	370	58	9	644	73	8	912	127	9	1,411
20,000	37	17	217	47	11	427	42	10	420	67	9	744	81	6	1,350	134	7	1,914
25,000	43	15	286	56	10	560	47	9	522	81	7	1,157	89	6	1,483	152	6	2,533
30,000	49	15	326	67	10	670	56	8	700	89	6	1,483	102	4	2,550	159	6	2,650
40,000	53	13	407	75	9	833	63	9	700	95	6	1,583	116	3	3,866	168	6	2,800
50,000	59	12	491	82	7	1,171	79	7	1,128	103	5	2,060	128	4	3,200	192	5	3,840
100,000	72	11	654	94	7	1,342	98	7	1,400	118	4	2,950	178	3	5,933	253	4	6,325

A robust temperature sweet-spot clearly exists for our scalable benchmarks, suggesting that the quest for robust and generic good cooling schedules is worth pursuing.

3.2   Impact of Population Size for Genetic Programming

One of the important parameters of genetic programming is the initial population size; another parameter worth tuning is the number of candidates η selected for mating at each iteration of the algorithm. In order to investigate their effects on our synthesis approach and evaluate the actual cost of large population sizes, we defined several setups with various values for the population size $|P|$ and amount of mating candidates η. We then performed 100 executions of our GP-based algorithms with each of these setups for the 2 shared bits mutual exclusion and 2-Tasks problems.

We show the results in Tables 3 and 4. As expected, increasing the size of the initial population also dramatically increases the cost of finding a good solution. Broadly speaking, increasing the population size reduces the number of iterations and increases the success rate, but it also increases the computation time required at each individual iteration. Smaller population sizes appear to benefit individual running times more than they harm success rates.

The impact of η on performance appears very limited on the range we have investigated.

Table 3. Impact of population size ($|P|$) for Program Synthesis (2 shared bits mutual exclusion only)

$|P|$	η	Rigid GP						Safety-first GP					
		w/o crossover			with crossover			w/o crossover			with crossover		
		$\bar{t}$	%	T	$\bar{t}$	%	T	$\bar{t}$	%	T	$\bar{t}$	%	T
150	5	583	7	8,328	589	9	6,544	113	31	364	115	33	348
	7	583	7	8,328	589	9	6,544	113	31	364	115	33	348
	9	584	7	8,342	588	9	6,533	113	31	364	114	33	345
250	5	1,024	12	8,533	1,057	15	7,046	230	46	500	245	49	500
	7	1,024	12	8,533	1,057	15	7,046	230	46	500	245	49	500
	9	1,024	12	8,533	1,057	15	7,046	231	46	502	245	49	500
350	5	1,435	15	9,566	1,451	18	8,061	325	63	515	367	67	547
	7	1,435	15	9,566	1,451	18	8,061	325	63	515	366	67	546
	9	1,435	15	9,566	1,451	19	7,636	325	64	507	367	67	547

Table 4. Impact of population size ($|P|$) for DCS (2-Tasks only)

$|P|$	η	Rigid GP						Safety-first GP					
		w/o crossover			with crossover			w/o crossover			with crossover		
		$\bar{t}$	%	T	$\bar{t}$	%	T	$\bar{t}$	%	T	$\bar{t}$	%	T
150	5	463	3	15,433	484	4	12,100	132	13	1,015	138	15	920
	7	463	3	15,433	485	4	12,125	132	13	1,015	139	15	926
	9	464	3	15,466	485	4	12,125	131	13	1,007	139	14	992
250	5	943	5	18,860	969	7	13,842	241	18	1,338	218	19	1,147
	7	943	5	18,860	969	7	13,842	241	18	1,338	218	19	1,147
	9	943	5	18,860	969	7	13,842	242	18	1,344	218	19	1,147
350	5	1,517	9	16,855	1,557	10	15,570	403	24	1,679	340	24	1,416
	7	1,517	9	16,855	1,557	10	15,570	403	24	1,679	340	24	1,416
	9	1,518	9	16,866	1,557	10	15,570	403	24	1,679	340	24	1,416

3.3   Impact of Crossover Ratio for Genetic Programming

Finally, we have also studied the effect of changing the share between crossover and mutation in genetic programming.

We report our results in Tables 5 and 6. Interestingly, the running time per instance increased with the share of crossovers, which might point to a production of more complex candidate solutions. Regarding expected running times, the results also indicate the existence of a sweet-spot for the crossover ratio at around 20% for both Rigid and Safety-first variants of the algorithm.

Table 5. Impact of crossover ratio (ρ, in percent) for Program Synthesis with Rigid and Safety-first GP

		Rigid GP			Safety-first GP		
	ρ	$\bar{t}$	%	T	$\bar{t}$	%	T
2 shared bits	0	583	7	8,328	113	31	364
	20	589	9	6,544	115	33	348
	40	602	9	6,688	123	33	372
	60	614	8	7,657	134	33	406
	80	613	8	7,662	142	21	676
	100	652	2	32,600	151	5	3,020
3 shared bits	0	615	7	8,785	171	17	1,005
	20	620	9	6,888	175	19	921
	40	637	9	7,077	187	19	984
	60	658	8	8,225	196	19	1,031
	80	669	4	16,725	207	11	1,881
	100	682	2	34,100	223	3	7,433
3 nodes	0	1,120	3	37,333	418	15	2,786
	20	1,123	6	18,716	421	16	2,631
	40	1,137	5	22,740	427	16	2,668
	60	1,149	5	22,980	453	13	3,484
	80	1,154	3	38,466	469	9	5,211
	100	1,167	2	58,350	487	4	12,175
4 nodes	0	1,311	3	43,700	536	11	4,872
	20	1,314	5	26,280	541	14	3,864
	40	1,325	4	33,125	557	13	4,284
	60	1,336	3	44,533	569	13	4,376
	80	1,345	3	44,833	581	9	6,455
	100	1,353	2	67,650	593	3	17,966

Table 6. Impact of crossover ratio (ρ, in percent) for DCS with Rigid and Safety-first GP

		Rigid GP			Safety-first GP		
	ρ	$\bar{t}$	%	T	$\bar{t}$	%	T
1-Task	0	378	4	9,450	89	17	523
	20	385	5	7,700	94	20	470
	40	403	5	8,060	101	19	531
	60	418	4	10,450	109	19	573
	80	425	3	14,166	116	12	966
	100	438	1	43,800	124	5	2,480

(continued)

Table 6. (*Continued*)

	ρ	Rigid GP $\bar{t}$	%	T	Safety-first GP $\bar{t}$	%	T
2-Tasks	0	475	3	15,833	127	13	976
	20	484	4	12,100	138	15	920
	40	491	4	12,275	146	15	973
	60	501	3	16,700	158	13	1,215
	80	509	2	25,450	169	11	1,536
	100	521	1	52,100	181	4	4,525
3-Tasks	0	571	3	19,033	189	9	2,100
	20	589	4	14,725	201	11	1,827
	40	597	3	19,900	209	11	1,900
	60	606	3	20,200	217	8	2,712
	80	613	1	61,300	225	7	3,214
	100	627	1	62,700	239	3	7,966
4-Tasks	0	658	3	21,933	288	9	3,200
	20	664	4	16,600	296	12	2,466
	40	679	4	16,975	303	11	2,754
	60	687	3	22,900	313	10	3,130
	80	693	2	34,650	321	8	4,012
	100	711	1	71,100	333	4	8,325
5-Tasks	0	776	1	77,600	438	7	6,257
	20	787	3	26,233	445	11	4,045
	40	792	3	26,400	451	8	5,637
	60	799	2	39,950	459	7	6,557
	80	804	2	40,200	467	5	9,340
	100	815	1	81,500	479	2	23,950
6-Tasks	0	961	2	48,050	659	6	10,983
	20	972	3	32,400	673	10	6,730
	40	981	2	49,050	679	10	6,790
	60	989	2	49,450	695	7	9,928
	80	997	2	49,850	703	4	17,575
	100	1,011	1	101,100	718	2	35,900

4   Conclusion

Together with our extensive exploration of the parameter space, the evaluation of PranCS indicates that simulated annealing is faster than genetic programming (we report some synthesis times with the best parameters observed using simulated annealing in Table 7), and that some temperature ranges are more useful than others. Additional information about the tool can be found at: https://cgi. csc.liv.ac.uk/~idresshu/index2.html.

Table 7. Synthesis times with the best parameters observed for Simulated Annealing with linearly decreasing cooling schedule applied to our DCS benchmarks; results for row "2-Tasks" should be compared with best results reported in Table 4 for solving the same DCS benchmark problem using GP-based algorithms.

	Rigid SA			Safety-first SA		
	$\bar{t}$	%	T	$\bar{t}$	%	T
1-Task	20	13	153	19	16	**118**
2-Tasks	25	10	250	24	13	**184**
3-Tasks	33	9	366	29	10	**290**
4-Tasks	47	9	522	43	9	**477**
5-Tasks	76	8	950	70	9	**777**
6-Tasks	119	7	1,700	106	7	**1,514**

In order to integrate this result into the cooling schedule we plan to use an adaptive cooling schedule, in which the decrements of the temperature depends on the improvement of the fitness.

Appendix A Pseud-Code to NuSMV Translation Example

To evaluate the fitness of the produced program, it is first translated into the language of the model checker NuSMV [6]. We have used the translation method suggested by Clark and Jacob [7].

In this translation, the program is converted into very simple statements, similar to assembly language. To simplify the translation, the program lines

```
1: process me
2: while (true) do
3:    noncritical section
4:    while (turn==me) do
5:       skip
6:    end while
7:    critical section
8:    turn=other
9: end while
```

'me' and 'other' denote (different) variable valuations, in this example implemented as boolean variables. In other instances, they might be have a different (finite) datatype.

```
1:  MODULE p(turn)
2:  VAR
3:    pc: {11, 12, 14,15};
4:  ASSIGN
5:    init(pc) := 11;
6:    next(pc) := case
7:     (pc=11) : {11, 12};
8:     (pc=12)&(turn=me) : 14;
9:     (pc=14) : 15;
10:    (pc=15) : 11;
11:    TRUE: pc;
12:  esac;
13:  next(turn):= case
14:   (pc=15): other;
15:   TRUE :turn;
16:  esac;
```

Fig. 3. Translation example – source pseudo-code (left) and target NuSMV (right)

are first labeled, and this label is then used as a pointer that represents the program counter *(PC)*. From this intermediate language, the NuSMV model is built by creating *(case)* and *(next)* statements that use the *PC*. Figure 3 shows the translation of a mutual exclusion algorithm.

References

1. Altisen, K., Clodic, A., Maraninchi, F., Rutten, E.: Using controller-synthesis techniques to build property-enforcing layers. In: Degano, P. (ed.) ESOP 2003. LNCS, vol. 2618, pp. 174–188. Springer, Heidelberg (2003). doi:10.1007/3-540-36575-3_13
2. Asarin, E., Maler, O., Pnueli, A.: Symbolic controller synthesis for discrete and timed systems. In: Antsaklis, P., Kohn, W., Nerode, A., Sastry, S. (eds.) HS 1994. LNCS, vol. 999, pp. 1–20. Springer, Heidelberg (1995). doi:10.1007/3-540-60472-3_1
3. Berthier, N., Maraninchi, F., Mounier, L.: Synchronous Programming of Device Drivers for Global Resource Control in Embedded Operating Systems. ACM Trans. Embed. Comput. Syst. 12(1s), 39: 1–39: 26., March 2013
4. Berthier, N., Marchand, H.: Discrete controller synthesis for infinite state systems with ReaX. In: 12th Internation Workshop on Discrete Event Systems. WODES 20114, IFAC, pp. 46–53, May 2014
5. Burch, J.R., Clarke, E.M., McMillan, K.L., Dill, D.L., Hwang, L.J.: Symbolic model checking: 10^{20} states and beyond. Inf. Comput. **98**(2), 142–170 (1992)
6. Cimatti, A., Clarke, E., Giunchiglia, E., Giunchiglia, F., Pistore, M., Roveri, M., Sebastiani, R., Tacchella, A.: NuSMV 2: an opensource tool for symbolic model checking. In: Brinksma, E., Larsen, K.G. (eds.) CAV 2002. LNCS, vol. 2404, pp. 359–364. Springer, Heidelberg (2002). doi:10.1007/3-540-45657-0_29
7. Clark, J.A., Jacob, J.L.: Protocols are programs too: the meta-heuristic search for security protocols. Inf. Softw. Technol. **43**, 891–904 (2001)
8. Clarke, E.M., Grumberg, O., Peled, D.A.: Model Checking. MIT Press, Cambridge (1999)
9. Connolly, D.: An improved annealing scheme for the qap. Eur. J. Oper. Res. **46**, 93–100 (1990)
10. Cury, J.E., Krogh, B.H., Niinomi, T.: Synthesis of supervisory controllers for hybrid systems based on approximating automata. IEEE Trans. Autom. Control **43**(4), 564–568 (1998)
11. Girault, A., Rutten, É.: Automating the addition of fault tolerance with discrete controller synthesis. Formal Methods Syst. Des. **35**(2), 190 (2009)
12. Henderson, D., Jacobson, S.H., Johnson, A.W.: The theory and practice of simulated annealing. In: Glover, F., Kochenberger, G.A. (eds.) Handbook of Metaheuristics, International Series in Operations Research & Management Science, vol. 57, pp. 287–319. Springer, Boston (2003). doi:10.1007/0-306-48056-5_10
13. Husien, I., Berthier, N., Schewe, S.: A hot method for synthesising cool controllers. In: Proceedings of the 24th ACM SIGSOFT International SPIN Symposium on Model Checking of Software. SPIN 2017, pp. 122–131. ACM, New York (2017)
14. Husien, I., Schewe, S.: Program generation using simulated annealing and model checking. In: De Nicola, R., Kühn, E. (eds.) SEFM 2016. LNCS, vol. 9763, pp. 155–171. Springer, Cham (2016). doi:10.1007/978-3-319-41591-8_11

15. Johnson, C.G.: Genetic programming with fitness based on model checking. In: Ebner, M., O'Neill, M., Ekárt, A., Vanneschi, L., Esparcia-Alcázar, A.I. (eds.) EuroGP 2007. LNCS, vol. 4445, pp. 114–124. Springer, Heidelberg (2007). doi:10.1007/978-3-540-71605-1_11

16. Katz, G., Peled, D.: Model checking-based genetic programming with an application to mutual exclusion. In: Ramakrishnan, C.R., Rehof, J. (eds.) TACAS 2008. LNCS, vol. 4963, pp. 141–156. Springer, Heidelberg (2008). doi:10.1007/978-3-540-78800-3_11

17. Katz, G., Peled, D.: Model checking driven heuristic search for correct programs. In: Peled, D.A., Wooldridge, M.J. (eds.) MoChArt 2008. LNCS (LNAI), vol. 5348, pp. 122–131. Springer, Heidelberg (2009). doi:10.1007/978-3-642-00431-5_8

18. Koza, J.R.: Genetic Programming: On the Programming of Computers by Means of Natural Selection. MIT Press, Cambridge (1992)

19. Krogh, B.H., Holloway, L.E.: Synthesis of feedback control logic for discrete manufacturing systems. Automatica $27(4)$, 641–651 (1991)

20. Marchand, H., Bournai, P., Le Borgne, M., Le Guernic, P.: Synthesis of discrete-event controllers based on the signal environment. Discrete Event Dynamic Syst. Theory Appl. $10(4)$, 325–346 (2000)

21. Pnueli, A., Rosner, R.: On the synthesis of a reactive module. In: Proceedings of the 16th ACM SIGPLAN-SIGACT Symposium on Principles of Programming Languages. POPL 1989. pp. 179–190. ACM, New York (1989)

22. Ramadge, P., Wonham, W.: The control of discrete event systems. Proc. IEEE Spec. Issue Dyn. Discr. Event Syst. $77(1)$, 81–98 (1989)

23. Wang, Y., Lafortune, S., Kelly, T., Kudlur, M., Mahlke, S.: The theory of deadlock avoidance via discrete control. In: Proceedings of the 36th Annual ACM SIGPLAN-SIGACT Symposium on Principles of Programming Languages, pp. 252–263. POPL 2009. ACM, New York (2009)

24. Zhou, M., DiCesare, F.: Petri Net Synthesis for Discrete Event Control of Manufacturing Systems, vol. 204. Springer Science & Business Media, Heidelberg (2012). doi:10.1007/978-1-4615-3126-5

Author Index

Printed in the United States
By Bookmasters

Printed in the United States
By Bookmasters